Highway GEOMETRIC Design

First Edition

Application of Design Standards in InRoads

Xudong Jia, Ph.D., P.E.
Professor of Civil Engineering
California State Polytecnic University, Pomona

Wen Cheng, Ph.D., P.E.
Assistant Professor of Civil Engineering
California State Polytecnic University, Pomona

Ming Guan, P.E.
Highway Engineer
KOA Corporation, Ontario

Kendall Hunt
publishing company

Kendall Hunt
publishing company

www.kendallhunt.com
Send all inquiries to:
4050 Westmark Drive
Dubuque, IA 52004-1840

ISBN 978-1-4652-0964-1

Printed in the United States of America
10 9 8 7 6 5 4 3 2 1

CONTENTS

CHAPTER 2

CHAPTER 3

CHAPTER 4

CHAPTER 5

CHAPTER 6

CHAPTER 7

CHAPTER 8

CHAPTER 9

CHAPTER 10

CHAPTER 11

CHAPTER 12

CHAPTER 13

CHAPTER 1

CHAPTER 7

CHAPTER 9

CHAPTER 10

CHAPTER 11

CHAPTER 12

PREFACE

Highway Geometric Design - Application of Design Standards in InRoads is primarily intended for use as a textbook in university-level courses that deal with roadway design. This book is aimed to apply geometric design standards in InRoads to design a typical highway interchange.

This is the first edition from the draft manuscripts that have been extensively tested in the classrooms at California State Polytechnic University, Pomona (or Cal Poly Pomona). The chapters in this book have been structured to reflect students' learning process. Key features of this book include:

» Highway geometric design is a safety-centered design. This book places more focuses on how to apply standards, control criteria and project constraints in horizontal alignment, vertical alignment, and cross section design. Different from other books that are tutorials of design software, this book describes the process and procedures on how to use design standards and controls, step-by-step, in the design of a freeway, ramps and ramp terminal intersections. This book considers the geometric requirements of an undercrossing bridge and its falsework, as well as culverts under the freeway and ramps.

» The book extensively uses design standards from California State Department of Transportation (Caltrans). The process and procedures employed in this book can be useful for students who are interested in the design standards and controls of other state Departments of Transportation (such as New York Department of Transportation (DOT), Georgia DOT) for the interchange design.

» Design software InRoads from Bentley is selected in this book. The design process and procedures described in this book can be implemented in other design software such as AutoCAD Civil 3D.

We are indebted to a number of faculty members at the Department of Civil Engineering, Cal Poly Pomona who made valuable suggestions and contributions to this book. Professor Peter Clark helped the authors outline the key points in Chapter 2 on transportation project development process in Caltrans and made constructive comments to the book. Without the positive support from him and other faculty members, we would never have completed this book.

Many others also made significant contributions to the book. Students at Cal Poly Pomona who used various draft manuscripts of this book from 1998 to 2011 have provided constructive comments and suggestions. Special thanks go to Mr. Tran Dinh Huy, Mr. Nicholas Broskoff and Mr. Jose Carlos Herrera who ran through the design procedures step-by-step and provided insightful comments and suggestions.

A special note of thanks goes to Ryan L. Schrodt, Shelly Walia, and the team at Kendall Hunt Publishing for their wonderful help, advice, and support in producing this book.

The book cannot be error free. We would be especially grateful for suggestions, criticisms, and corrections that might improve this book.

Xudong Jia • Wen Cheng • Ming Guan

Pomona, CA
September, 2012

Chapter 1 Introduction

LEARNING OBJECTIVES

After completing this chapter you should know:

1. The relationships between the highway geometric design and construction;
2. Highway geometric design standards and the relationship between national standards and state-level standards;
3. Highway design controls and criteria.

SECTIONS

Highways are three-dimensional linear features that provide mobility, safety, drainage, and traffic controls to their users. Highways are classified into the following categories in response to the operational requirements (speed and volume) of drivers, pedestrians and vehicles and the environments the highways traverse:

- Principal arterials including freeways and expressways
- Minor arterials
- Collectors
- Local roads and streets

Principal arterials serve the major activity centers of urban areas and consist mainly of the corridors by which high traffic volumes are carried between activity centers. *Minor arterials* are the highways that interconnect with and supplement the principal arterials. They provide land access that cannot be made available by the principal arterials. Major and minor *collectors* are the highways that collect traffic from local streets in residential areas and local communities and carry that traffic to the minor and principal arterials. *Local highways* are streets that are not classified as principal/minor arterials and major/minor collectors.

To build a new highway or improve an existing highway, highway engineering contractors need to clear the project area, prepare appropriate grades for the roadway bed, and overlay aggregates and asphalt/Portland cement concretes onto the roadway bed. During the grading and overlaying processes, contractors conduct construction surveys to establish the highway centerline and the construction limits of the highway project, create bench marks outside of the project limits, and set up stakes for the guidance and control of construction operation.

Highway geometric design is a process that produces construction detail plans to guide and control how a highway is graded and construction stakes are established. Construction detail plans describe horizontal alignments, vertical alignments, and cross sections as well as specifications and construction notes with which the cross sections are accurately superimposed or cast along the horizontal and vertical alignments and the highway is formed on the existing ground.

1.1 Evolution of Highway Geometric Design Practice

Highway geometric design has gone through three design periods: hand-based design, computer-aided design (CAD), and global positioning system (GPS)-based design. The hand-based geometric design was a common practice prior to the invention of computers. During the hand-based period, highway designers used pencils/pens and engineering papers to produce horizontal alignments, vertical alignments, and cross sections. Highway designers typically relied on contour maps to understand elevation changes of highway project areas. Horizontal alignments were overlaid directly onto contour maps. Stations and key design parameters were calculated manually, such as point of intersection (PI), beginning of curve (BC), ending of curve (EC), and curve radius (R).

Hand-based geometric design was a challenging and time-consuming task. It required highway designers to possess both engineering design capabilities and drafting skills. Typically, it could take several days to make a horizontal alignment or a vertical alignment for a highway project. When design mistakes were found, highway engineers were required to redo the design from scratch.

Computer-aided design (CAD) technologies, since their emergence in the 1980s, have significantly changed how highways are designed. Highway designers relied on design software (such as AutoCAD from Autodesk, MicroStation from Bentley, and others) to facilitate geometric design procedures. At CAD's early stage (1980s–90s), highway designers used design software to help them draft highway elements (horizontal alignments, vertical alignments, and cross sections). Features related to earthwork calculation were limited.

In the late 90s, CAD technologies were improved significantly by incorporating various algorithms to represent digital surfaces or to create digital terrain models for highway projects. One of the revolutionary algorithms was the triangulated irregular network (TIN), which uses irregularly distributed nodes and lines with three-dimensional coordinates (x, y, and z) to form a network of non-overlapping triangles for a surface. An advantage of using a TIN for surface mapping and analysis is that points required for a TIN can be randomly distributed geographically, making data inputs flexible. The TIN algorithm determines which points are most necessary for the accurate representation of terrains or surfaces.

The design software in the CAD-based period has incorporated TIN technologies and provided a set of digital terrain model (DTM) features to help highway designers create 3-D surfaces for various types of highways. Besides DTM features, some design software such as AutoCAD Civil 3D has incorporated additional features that automatically check design standards and criteria, relieving the burden on highway designers to ensure conformance of their design to standards and criteria.

The CAD-based geometric design is the current practice in producing construction detail plans and estimating costs for highway projects. Compared to hand-based design, CAD significantly reduces the design time for a highway project. Using 2-D design drawings from CAD-based design software, highway contractors can easily lay out highway centerlines, cross sections, and project limits through various types of stakes.

Recently CAD-based geometric design has been challenged by GPS technologies. Since 2005, GPS technologies have been applied in highway construction, thanks to their abilities to get high-accuracy elevation measurements. GPS-based staking has become widespread, significantly reducing the time required to implement construction details (that are described by paper plans) on physical ground. It is anticipated in the near future that construction survey will eliminate paper media. Construction plans produced from CAD-based design software can be directly downloaded to GPS-based construction equipment for grading and staking.

GPS-based 3-D laser scanning technology has imposed a new challenge for geometric design practice. Three dimensional laser scanning now becomes a new way to create accurate survey measurements in complicated highway environments. With several "scan" setups, a complete 3-D model can be made of existing highway conditions. The model is then used to create 2-D design drawings, 3-D computer models, and final survey documents.

It is anticipated that the GPS-based 3-D laser scanning technology will change not only highway construction procedures (by replacing traditional staking process), but also the way we design highways. The current 2-D-based geometric design methodology could eventually be replaced by a new 3-D approach that automates design drawings into construction equipment for three-dimensional highways.

1.2 Highway Geometric Design Standards

Three-dimensional highway design requires a good understanding of the interactive relationships among drivers, pedestrians, vehicles, and highways. The characteristics of drivers, pedestrians, and vehicles serve as the controls and criteria for determining the highways' physical dimensions.

The American Association of State Highway and Transportation Officials (AASHTO) provides a national policy on geometric design of highways and streets (2004 version) that defines comprehensive characteristics of drivers, pedestrians, and vehicles, and provides design guidance based on established practice nationwide. This policy is referred to as the *AASHTO Green Book*. The purpose of this policy is to help highway designers consider the following factors when they design highways:

- Need for a safe and efficient highway system,
- Needs of users, in particular, the low mobility and disadvantaged groups,
- Needs for maintaining the integrity of the environment,
- Adverse effects on natural resources, environmental values, public services, aesthetic values, and community and individual integrity that should be minimized.

The design guidelines included in the *Green Book* consist of recommended practices, design controls, and criteria for freeways, arterials, collectors, and local roads, in both urban and rural settings.

Some state departments of transportation (DOTs) use the *Green Book* to establish their standards for state highways. However, most state DOTs have developed their own standards for highway geometric design. For example, the California State Department of Transportation (Caltrans) provides its own *Highway Design Manual (HDM)* for highway projects in California (see www.dot.ca.gov/hq/oppd/hdm/hdmtoc.htm).

The highway design criteria and policies in the Caltrans *HDM* provide a guide for engineers who do business in California on state highways to exercise sound judgment in applying standards in the design of transportation projects. This manual allows some flexibility in applying design standards to take into consideration the context of the project settings. *HDM* enables designers to create context-sensitive solutions or alternatives, as appropriate, for specific circumstances while maintaining environmental safety and integrity.

The design standards described in the *HDM* are the minimum requirements. All projects shall be designed, to the maximum extent feasible, to meet or exceed these minimum requirements, while taking into account costs, traffic volumes, traffic and safety benefits, right of way, socioeconomic and environmental impacts, maintenance, etc.

The Caltrans *HDM* defines standards in the following order of importance in developing a safe state highway.

1. **Mandatory Standards.** Mandatory design standards are those considered most essential to achieve the overall design objectives. Many standards pertain to legal requirements or regulations such as those embodied in the Federal Highway Administration (FHWA) controlling criteria. Mandatory standards use the word *shall* and are printed in **boldface** type. For example, **Table 201.1 shows the standards for stopping sight distance related to design speed, and these shall be the minimum values used in design.**
2. **Advisory Standards.** Advisory design standards are also important. They allow greater flexibility in projects to accommodate design constraints or to be compatible with local conditions. Advisory standards use the word should and are indicated by underlining. For example, <u>the stopping sight distances in Table 201.1 should be increased by 20 percent on sustained downgrades steeper than 3 percent and longer than one mile.</u>
3. **Permissive Standards.** All standards other than mandatory or advisory, whether indicated by the use of *should* or *may*, are permissive with no requirement for application intended. For example, the minimum stopping sight distance is the distance required by the driver of a vehicle, traveling at a given speed, to bring the vehicle to a stop after an object on the road becomes visible. Stopping sight distance is measured from the driver's eyes, which are assumed to be 3½ feet above the pavement surface, to an object ½-foot high on the road.

The standards in Caltrans HDM generally conform to the standards and policies set forth in the AASHTO *Green Book*. It is important to note that AASHTO policies and standards, which are established as nationwide standards, do not always satisfy California conditions since Caltrans standards in many cases are more stringent than those in the Green Book. When standards differ, Caltrans standards govern, except when necessary for FHWA project approval.

1.3 Highway Design Controls and Criteria

Before highway designers design a highway, they shall have a good understanding of controls and criteria associated with the subject project. Generally speaking, there are two types of control and criteria to be considered: 1) general controls and criteria and 2) project-specific controls and criteria.

General Controls and Criteria

1. Functional Classifications

Highway designers shall know the type of highway facility (freeway, ramp, expressway, 2-lane conventional highway, multilane conventional highway, or frontage road) to be designed. Different highway facilities require different sets of controls and criteria. For example, the design speed for an urban freeway in California is in the range of 55–80 mph, while the design speed for an urban arterial street is 40–60 mph. In practice, highway facility type is provided as an input to highway designers when a project is initiated.

2. Design Speed

The choice of design speed for a highway project is influenced principally by facility type, terrain characteristics, economic considerations, environmental factors, traffic volume, and whether the project area is rural or urban. The design speed for a highway in level or rolling terrain can be justified as higher than one in mountainous terrain. A rural highway justifies a higher design speed than an urban highway. Typically design speed is provided to highway designers prior to the design of a highway.

3. Design Vehicles

The design vehicle should be determined when a highway project involves the design of intersections, ramp terminals, or roundabouts. *AASHTO Green Book* recommends that: 1) a passenger car to be selected when a main traffic generator is a parking lot or a series of parking lots; 2) a single-unit truck for intersections of residential streets and park roads; 3) a city transit bus for state highway intersections with city streets that are designated bus routes and that have relatively few large trucks using them; 4) a large school bus or conventional school bus for intersections with low-volume county highways and township/local roads under 400 average daily traffic (ADT) volume and for subdivision street intersections; and 5) WB-20 [WB-65 or -67] truck for freeway ramp terminals and intersections on state highways and industrialized streets that carry high traffic volumes and/or that provide local access for large trucks.

Caltrans defines design vehicles by five groups:

a. STAA[1] Design Vehicle, STAA Vehicle Long Tractor STAA Vehicle 53-FT Trailer

These vehicle types can only be allowed to operate on the STAA Truck Network. They cannot be operated on any other California Legal Network or routes. The STAA Truck and California Legal Networks in Caltrans District 7 can be found at www.dot.ca.gov/hq/traffops/trucks/truckmap/truckmap-d07.pdf.

b. California Legal Vehicles

California Legal Vehicles can be used in the design of all interchanges and intersections on California Legal routes and California Legal KPRA[2] Advisory routes for both new construction and rehabilitation projects.

c. 40-Foot Bus Design Vehicles

Forty-foot bus design vehicles are considered at intersections where truck volumes are light or where truck traffic consists of mostly three-axle units.

d. 45-Foot Bus and Motorhome Design Vehicles

Forty-five-foot bus and motorhome design vehicles are allowed only on the 45-foot bus and motorhome routes.

e. 60-Foot Articulated Bus Design Vehicles

Sixty-foot articulated bus design vehicles are allowed only on the 60-foot articulated bus routes. Caltrans does not provide a list of such bus routes. Local transit agencies should be consulted to obtain the 60-foot bus routes for a proposed project.

---

[1] *The Surface Transportation Assistance Act of 1982 (STAA)*
[2] *Kingpin-to-rear-axle distance*

4. Design Period

Each highway project has its own service life. Caltrans defines *service life* as 20 years after completion of construction of a new highway. For example, a new project is scheduled to finish its construction in 2014. The service life for this project is 20 years from 2014 to 2034.

5. Traffic Volumes

Highways are designed to accommodate design hourly volume (DHV), truck needs, and directional distribution of future year (or 20 years after the completion of construction). Therefore an understanding of future year traffic pattern is required before highway designers conduct geometric design.

The future year DHV is the projected hourly volume used to define the capacity (or the number of lanes) of a highway to be built. In practice, the future year DHV is difficult to estimate directly. Often we rely on the future average daily traffic (ADT) volume and the equation below to estimate the future DHV indirectly:

$$\text{Future Year DHV} = \text{Future Year ADT} \times K \qquad (1\text{-}1)$$

Where	Future Year DHV	Vehicles/Hour
	Future Year ADT	Vehicles/Day
	K	Ratio of DHV to ADT

The future year ADT volume is estimated through either a growth factor method or a travel demand forecasting model. The growth factor method assumes that ADT volume or traffic demand increases year by year at a fixed growth rate within the design period of a highway. With ADT volume or traffic counts that can be directly observed on an existing highway (which is similar to the new highway to be built), the future year ADT volume for the new highway can be estimated as:

$$\text{Future Year ADT} = \text{Existing ADT at the Current Year} \times (1 + r\%)^{20+3} \qquad (1\text{-}2)$$

Where	r	Growth rate
	n	Number of years taken for design and construction of a highway project

The current year is defined as the year when a project is initiated.

For example, a highway project is initiated in 2011. It is anticipated that the project requires three years for the highway design and construction. In other words, construction will be completed in 2014. Therefore, the future year will be 2034. Also assume the ADT volume on a similar, existing highway is observed to be 6,000 vehicles/day, the growth rate is 1%, then the future year ADT volume will be:

$$\begin{aligned} ADT_{2034} &= ADT_{2011} \times (1 + 1\%)^{20+3} \\ &= 6000 \times (1 + 1\%)^{23} \\ &= 7543 \; \textit{Veh/Day} \end{aligned}$$

The travel demand forecasting model, which involves trip generation, trip distribution, modal split, and trip assignment, can predict the future year ADT volume directly. This method is often used when a highway project has sufficient financial resources to support the data collection and analysis required for the model. For example, the travel demand forecasting model recently developed by the Southern California Association of Government is used to estimate the future traffic demand of the High Desert corridor, a 63-mile freeway, which will connect the cities of Palmdale and Victorville, California.

K-factor represents the DHV as a percentage of ADT. Experience has shown that design hourly volume (or the 30th highest hourly volume in a year) is typically between 12% and 18%, with the average being 15%. In practice, K-factor is provided to highway designers when a highway project is initiated. The K-factor, once determined, does not change from year to year within the design period.

Truck volumes should be considered in geometric design since trucks have critical impacts on the capacity of highway facilities. Unbalanced traffic in different directions, which is represented by directional factor (D), should also be considered in geometric design. A directional factor (D) of 60% indicates that traffic in the heavier direction is 60% of the total traffic of both directions.

Design speed, traffic volumes, truck percentage in the traffic flow, and directional factor together form the design designation that controls the highway design. Here is an example of the design designation for a highway:

ADT (2011) = 6,000 veh/day

ADT (2034) = 7,543 veh/day

DHV = 2,500 veh/hr

Directional Factor, D = 60%

Truck Percentage in the flow, T = 12%

Design Speed, V = 70 mph

6. Design Level of Service

Level of Service (LOS) is a performance indicator to descriptively assess highway effectiveness in serving traffic. LOS A describes a free-flow operation. Drivers have a sense of full mobility and comfort between lanes. An example of LOS A can be found late at night in both urban and rural areas.

LOS B describes a reasonably free-flow operation. Free-flow speed is maintained and maneuverability within the traffic flow is slightly restricted. Drivers have a high level of physical and psychological comfort.

LOS C describes an at-or-near tolerable free-flow operation. Changes among lanes are noticeably restricted. Drivers feel comfortable. Traffic flow reaches close to the capacity of the facility.

LOS D describes a traffic flow operation in a decreasing level of comfort. Maneuverability within the traffic flow is more restricted and driver comfort levels decrease.

LOS E describes a traffic operation at capacity. Traffic becomes irregular, flow speed changes rapidly because sufficient gaps to maneuver in traffic stream are barely found, and flow speed normally is below the posted limit.

LOS F describes a breakdown in traffic flow. Flow is forced into stop-go patterns. An example of LOS F is a road in a constant traffic jam.

LOS of a highway varies from time to time. During the AM and PM peak hours, the highway might operate at LOS D, while during the nonpeak hours the LOS could be B or C. Which LOS should be used in geometric design? If LOS A is selected for design, the highway would be overdesigned. If LOS E or F is selected the highway could be underdesigned. AASHTO provides the following reasonable recommendations to guide the selection of design LOS for freeway, arterial, collector, and local roads:

B, B, C, D within the rural level areas,

B, B, C, D within rural rolling areas,

C, C, D, D within rural mountainous areas, and

C, C, D, D within urban and suburban areas, respectively.

Project-Specific Controls and Criteria

Project-specific controls and criteria vary, depending on location and nature of highway projects. For example, the project used in later chapters of this text requires consideration of an undercrossing bridge for the Route 8 freeway that runs on top of a local road called Laguna Canyon Road. Therefore, the vertical bridge clearance in relation to the local road should be considered. Also the project needs to consider lane width, presence or absence of shoulders, grades, lateral clearance, etc.

1.4 Software Use in Highway Geometric Design

Use of software continues to revolutionize the way of doing geometric design. One of the benefits of using design software is enhancement of a highway designer's ability to automate design tasks (such as horizontal alignment design, profile creation, vertical alignment design, cross section design, quantity calculation, and drawings production), implement design controls and criteria through software functions, and prepare information needed in construction drawings.

Table 1-1 provides a review of design software currently used by all 50 states in 2010[3]. Note in the table that InRoads and GeoPAK from Bentley are the official software packages selected by many state DOTs for geometric design, while four state DOTs selected Autodesk's AutoCAD Civil 3D for their geometric design.

[3] *The CAD platform and design software for each state can change from time to time. The list in this book was referenced to the design practice in 2010.*

Name____ Date________________

1.5 Questions

Q1.1 List the three periods the highway geometric design has experienced. Provide a summary of these three periods.

Q1.2 GPS-based staking is popular. Do an Internet review on GPS-based staking practice in the U.S. and write a review on the difference between the GPS-based staking and conventional staking.

Q1.3 The Caltrans *HDM* defines standards in the specific order of importance in developing a safe state highway. Describe this order by providing an example standard for each type of standard.

Q1.4 List and describe general controls and criteria used for highway geometric design.

Q1.5 Assume the current traffic volume (in 2012) is 10,000 veh/day; the growth factor is 2%; and the project service life is 23 years, what is the future year? What is the traffic volume of the future year?

Q1.6 Given the 30th-highest hourly volume is 300 veh/hr and the average daily traffic volume is 2,000 veh/day for an existing highway, what is the K-factor for this highway?

Q1.7 What is the design LOS for a freeway in a suburban area according to AASHTO recommendations?

Q1.8 What is the current CAD platform and design software adopted by California State Department of Transportation (Caltrans) for highway geometric design?

Table 1-1 Software Use in State DOTs for Geometric Design

State	CAD Platform	Design Platform	Design Standards
Alabama	Microstation	InRoads	ALDOT Standard Specifications for Highway Construction
Alaska	AutoCAD	Civil 3D	ALDOT Standard Specifications for Highway Construction
Arizona	Microstation	InRoads	ADOT Roadway Design Guidelines
Arkansas	Microstation	InRoads	Arkansas State Highway and Transportation Roadway Design Plan Development Guidelines
California	Microstation	CAiCE	Caltrans HDM
Colorado	Microstation	InRoads	Colorado DOT HDM
Connecticut	Microstation	InRoads	ConnDOT HDM
Delaware	Microstation	InRoads	Civil 3D
District of Columbia	Microstation	GeoPAK	DDOT HDM
Florida	AutoCAD	CAICE	FDOT Roadway Design Standards
Georgia	Microstation	InRoads	State Roadway Design Manual
Hawaii	AutoCAD	Civil 3D	AASHTO Green Book
Idaho	Microstation	InRoads	State Roadway Design Manual
Illinois	Microstation	GeoPAK	Bureau of Design & Environment Manual
Indiana	Microstation	MX/InRoads	Indiana Design Manual
Iowa	Microstation	GeoPAK	Iowa DOT Design Manual
Kansas	Microstation	GeoPAK	KDOT Road Design Manual
Kentucky	Microstation	InRoads	Kentucky DOT Highway Design Manual
Louisiana	Microstation	InRoads	Louisiana Road Design Manual
Maine	Microstation	MX	Maine DOT Highway Design Guide
Maryland	Microstation	InRoads	AASHTO
Massachusetts	AutoCAD	Civil 3D	Mass DOT Project Development & Design Guide
Michigan	Microstation	GeoPAK	Michigan DOT Road Design Manual
Minnesota	Microstation	GeoPAK	Minnesota Road Design Manual
Mississippi	Microstation	GeoPAK	Mississippi DOT Roadway Design Manual

Missouri	Microstation	GeoPAK	MoDoT
Montana	Microstation	GeoPAK	MDT Design Manual
Nebraska	Microstation	GeoPAK	Nebraska DOT Roadway Design Manual
Nevada	Microstation	InRoads	Nevada DOT Design Standards
New Hampshire	Microstation	InRoads	NHDOT Standard Specification 2010
New Jersey	Microstation	InRoads	NJDOT Roadway Design Manual
New Mexico	Microstation	InRoads	AASHTO Green Book
New York	Microstation	InRoads	AASHTO Green Book
North Carolina	Microstation	InRoads	The New York Street Design Manual
North Dakota	Microstation	InRoads	AASHTO Strategic Highway Safety Plan
Ohio	Microstation	GeoPAK	Ohio DOT HDM
Oklahoma	Microstation	GeoPAK	AASHTO Green Book
Oregon	Microstation	InRoads & GeoPAK	Oregon DOT HDM
Pennsylvania	Microstation	InRoads	PennDOT Design Standards
Rhode Island	AutoCAD	Civil 3D	SCDOT Design Standards
South Carolina	Microstation	GeoPAK	SCDOT Design Standards
South Dakota	Microstation	InRoads	SDDOT Design Standards
Tennessee	Microstation	GeoPAK	Tennessee DOT Roadway Design Guidelines
Texas	Microstation	GeoPAK	Texas DOT Roadway Design Manual
Utah	Microstation	InRoads	Roadway Design Manual of Instruction
Vermont	Microstation	InRoads	Vermont State Design Standards
Virginia	Microstation	InRoads	VDOT Design Manual
Washington	Microstation	GeoPAK	WSDOT Design Manual
West Virginia	Microstation	InRoads	AASHTO Green Book
Wisconsin	AutoCAD	Civil 3D	WIDOT Facilities Design Manual
Wyoming	Microstation	GeoPAK	WYDOT Road Design Manual

Chapter 2

Transportation Project Development Process

LEARNING OBJECTIVES

After completing this chapter you should know:

1. How a transportation project is developed from its feasibility study to the completion of construction;
2. How engineering requirements are met by a transportation project;
3. How environmental laws and regulations are applied in a transportation project;
4. How management approval procedures are applied in Caltrans for a transportation project.

SECTIONS

How do we build transportation projects? Elected officials and legislators are interested in this question since they make policies and regulations with which transportation projects must comply. Transportation agencies and governmental officials need to know how transportation projects are developed so they can allocate resources and engineering techniques to solve transportation issues. Transportation engineers and consultants pay close attention to transportation projects so they can be actively involved in the project development process. The general public is interested in the benefits and environmental impacts of transportation projects.

2.1 Overview of Transportation Project Development Process

There are many project development processes and procedures that describe how transportation projects are built. These processes and procedures differ in format and content among transportation agencies. However, they all share the following key phases:

» Needs identification and assessment
» Project programming (or resource allocation to potential transportation projects)
» Project initiation
» Preliminary engineering and environmental analysis
» Final engineering and environmental analysis
» Project procurement
» Project construction
» Project close-out

For example, the Pennsylvania State Department of Transportation (PennDOT) has its own transportation development process that consists of:

1. **Project scoping and needs analysis.** The project scoping and needs analysis phase monitors transportation systems, identifies existing or potential deficiencies, and assesses transportation improvements needs.
2. **Prioritization and programming.** The prioritization and programming phase requires PennDOT to coordinate with regional/local planning organizations to review and evaluate planning data, assess resources available to fund planned projects, and prioritize and program projects for implementation.
3. **Preliminary design.** The preliminary design phase involves a task by which engineering and environmental studies are conducted to determine alternatives to a transportation improvement project or a new project. Environmental impacts are avoided, minimized, or mitigated through design.
4. **Final design.** The final design phase consists of detailed plans, specifications, and cost estimates for the best project alternative. The best alternative is further developed and approved, specific environmental permits are obtained, and mitigation commitments are incorporated into the best alternative.
5. **Construction.** Selected contractors in the construction phase construct the best alternative, along with required mitigation measures. New or improved transportation facilities are then opened to the public after the completion of the project alternative and a thorough final inspection.
6. **Maintenance.** In the maintenance phase, PennDOT monitors the operational performance of transportation facilities and conducts maintenance activities as needed to ensure the safe and efficient operation of the transportation facilities.

In this chapter, we use the Caltrans project development process (instead of the PennDOT process) to explain the key phases involved in transportation project development. The reason for selecting the Caltrans process is that the remaining chapters follow Caltrans design standards and practices, and the tutorial project is located in California.

2.2 Caltrans Transportation Project Development Process

Caltrans transportation project development process implements Caltrans' mission of improving mobility across California and, for each project, the process defines activities that commence with the project initiation and end with the assembly of the final project records after the project construction is completed. Different from other DOTs, Caltrans ties the project development process to the legal requirements of environmental laws and regulations. It incorporates engineering requirements and management approval steps with an environmental process.

The current Caltrans project development process is an implementation of the conventional design-bid-build method for construction of new highways or improvement of existing highway facilities. It consists of seven phases: System Planning, Project Initiation, Project Approval/Environmental Document (PA/ED), Plans, Specifications and Cost Estimates (PS&E), Contract Procurement, Construction, and Project Close-Out (see Figure 2-1).

Needs Assessment and System Planning

Before a project is initiated as a solution to a transportation problem, Caltrans must have a complete needs assessment of the state highway system, which consists of about 15,000 centerline miles and more than 49,000 lane miles. The assessment, being the critical task in transportation system planning, includes the needs evaluation for highway system preservation and capacity enhancement, as well as modal connectivity, accessibility, and safety improvements.

1. State Highway System Preservation

Caltrans performs an Automated Pavement Condition Survey (APCS) every fiscal year to evaluate severities and the extent of pavement distress for the entire state highway network. The APCS collects raw data, including longitudinal and transverse profiles, images of pavement surfaces, and right-of-way images using van-mounted GPS, inertial referencing systems, and distance measuring instruments. APCS provides a clear picture of the serving conditions of the state highway system and establishes an inventory of existing surface conditions on all the state routes.

Caltrans uses Ground Penetrating Radar (GPR) technology to create a pavement structure inventory that records pavement layer thicknesses and material types continuously along highways. The inventory also stores geographic coordinates of nonpavement assets such as bridges, ramps, county lines, etc. Additionally, Caltrans uses Global Navigation Satellite System (GNSS) technology as the primary tool for data collection of geographic locations.

With data collected from APCS and GPR, Caltrans uses a pavement management system called PaveM to develop and implement a proactive approach to prioritizing, preserving, rehabilitating, and maintaining existing highway pavement.

PaveM employs algorithms to develop pavement management strategies and create annual reports of pavement conditions for the California Transportation Commission (CTC). PaveM also provides information for the State Highway Operations and Protection Program (SHOPP).

2. Capacity Enhancement, Accessibility Improvement, and Connectivity Development

Caltrans conducts a set of system planning activities to assess travel demand and the capacity of the existing highway system. It also identifies needs for capacity enhancement, accessibility improvement, and connectivity development. The result of this assessment is a list of "planning concept and scope" documents that identify and clarify transportation system problems and issues and that recommend practical solutions or improvement concepts to achieve Caltrans' mobility goals.

3. Safety Improvement

Caltrans implements the Highway Safety Improvements Program (HSIP) to identify safety needs and reduce fatalities, serious injuries, and property damages on all public roads.

Caltrans compiles all the needs that resulted from the above assessment process, develops project concepts or strategies to address the needs, and defines the scope, schedule, and cost of potential projects.

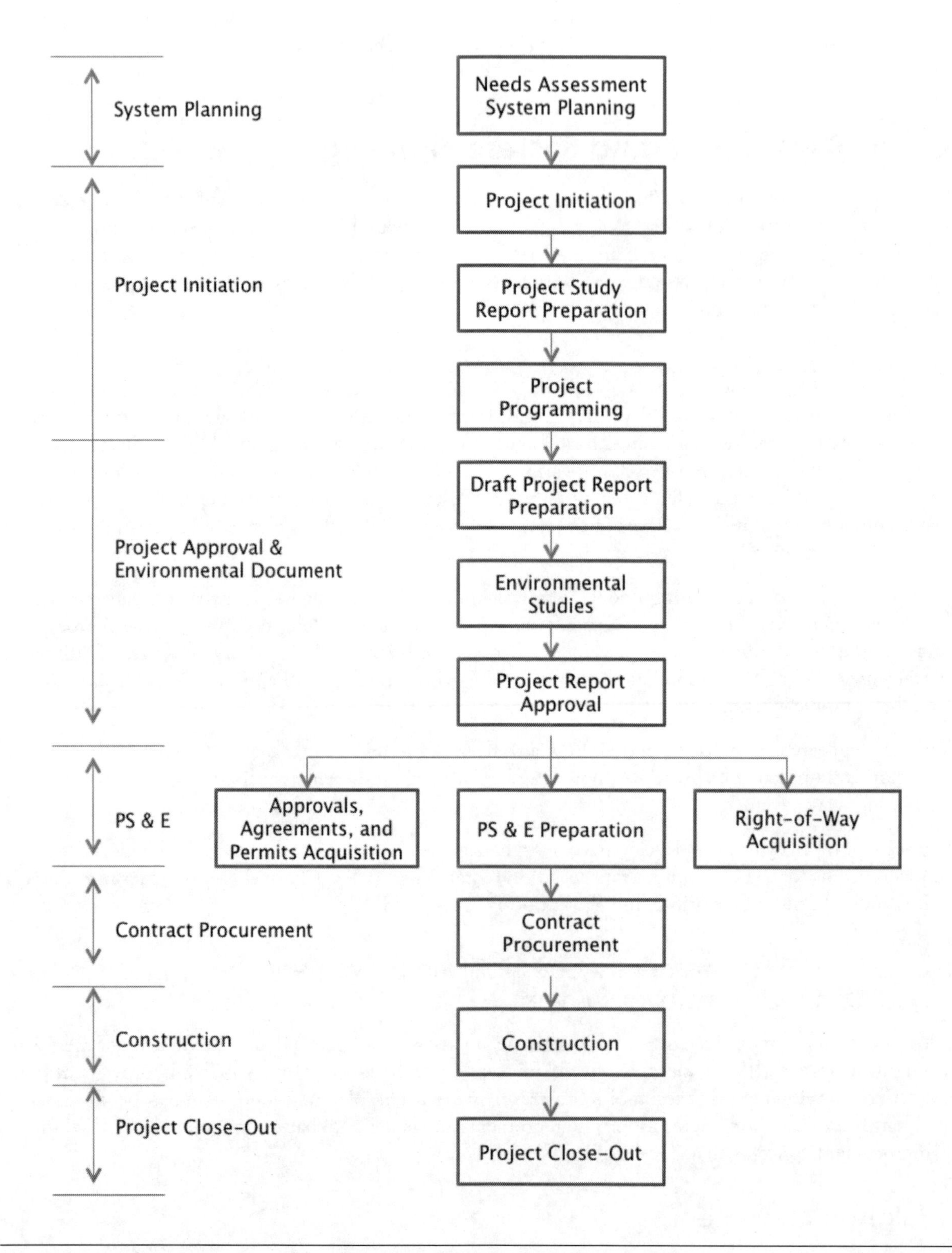

Figure 2–1 Caltrans Transportation Project Development Process

Project Initiation

The project initiation phase, following the needs assessment and system planning phase, produces a Project Initiation Document (PID) for each transportation project.

When a transportation project is initiated, a project manager (PM) should be designated. The PM is responsible for securing an expenditure authorization (EA), organizing a project development team (PDT), preparing a project work plan, and developing a Project Study Report (PSR[1]) or a Project Initiation Document.

To develop the PSR or the PID for a project, the Project Development Team (PDT) must develop viable alternatives (including a "no-build" alternative) that meet the project's purposes and needs. The alternatives should be context sensitive in avoiding, minimizing, or mitigating adverse environmental impacts.

Caltrans is mandated by federal law to conduct a Value Analysis (VA) study on all federal-aided projects with a total project cost of $25 million or more. The purpose of the value analysis is to identify the alternatives with the best value for the majority of the project stakeholders. The VA study provides a quantitative and qualitative assessment on competing alternatives for a federal project.

The PDT must conduct safety and constructability reviews on all alternatives before the PSR can be considered for approval.

When a PSR is approved, the project is eligible to compete for state and federal funds.

Project Approval/Environmental Document (PA/ED)

The PA/ED phase consists of three tasks: 1) Draft Project Report (DPR), 2) Public Hearing, and 3) Project Report (PR).

1. Draft Project Report

The DPR task involves engineering studies and environmental studies, as well as DPR preparation, circulation, and approval.

Engineering studies concentrate on the use of design standards in evaluating the alternatives identified in the PID phase. The purpose of the engineering studies is to support the environmental evaluation and the project approval.

The PDT first identifies physical features such as terrain (flat, hilly, mountainous), material (dirt, sand, rock), improvements (buildings, drainage, structures, and utilities), and environmental concerns (wetlands, archeological sites, etc.) and it establishes physical controls and constraints. The PDT also needs to scope the project features such as project period, design LOS, design speed, design hourly volume, truck percentage in the vehicle stream, etc. The PDT refines the geometric drawings (developed in the PID phase) in greater detail for the environmental evaluation of the project.

Environmental studies, which are parallel to engineering studies, focus on the environmental impacts of the project alternatives. They address environmental issues (such as air and water quality, noise impacts, wetlands, coastal zone infringement, floodplains, wildlife and plants, historic and cultural resources, social and economic changes, park lands and recreational areas, hazardous waste, energy, and visual effects) and they evaluate how environmental, social, and economic impacts can be minimized or mitigated.

---

[1] *PSR is a special PID that defines alternatives and provides cost estimates and a project schedule for project alternatives.*

There are three levels of environmental impact analyses the PDT may determine for each project alternative: 1) Finding of No Significant Impact (FONSI), 2) Environmental Assessment (EA), and 3) Environmental Impact Study (EIS)/Environmental Impact Report (EIR). When a FONSI is concluded for all the project alternatives, the project can proceed without having to complete a draft EIS/EIR. When environmental impacts are not easily identified as significant, Caltrans must prepare a smaller, shorter document called the Environmental Assessment (EA). The finding of the Environmental Assessment determines whether an EIS/EIR is required. If the Environmental Assessment indicates that no significant impact is likely, then the PDT proceeds with a FONSI. Otherwise, Caltrans must conduct a full-scale EIS/EIR. The EIS/EIR describes all the negative environmental impacts of alternatives and provides measures to avoid, reduce, or mitigate these impacts for the project.

The results of the engineering studies and environmental studies are documented in the Draft Project Report (DPR). The results include the project's needs, the summary of key points in the Draft Environmental Document (DED), and the findings of scope, cost, and overall impacts of alternatives. The DPR is used to decide if the project will proceed to the public hearing task.

2. Public Hearing

Public hearings are required for a transportation project with significant impacts. The hearings provide a forum that allows Caltrans to present to concerned communities and stakeholders the project's purposes and needs, major issues, alternative locations, design features, and potential social, economic, and environmental effects. The hearings help Caltrans obtain public comments and ensure that transportation decisions are consistent with the goals and objectives of federal, state, and local entities.

All comments and recommendations are documented and addressed in the Final Environmental Document (FED).

3. Project Report

A project report is created by updating the DPR after comments and recommendations from public hearings are incorporated into alternatives and a Least Environmentally Damaging Practicable Alternative (LEDPA) is selected as the project's preferred alternative. The Project Report (PR) is a decision document.

Plans, Specifications, and Cost Estimates (PS & E)

The Plans, Specifications, and Cost Estimates phase involves a set of design activities for the preferred alternative. These activities include: 1) preliminary plan development, 2) final plan preparation, and 3) preparation of a PS & E package for approval.

1. Development of Preliminary Plans

Preliminary plans consist of: 1) geometric base maps, 2) cross sections, 3) bridge site data, 4) right-of-way (R/W), and 5) skeleton layouts. To develop these plans, the PDT must collect such information as mapping and surveys data, materials report, drainage report, and traffic data. The mapping and surveys data are critical to the geometric base maps that show existing topography and proposed engineering features. The materials and drainage reports help the PDT determine right-of-way needs and design drainage facilities. The traffic data defines the nature of traffic plans.

The PDT must also finalize horizontal and vertical alignments as well as interchange and intersection details developed during the PID and PA/ED phases. The final design drawings provide information to help the PDT develop cross sections for earthwork calculation. The design drawings show the structure site information for bridges, pumping plants, pedestrian structures, retaining walls, noise barriers, culverts, and other highway- and transit-related facilities.

The PDT compiles geometric base maps, cross sections, and right-of-way details into skeleton layouts for use in developing final plans and a PS & E package.

2. Preparation of Final Plans

Final plans consist of final project plans, maps for right-of-way acquisition, bridge general plans, and quantities.

The project plans include layout sheets, typical cross sections, profile sheets, construction details, drainage sheets, and quantity summary sheets. The maps for right-of-way acquisition include right-of-way requirements needed to produce appraisal maps. The bridge general plans provide a description of bridge type, dimensions, aesthetics treatment, and cost estimates. The plans contain design detail information needed for the development of a bridge PS & E.

Project cost estimates represent the total costs for construction of the preferred alternative. Determining the costs requires unit price analysis for individual contract items.

3. Preparation of a PS & E Package

The Project Engineer (PE) of the PDT works with other Caltrans units to use the final plans to prepare a PS & E package. After the PS & E package has been reviewed (in terms of safety, constructability, standards, funding availability, etc.) and approved, a right-of-way certificate can be issued and contract documents can be prepared.

Contract Procurement

The Division of Engineering Services Office of Engineer (DES-OE) receives the PS & E package from the PDT, develops bidding documents, and requests for proposals through Caltrans' Centralized Bidder Inquiry System. Before the Request for Proposal (RFP) is advertised, the DES-OE works with the PDT to establish a set of criteria and a scoring or ranking system for proposal evaluation.

Civil engineering contractors doing business in California can submit their proposals to respond to the call for submission. One of the contractors wins the construction project. The project then proceeds to the Construction phase.

Construction

The winning contractor executes the contract and works on the project construction. Construction activities may include fencing, dust control, mobilization, construction area traffic control and management, existing highway facility removal, clearing and grubbing, watering, earthwork, erosion control and highway planting, pavement construction, and finishing roadways.

Caltrans provides resident engineers to oversee the construction process. When construction activities cannot be consistent with those specified with the contract documents (the project plans and specifications), change orders may be issued to change the construct scope of work.

Project Close-Out

The project moves into this phase when the construction work is completed and the project facilities have been accepted and approved for operation. The PDT archives all the project records and documents (including change orders) into the Caltrans project archiving system as "as-built" documents for later projects.

Name____ Date______________

2.3 Questions

Q2.1 List the common key phases used to develop a transportation project by transportation agencies.

Q2.2 Visit the PennDOT website and provide a summary of the activities required to develop a transportation project in Pennsylvania.

Q2.3 What is the Caltrans mission statement?

Q2.4 What are the unique features of the Caltrans transportation project development process, as compared with those of other state departments of transportation?

Q2.5 List the key phases in the development of a Caltrans transportation project.

Q2.6 Visit the Caltrans website to summarize the collection of pavement data.

Q2.7 Describe the features of the Caltrans pavement management system (PaveM).

Q2.8 What are the key types of needs that should be identified for the implementation of Caltrans mission?

Q2.9 What is a Project Study Report (PSR)? Is a PSR one type of Project Initiation Document (PID)?

Q2.10 Visit the Caltrans website and provide a summary of steps required to develop a PSR for a transportation project.

Q2.11 What is the PA & ED phase?

Q2.12 Describe the key tasks of the PA & ED phase.

Q2.13 List three levels of environmental impact analyses.

Q2.14 Why are public hearings required when a project needs an EIR/EIS document?

Q2.15 What is PS & E?

Q2.16 Which Caltrans unit is responsible for the contract procurement?

Q2.17 Why must Caltrans archive project documents when the project construction is completed?

Chapter 3 Highway Geometric Design Process

LEARNING OBJECTIVES

After completing this chapter you should know:

1. The highway geometric design process and basic design elements that include project site surface, horizontal alignment, profile and vertical alignment, typical cross section, roadway design, and earthwork calculation;
2. Interactions among design elements that ensure a highway (to be designed) meets or exceeds design criteria, controls, and standards;
3. Prerequisite requirements for each design element.

SECTIONS

Geometric design is critical to the transportation project development described in Chapter 2. It is one of the most important tasks in the Project Initiation Document (PID), the Project Approval/Environmental Document (PA/ED), and the Plans, Specifications, and Cost Estimates (PS & E) phases of the Caltrans project development process.

3.1 Highway Geometric Design Process

The geometric design process of a highway consists of basic design elements, including project site surface generation, horizontal alignment design, profile and vertical alignment design, typical cross section design, roadway design, and earthwork calculation (see Figure 3-1).

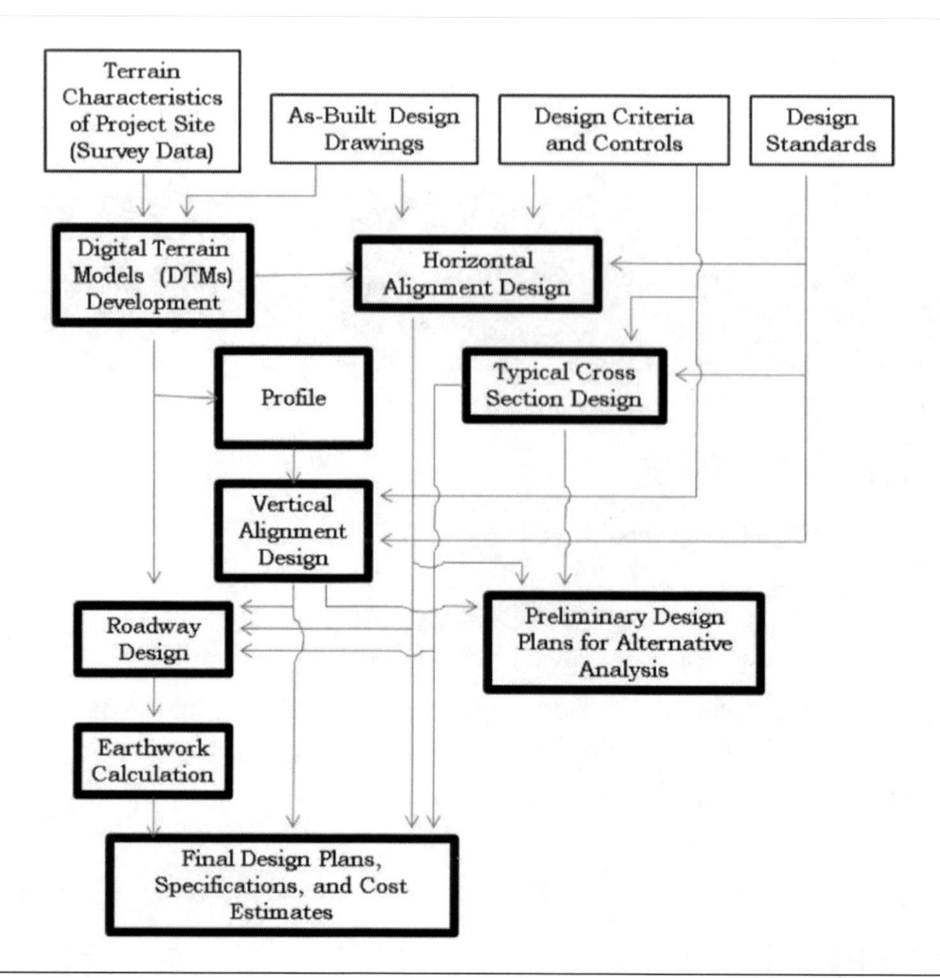

Figure 3-1 Highway Geometric Design Process

1. Project Site Surface Generation

Generating the surface of a project site is the first step in the highway geometric design process. Using computers, designers build digital surfaces that represent the existing ground at project sites. Digital surfaces assist designers in understanding the terrain characteristics of project sites through three-dimensional mathematical models.

There are two types of data required to generate the project site surfaces: engineering survey data and as-built drawings. Engineering survey uses total stations and Global Positioning System (GPS) technologies to measure existing facilities at project sites. These measurements, represented by a series of (x, y, z) points, are entered into highway design software packages such as Bentley's InRoads and Autodesk's AutoCAD Civil 3D to create digital surfaces.

As-built drawings supplement engineering survey data to further define the types of existing facilities (highways, culverts, bridges, water streams, etc.). These drawings, overlaid with survey data, help designers refine digital surfaces by incorporating existing terrain features such as breaklines (water streams, mountain ridges, existing highways, and water channels) and boundaries (water basins and subbasins, buildings, and existing roadside facilities) into the surfaces.

Digital surfaces provide basic terrain characteristics of project sites for horizontal alignment design, profile and vertical alignment design, and roadway design.

2. Typical Cross Section Design

Typical cross section design is an independent step that parallels or follows the project site surface generation step. It involves the determination of traveled ways, medians, curbs, and gutters, as well as side slopes on tangent segments of a highway (see Figure 3-2).

Design criteria, controls, and design standards govern the design of cross section layouts. Designers need to determine the number of lanes before they design traveled ways and shoulders. Designers also determine the width of medians based on terrain characteristics, project site location (urban, suburban, or rural), land availability, right-of-way, and future expansion to accommodate other transportation facilities, including High Occupancy Vehicle (HOV) lanes, High Occupancy Toll (HOT) lanes, mass transit facilities, etc.

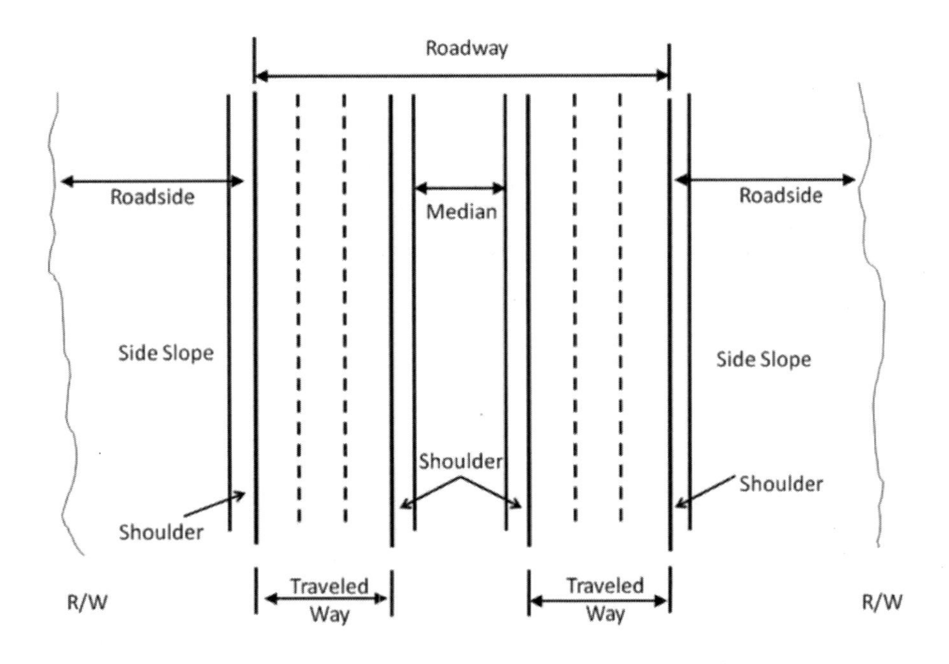

Figure 3-2 Elements of Cross Section Design

Side slope and its width are important to roadside safety, erosion control, and right-of-way (R/W) acquisition of a project. Designers shall make side slopes as flat as possible and ensure roadsides are safe recovery areas for errant drivers.

Cross section template and layout is one of the required design elements in the engineering preliminary study (or the alternative analysis) of a project. They are the inputs to the roadway design step that produces final design plans and construction details for PS & E packages.

3. Horizontal Alignment Design

Horizontal alignment design is a step that follows the project site surface generation step. It defines the highway's centerline or the station line by a series of tangent lines and curves. One of the challenging tasks in horizontal alignment design is to determine superelevation transition from tangent to curve and from curve to tangent.

Horizontal alignment design is a repetitive, trial and error process. It involves tasks to identify alternatives that meet the requirements of design criteria, controls, and standards.

Designers select an engineering feasible and environmentally sound alternative as the best choice. The best alternative shall be well coordinated with the vertical and cross-sectional features of the planned highway.

4. Profile and Vertical Alignment Design

Creating the profile and vertical alignment(s) is the step after the highway's horizontal alignment and project surface are created.

A profile or a longitudinal profile is the result of cutting a project site surface along a horizontal alignment with a saw. It represents the existing elevation of the horizontal alignment.

The vertical alignment design establishes a series of grade lines connected by parabolic curves. There are many vertical alignment alternatives for a highway. Designers attempt to lay out vertical alternatives that meet the design criteria, controls, and standards, while considering the context of project site surfaces and as-built drawings. Vertical alignments cannot be finalized until they have been well coordinated with their corresponding horizontal alignments and cross-sectional highway features.

5. Roadway Design

Roadway design is a step that assembles cross-section template(s), horizontal alignment(s), and vertical alignment(s) to create a three-dimensional corridor for the highway. The corridor is a 3-D design surface that rolls the cross-section templates along the horizontal and vertical alignments.

Designers rely on design software (InRoads or AutoCAD Civil 3D) to create the highway corridors. No design criteria, design controls, and design standards are needed for these corridors.

Once a corridor is created, a series of transverse profiles, which show the results of cutting the existing and design surfaces along the cross section of each station, can be established. The transverse profiles consist of elevations of the design and existing surfaces.

Designers use transverse profiles to determine earthwork. If an existing surface profile line in a transverse profile is below a design surface profile line (see Figure 3-3), the excavation or cut is required. If an existing surface profile line is above a design surface profile line, the embankment or fill is required.

6. Earthwork Calculation

With transverse profiles, designers can calculate earthwork between two consecutive stations using the average end area method. Designers also can create mass hauling diagrams to facilitate the determination of cut/fill sections, earthwork balance lines, hauling direction, distance, and volume, as well as the amount of waste/borrow.

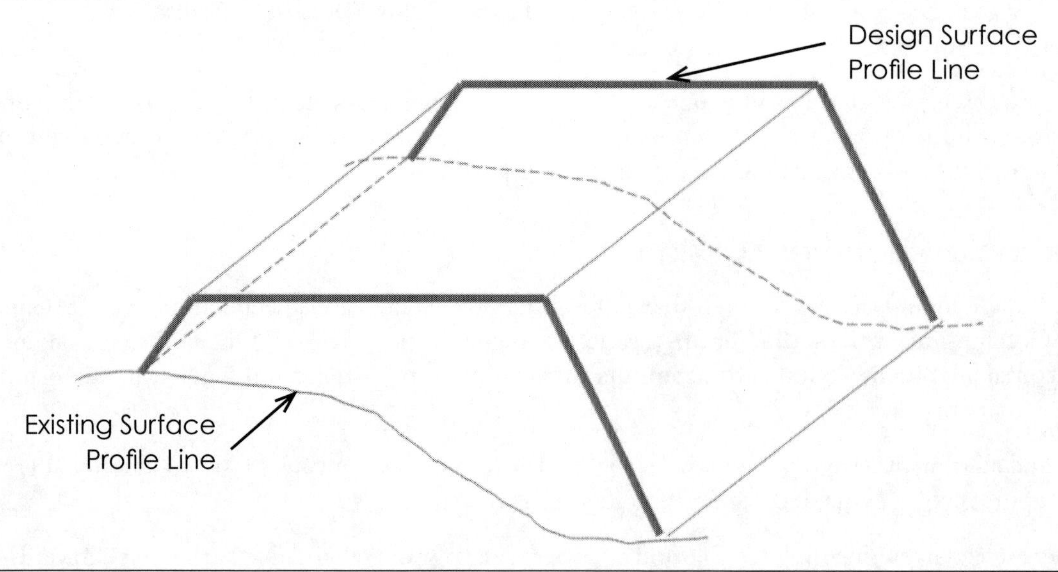

Figure 3-3 Transverse Profile

Name____ Date______________

3.2 Questions

Q3.1 Which types of data are required before the geometric design for a highway is conducted?

Q3.2 Which step of the highway geometric design process should be conducted first?

Q3.3 List the types of data required for the horizontal alignment design.

Q3.4 Can we design a vertical alignment without having horizontal alignments created?

Q3.5 Illustrate the roadway and roadside parts of a highway.

Q3.6 What is a longitudinal profile?

Q3.7 What is the difference between a longitudinal profile and a transverse profile?

Q3.8 List the types of inputs that should be prepared to create a corridor.

Q3.9 Is excavation work needed when an existing surface profile line is above a design surface profile line between two consecutive stations?

Chapter 4
Tutorial Project

LEARNING OBJECTIVES

After completing this chapter you should know:

1. The scope of the tutorial project;
2. The data and the design requirements related to the tutorial project.

SECTIONS

Caltrans, in partnership with the Federal Highway Administration (FHWA) and the Riverside County Transportation Commission (RCTC), proposes to improve accessibility and traffic conditions along Dillon Road, the corridor between the cities of Palm Springs and Coachella. A new state highway (Route 8[1]), parallel to Dillon Road and approximately 10 miles north of I-10, is planned.

Along Route 8 there are three interchanges planned for the corridor. One of the interchanges is Route 8 with Laguna Canyon Road. This interchange is located within the City of Coachella (see Figure 4-1).

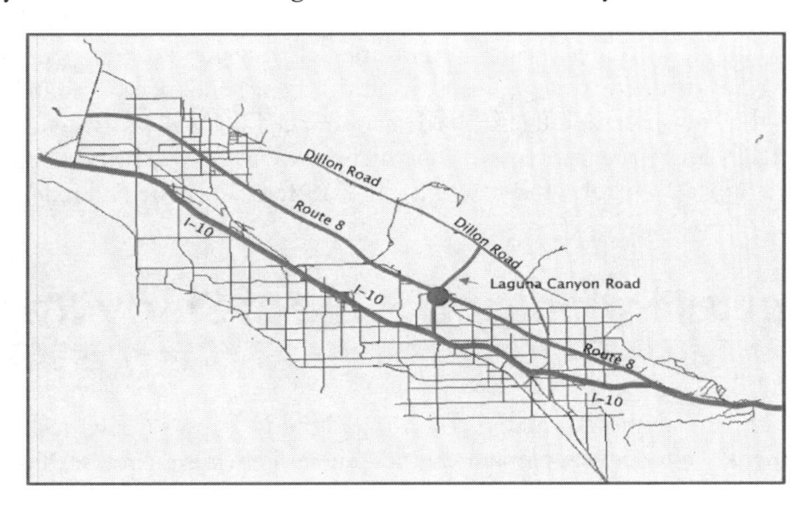

Figure 4-1 Route 8 Project Area

[1] *The tutorial project (or the Route 8 project) is a fictitious project that mimics a real Caltrans project for accessibility improvement and traffic congestion reduction of a corridor. The Dillon corridor is selected subjectively to show that the geometric design procedures and steps involved in this textbook can be applied to any other interchange project.*

4.1 Purpose and Need

State Highway 8 (or Route 8) is a proposed 20-mile transportation corridor that will relieve the congestion of the east-west traffic in Riverside County between the San Jacinto and Coachella Valley areas and will help meet transportation needs through 2035.

By 2035, the population in the regions of San Jacinto and Coachella Valley is expected to reach 1 million. The growing population and development will increase traffic volumes to a level that could cause the overall mobility and traffic flow to decline substantially. All the major transportation routes (particularly I-10) in the region will experience significant congestion.

Route 8 will play a critical role in the regional economy of Coachella Valley and Palm Springs. It will carry heavy commuter and truck traffic inside and outside the regions. It will help maintain and enhance the quality of life in western Riverside County.

The Route 8 project will:

1. Improve I-10 traffic flow and mobility within the regions of Coachella Valley and Palm Springs by diverting traffic onto Route 8.
2. Improve accessibility within the region through three interchanges on Route 8.
3. Provide an "energy" corridor for easy access to the solar industry that will be established around the Joshua Tree National Park.
4. Enhance the operational efficiency of the highway facilities (such as Dillon Road).

4.2 Project Programming

Programming is the process by which Caltrans identifies funds for a project, based on a projection of revenues expected to be available at a specific time in the future.

Caltrans has established a cooperative agreement with the RCTC and programmed the Route 8 project with sufficient funds through the State/Local Transportation Partnership Program (SLTPP). The preliminary design (including the geometric design illustrated in this book) and environmental studies for Route 8 are funded through the Transportation Uniform Mitigation Fee (TUMF) of Riverside County. The final engineering design, right-of-way acquisition, and construction will be funded by a combination of Measure A (Riverside County's ½¢ sales tax for transportation), TUMF fees, and state and federal dollars.

4.3 Project Initiation and Project Study Report

The Route 8 project has already been initiated. The Project Study Report (PSR) task for this project is being conducted to identify alternatives, prepare engineering and environmental studies for each alternative, and document the analysis results in the Project Initiation Document (PID).

There are four alternatives (Route 8 alignment and interchange layouts) that have been considered for PSR approval. Figure 4-1 illustrates one of the alternatives. In this alternative, the Laguna Canyon Road is designed to go under Route 8.

The spread diamond layout (or Type L-2 as defined in Caltrans *HDM*) is selected in this alternative for the Route 8/Laguna Canyon Road interchange. This alternative provides ramp grades and a left-turn storage, and is flexible enough for constructing loop ramps if required in the future.

This textbook focuses on the engineering study of the spread diamond interchange.

4.4 Project-Related Data

The project data used in this book can be downloaded from the website www.csupomona.edu/~xjia/GeometricDesign/. The **Book.zip** file on the website contains all the files, chapter by chapter, for the book. The website also contains data files required for each chapter. For example, www.csupomona.edu/~xjia/GeometricDesign/Chp4 contains two files, **Tutorial.dgn** and **Tutorial2D.dgn**, in the **Master Files** folder. The website is later referenced to the book website.

You are encouraged to create two project folders called **Book** and **Tutorial** in **C:/Temp/** folder of your computer. You need to download all the files on the book website into the **C:/Temp/Book**[2] folder with same subfolder structure. Your **C:/Temp/Book** folder should contain the following files:

Chp4 **.../Chp4/Master Files/** Tutorial.dgn, Tutorial2D.dgn

Chp5 **.../Chp5/Master Files/** Tutorial.dgn, Tutorial2D.dgn

Chp6 **.../Chp6/Master Files/** Tutorial.dgn, OG.dtm

Chp7 **.../Chp7/Master Files/** Tutorial2D.dgn, OG.dtm, Tutorial.alg, Turorial_2D_HA.dgn

Chp8 **.../Chp8/Master Files/** Turorial.dgn, Tutorial2D.dgn, OG.dtm, Tutorial.alg, Turorial_2D_HA.dgn, Tutorial_2D_Profile.dgn

Chp9 **.../Chp9/Master Files/** Turorial.dgn, Tutorial2D.dgn, OG.dtm, Tutorial.alg, Turorial_2D_HA.dgn, Tutorial_2D_Profile.dgn, Tutorial_2D_VA.dgn

Chp10 **.../Chp10/Master Files/** Turorial.dgn, Tutorial2D.dgn, OG.dtm, Tutorial.alg, Turorial_2D_HA.dgn, Tutorial_2D_Profile.dgn, Tutorial_2D_VA.dgn, Dike_E.dgn, Tutorial.itl

Chp11 **.../Chp11/Master Files/** Turorial.dgn, Tutorial2D.dgn, OG.dtm, Tutorial.alg, Turorial_2D_HA.dgn, Tutorial_2D_Profile.dgn, Tutorial_2D_VA.dgn, Dike_E.dgn, Tutorial_Chapter11.itl, CaltransRamps.sup, CaltransMultiLane.sup, Turorial.ird, Route8.dtm, LagunaCanyonRoad.dtm, EBOFF.dtm, EBON.dtm, WBOFF.dtm, WBON.dtm

Chp12 **.../Chp12/Master Files/** Turorial.dgn, Tutorial2D.dgn, OG.dtm, Tutorial.alg, Turorial_2D_HA.dgn, Tutorial_2D_Profile, Tutorial_2D_VA.dgn, Dike_E.dgn, Tutorial_Chapter11.itl, CaltransRamps.sup, CaltransMultiLane.sup, Turorial.ird, Route8.dtm, LagunaCanyonRoad.dtm, EBOFF.dtm, EBON.dtm, WBOFF.dtm, WBON.dtm

Chp13 **.../Chp13/Master Files/** ctcellib.cel, V8eSeed.dgn, Turorial.dgn, Tutorial2D.dgn, OG.dtm, Tutorial.alg, Tutorial_2D_HA.dgn, Tutorial_2D_Profile, Tutorial_2D_VA.dgn, Dike_E.dgn, Tutorial_Chapter11.itl, CaltransRamps.sup, CaltransMultiLane.sup, Turorial.ird, Route8.dtm, LagunaCanyonRoad.dtm, EBOFF.dtm, EBON.dtm, WBOFF.dtm, WBON.dtm, Tutorial_CrossSection, FULPROF.dgn,

.../Chp13/Plan Sheets/ KeyMap.dgn, 01_Title_Sheet.dgn, 02_Typical_Xsec.dgn, 03_Key_Map.dgn, 04_Route8_Layout_1.dgn, ...,4_Route8_Layout_11, 05_Route8_Profile_1.dgn,... 05_Route8_Profile_11.dgn, etc.

---

[2] *C:/Temp/Book and C:\Temp\Book are interchangeable in this textbook.*

You need to create two subfolders, **Master Files** and **Plan Sheets**, in the **C:/Temp/Tutorial/** folder. The **Tutorial** folder and its subfolders do not have any files. You will place all the working design files (to be discussed in chapters 5–13 of the book) into the **Tutorial** folder.

Figure 4-2 shows the study area for the Route 8/Laguna Canyon interchange.

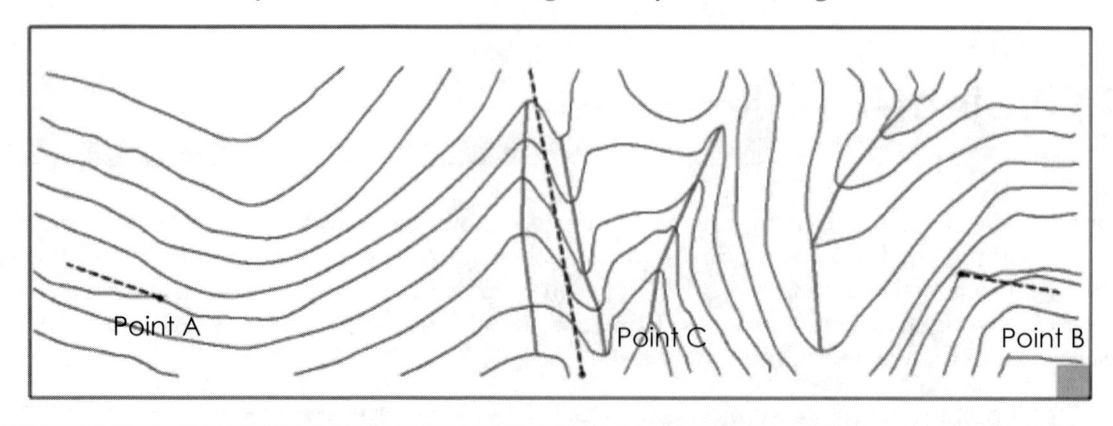

Figure 4.2 Study Area of the Laguna Canyon Interchange

Several design squads are assumed to work simultaneously for the Route 8 project. You are responsible only for the Route 8/Laguna Canyon interchange design between Points A and B.

Design Speed. In this tutorial project, the design speed is provided as follows:

Route 8	60 mph
Ramps	25–50 mph
Laguna Canyon Road	45 mph

Because the project site is in a suburban area with hilly terrain, it is reasonable to assume that traffic at the ramp termini makes a turning movement instead of a full stop. The design speed along the ramps varies depending on the location of vehicles on the ramps.

An acceptable approach is to set design speeds of 25 mph and 50 mph at the ramp terminus and the exit/entrance nose, respectively. The appropriate design speed for any intermediate point on the ramps is based on its location relative to ramp ends. When short-radius curves are considered in horizontal alignment design, they should be consistent with the design speeds related to the curves.

Number of Lanes. A traffic study done for the Route 8 project demonstrates that Route 8 should have two lanes in each direction, and each ramp should have one lane to meet future traffic needs. The Laguna Canyon Road has four lanes, two in each direction. The lane width for Route 8, ramps, and Laguna Canyon Road is 12 ft.

Geometric Control Points. There are three control points that govern the horizontal alignment design of Route 8 and Laguna Canyon Road. The direction and coordinates of these control points are provided below in English units:

Route 8: Point A

Direction: S 70° E[3]

Coordinates[4]: N 7545.9318 ft

E 6889.7638 ft

Elevation: 3083.99 ft

Grade: 2.0% (West to East)

Point B

Direction: S 80° E[5]

Coordinates: N 7874.0157 ft

E 17388.4514 ft

Elevation: 3051.18 ft

Grade: 1.5% (West to East)

Horizontal alignment between Points A and B is only a portion of the entire Route 8 project. It shall be consistent with alignments leading to Point A and coming out of Point B (see Figure 4-2). The directions at Points A and B are the design "agreement" among design teams for Route 8. Each design team shall comply with this agreement.

Points A and B can be used *only* as beginning of curve (BC), ending of curve (EC), or points on tangent segments outside of any curves.

Vertical alignment between Points A and B shall be consistent with vertical alignments leading to Point A and coming out of Point B. The grade leading to Points A (–2.0%) and the grade coming out of Point B (1.5%) are the design "agreement" among design teams for Route 8. Each design team or squad shall comply with this agreement.

Points A and B can be used *only* as beginning of vertical curve (BVC), ending of vertical curve (EVC), or points on grade lines outside of any vertical curves.

Laguna Canyon Road: Point C

Direction: N10°W

Coordinates: N 6561.6798 ft

E 12467.1916 ft

Elevation: 2969.16 ft

Grade: 4.00% (South to North)

---

[3] *S70°E: Bearing of the direction leading to Point A.*

[4] *The coordinates are obtained from the contour map shown in Figure 4-2. These coordinates are not the state plane coordinates normally required by Caltrans*

[5] *S80°E: Bearing of the direction coming out of Point B.*

Horizontal alignment for Laguna Canyon Road shall pass through Point C and shall be in the direction of N10°W.

Vertical alignment for Laguna Canyon Road shall pass through Point C and shall be in the grade of 4% up to the North.

Bridges. Two undercrossing bridges are considered on Route 8. One is on Route 8 Eastbound and the other is on Route 8 Westbound.

These two bridges will be constructed using the Cast-in-Place method. Therefore falsework for the bridge construction should be considered.

Horizontal alignment for Laguna Canyon Road shall pass through Point C and shall be in the direction of N10°W.

Vertical alignment for Laguna Canyon Road shall pass through Point C and shall be in the grade of 4% up to the North.

Culverts. Culverts should be considered to carry storm water across Route 8. A previous hydrology and hydraulic study recommended that 48-inch diameter pipes would be sufficient to accommodate a 50-year storm event.

A 2-ft minimum elevation difference between the top line of the pipes and the pavement surface of roadways (Route 8 and ramps) should be provided.

Side Slopes and Right-of-Way (R/W) Lines. Cut and fill slopes should be at least 2:1 (Horizontal: Vertical). If possible, the slopes should be 4:1 or flatter.

For ease of roadway construction and future maintenance, the R/W line should be located a minimum of 8 ft. from the catch point where the side slope intersects the existing ground.

4.5 Design Standards

All design standards should be based on the following order of references: (1) Caltrans Highway Design Manual, (2) A Policy on Geometric Design of Highways and Streets, American Association State Highway and Transportation Officials (AASHTO), 2004, and (3) current acceptable highway design practice.

State Highway 8 and Laguna Canyon Road are not the real roadways in Riverside County. They are created for this textbook only.

Name____ Date______________

4.6 Questions

Q4.1 What is the purpose of this tutorial design project?

Q4.2 Which transportation agencies provide funding for the design and construction of this project?

Q4.3 What is the website that provides the detailed information for this project design?

Q4.4 How many lanes do we need to consider for the Route 8 mainline and the ramps?

Q4.5 Describe the three control points in terms of their location and elevation.

Q4.6 What is a culvert?

Chapter 5
Introduction to MicroStation and InRoads

LEARNING OBJECTIVES

After completing this chapter you should know:

1. The relationship between InRoads and MicroStation;
2. The basic working environment of MicroStation;
3. The basic working environment of InRoads;
4. How to navigate and use InRoads interfaces.

SECTIONS

MicroStation® and InRoads® are two Computer Aided Design (CAD) software products of Bentley Systems, Inc. Roadway designers use these two products to 1) create digital surfaces that represent existing grounds of roadway project sites, 2) develop horizontal and vertical alignments of roadways, and 3) establish roadway design surfaces with which design plans can be prepared and earthwork costs can be estimated.

The objective of this chapter is to introduce the working environments of MicroStation and InRoads. The remainder of the chapter describes the relationships between InRoads and MicroStation, followed by a review of the general interfaces of MicroStation and InRoads and the associated file types. The chapter ends with hands-on procedures that describe how to start and close InRoads and MicroStation.

5.1 MicroStation and InRoads

MicroStation provides a platform for two- and three-dimensional (2-D/3-D) design and drafting. It creates 2-D/3-D vector graphic objects and elements. In this book, MicroStation V8i SELECT Series 2 is used to develop the tutorial project.

InRoads provides a platform for major civil engineering and transportation projects. It works with MicroStation to provide civil engineers with CAD design and drafting capabilities, powerful mapping tools, and roadway design automation. In this book, InRoads SELECT Series 1 is used for the design of the tutorial project.

MicroStation and InRoads are complementary products for roadway design (see Figure 5-1). InRoads relies on MicroStation to display CAD graphics, such as alignment layouts, 3-D terrain models, etc. Although InRoads can create graphic elements, its drafting capabilities are limited. InRoads imports CAD graphics mainly from MicroStation for roadway design. In this regard, InRoads can be viewed as the "data processor" for horizontal station calculation, vertical elevation determination, superelevation rate computation, and earthwork volume estimation, whereas MicroStation can be seen as the "data viewer and provider" of InRoads.

InRoads cannot run alone. It requires MicroStation to provide a supporting CAD environment within which InRoads performs calculations. When users activate InRoads, MicroStation is loaded first, then InRoads. Users can close InRoads and keep MicroStation running. However when users close MicroStation, InRoads is closed first.

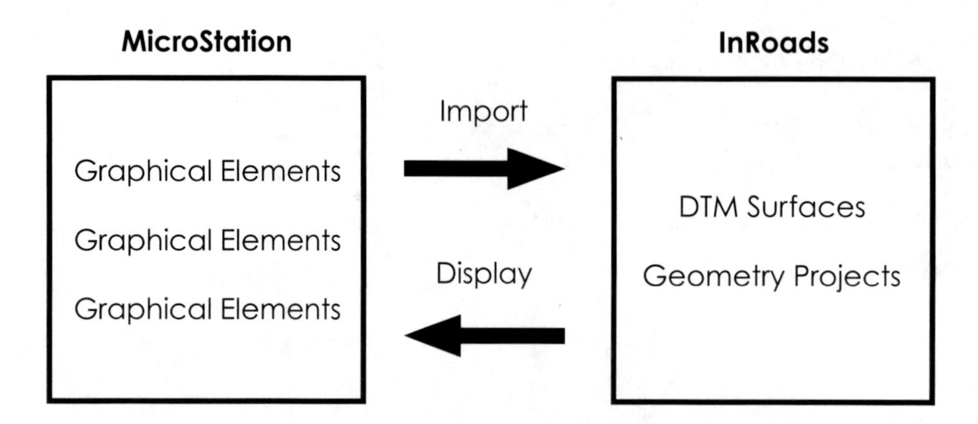

Figure 5-1 Relationships Between MicroStation and InRoads

MicroStation and InRoads are individual applications. Deleting and/or modifying CAD graphics in the MicroStation environment does not impact InRoads data. However, deleting or modifying elements in the InRoads environment makes the deleted elements disappear and the modified elements are reflected in the MicroStation environment.

5.2 MicroStation Environment

The MicroStation environment consists of a set of interfaces that help designers use drafting tools to create and manage graphic elements in 2-D or 3-D settings. These interfaces include:

1. Drafting and annotation tools for 2-D/3-D design productivity (for example, AccuDraw and AccuSnap);
2. Project explorer that manages and navigates files, models, saved views, and references;
3. 3-D tools for wireframe, surface, mesh and feature-based modeling, editing, and manipulation, as well as processing tools for carving and sculpting 3-D geometry;
4. Visualization and animation tools for realistic rendering (including ray-trace and particle trace solutions) and extensive material, texture, and lighting libraries;
5. Tools that attach drawings and images to a drawing environment.

The textbook assumes that readers have acquired a certain level of familiarity with MicroStation interfaces. This section presents only the general interface of MicroStation. For readers who want to know more about MicroStation, they can explore the details in MicroStation references.

The general interface of MicroStation, as shown in Figure 5-2, is composed of a Menu Bar, Tool Box, Tool Window, Tool Setting Window, View Window, Status Bar, and Key-in Window.

Menu Bar	The primary location that allows users to access all MicroStation commands.
Tool Box	Consists of various tools. The list of available tool boxes in MicroStation can be identified from the Tools Menu drop-down submenus.
Tool Window	Contains a set of drafting tools.
Tool Settings Window	Displays a number of specific parameters related to a tool when being invoked.
View Window	Eight view windows that can display an engineering drawing from different perspectives. Each view window has its own view control toolbox. Figure 5-2 shows one of the view windows (View 1—the Top View window). The other view windows can be invoked by going to **Window → Views.**
Key-in Window	Allows users to enter commands to control MicroStation. It can be invoked by selecting **Utilities → Key-in.**
Status Bar	Displays a variety of useful information that contains the name, prompts, and messages of the selected tool.

5.3 InRoads Environment

The InRoads environment consists of a set of interfaces that help designers use roadway design tools to create and manage geometric projects. The general interface as shown in Figure 5-3 contains five components: Menu Bar, Workspace Bar, Feedback Pane, Status Bar, and Tool Bar.

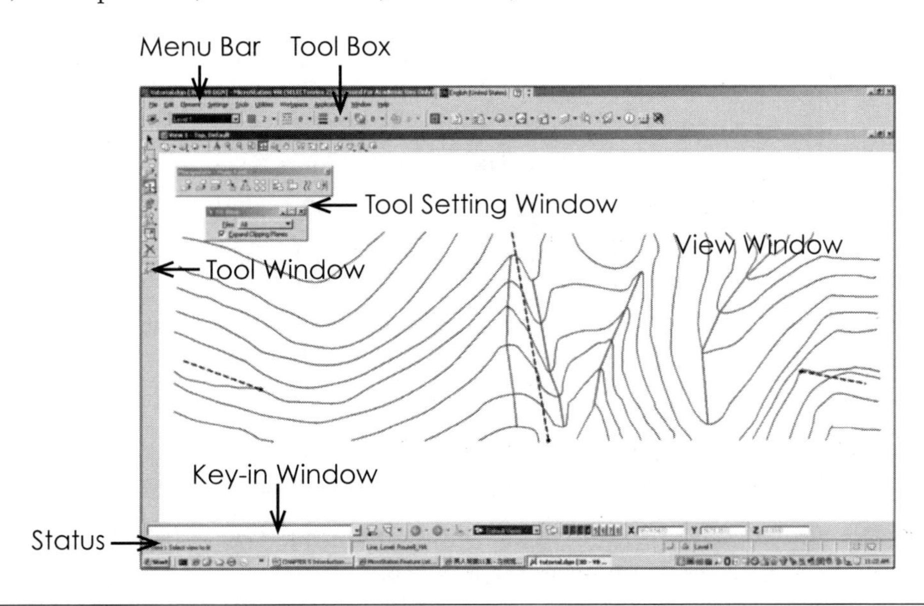

Figure 5-2 MicroStation Interfaces

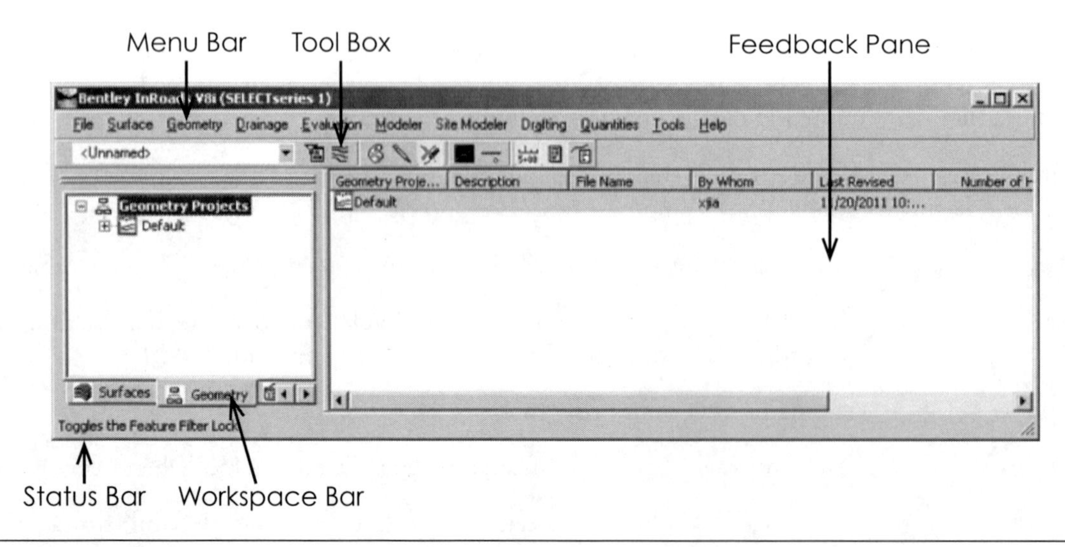

Figure 5-3 InRoads Environment

Menu Bar — Makes all InRoads commands accessible to users. If a menu item has ellipses (...) to the right, a dialogue box displays. Menu items that include a small arrow have a cascading submenu. From left to right, the menu bar contains File, Surface, Geometry, Drainage, Evaluation, Modeler, Drafting, Tools menus. The File, Drafting, and Tools menus are similar to other software packages.

Surface — Enables users to view surface contours, surface triangles, perimeters, and features.

Geometry — Allows users to open geometry projects, set and view horizontal alignments, and view alignment stationing.

Evaluation — Allows users to create profiles that are instrumental in viewing vertical alignments and associated annotations. Other commands are related with volume calculations, cross sections, mass haul diagrams, and so on.

Modeler — Assembles superelevation, templates, and corridors to create roadways.

Workspace Bar — The Workspace Bar (i.e., the InRoads Explorer Window, see Figure 5-3) is the area where users can interact with project-related data. Like the standard Windows Explorer, users can conveniently collapse and expand the data by clicking the plus (+) and minus (–) signs next to the data. At the bottom of this area are tabs, including Surface, Geometry, Preference, Drainage, Templates, and Corridors. These tabs allow users to easily switch among data files.

Feedback Pane — Provides additional information associated with project data (e.g., end station, type, author features, etc.; see Figure 5-3). The pane allows users to track projects and identify potential issues in a design. The information shown in the pane varies, depending on the type of data selected in the workspace bar. Figure 5-3 shows the feedback pane for a surface.

Status Bar — Presents users with information about commands being implemented. The Status Bar is located at the bottom left corner of the InRoads interface.

Tool Bar — Provides shortcuts that link to often-used commands. The display of tool bars can be controlled via **Tools → Customize → Toolbars** tab. (Figure 5-3 shows the Locks toolbar.)

The Workspace Bar and Feedback Pane menus display different tabs for different file types of InRoads data as described below:

a. Surface File (.dtm)

Stores 3-D surface models representative of either existing grounds or design conditions. The 3-D models consist of features such as points, breaklines, exterior boundaries, interior boundaries and contours.

Multiple .dtm files can be loaded in InRoads concurrently. Only one file can be set active at a given time.

b. Geometry Project File (.alg)

Stores geometric data, including horizontal and vertical geometry layouts as well as coordinate geometry (COGO) points. The layouts consist of various types of linear features, including tangent line, circular curve, parabolic curve, and spiral curve. Similar to .dtm files, multiple .alg files can be loaded into InRoads at the same time, while only one file can be set active at a given time.

c. Template Library File (.itl)

Stores typical cross sections and their associated components for a roadway or project. Different from .dtm and .alg files, only one template library file can be loaded into InRoads for a project.

d. Roadway Design File (.ird)

Stores and organizes information required to define a roadway design, including horizontal and vertical alignments, various templates, superelevation transition, etc. Similar to an .itl file, only one roadway design library file can be loaded into InRoads for a project.

e. Project File (.rwk)

Does not store any InRoads data. It is an ASCII file that outlines all the InRoads data required for a specific roadway project.

Note that some MicroStation and InRoads tools and commands are similar in name but different in function. Readers should pay attention to the differences when using the commands. In the remaining chapters of this text, the MicroStation commands or tools begin with **MicroStation →** while the InRoads commands or tools begin with **InRoads →.**

5.4 Getting Started with InRoads

This section describes the procedures to start and close MicroStation and InRoads.

1. Start MicroStation and InRoads.
    a. Locate the project file called ***tutorial.dgn*** from the **C:/Temp/Book/Chp5/Master Files/** folder and copy this file to **C:/Temp/Tutorial/Master Files** folder. You can also download it from the book website: www.csupomona.edu/~xjia/GeometricDesign/Chp5/ Master Files/.

b. Start InRoads.

   Go to **Start → Programs → Bentley → InRoads Group V8i (SELECTseries 1) → InRoads.** (This command path may be different if MicroStation and InRoads are not installed with default settings.)

c. Select ***tutorial.dgn*** in the MicroStation Manager Dialog window. Figure 5-4 shows the MicroStation design environment with the InRoads window on the top.

d. Minimize the InRoads window.

e. Click the **Fit View** icon in the view window of the MicroStation environment to fit the design file. Figure 5-2 shows the project area.

Note that this design file is a 3-D MicroStation file. The middle dash line (in the North-South direction) is Laguna Canyon Road. Two short dash line segments that run East-West define the alignments at each end of the project. These alignments are considered as design constraints.

There are also four solid breaklines that represent ridge or valley lines on the project site. More detailed information about the tutorial project is provided in later chapters.

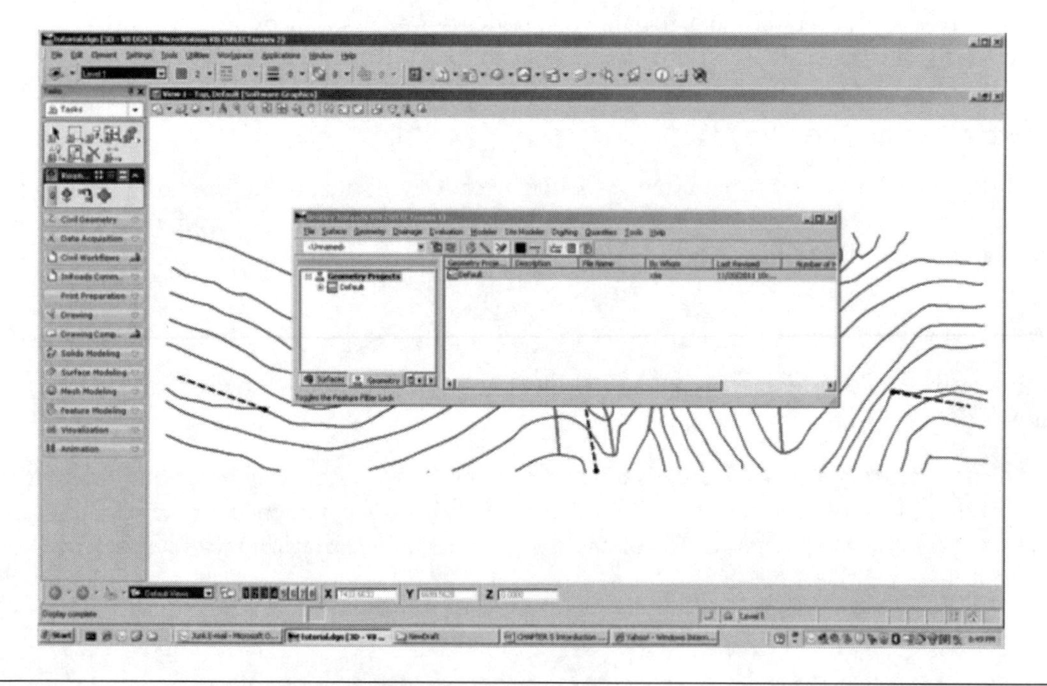

Figure 5-4 MicroStation and InRoads Environments

2. Close InRoads and MicroStation.

   a. To close the InRoads program, select **InRoads → File → Exit.** Note that the Exit command closes only InRoads; the MicroStation will still be running.

   b. To close MicroStation, select **MicroStation → File → Exit.**

Name____ Date________________

5.5 Questions

Q5.1 What is a Geometry Design File (.alg)?

Q5.2 Can multiple geometry project files be open for editing concurrently?

Q5.3 Can multiple surface files be open for editing concurrently?

Q5.4 Can multiple template library files be open for editing concurrently?

Q5.5 Can multiple roadway design files be open for editing concurrently?

Q5.6 What are the relationships between InRoads and MicroStation? Does deleting a MicroStation item affect InRoads data?

Q5.7 Locate two ridge lines and two valley lines in **tutorial.dgn** and explain your choice.

Q5.8 Can a Project File (.rwk) store data?

Q5.9 When saving design files, can you save them all at once? Explain.

Q5.10 Is the MicroStation design file (**tutorial.dgn**) a 3-D file? How can you verify this?

Chapter 6
Project Site Surface

LEARNING OBJECTIVES

After completing this chapter you should know:

1. How to create a new Digital Terrain Model (DTM) surface;
2. How to load graphics into a surface;
3. How to create a Triangulated Irregular Network (TIN) and triangulate loaded surface data;
4. How to save a DTM surface;
5. How to visualize a DTM surface.

SECTIONS

Existing ground surface plays a critical role in many highway design projects. This chapter discusses the use of Digital Terrain Model (DTM) technology to represent the existing ground surface of the project site used in this book.

A *DTM* is a 3-D model that captures and represents the ground surface of a highway project using triangulated irregular network (TIN) technologies (see Figure 6-1).

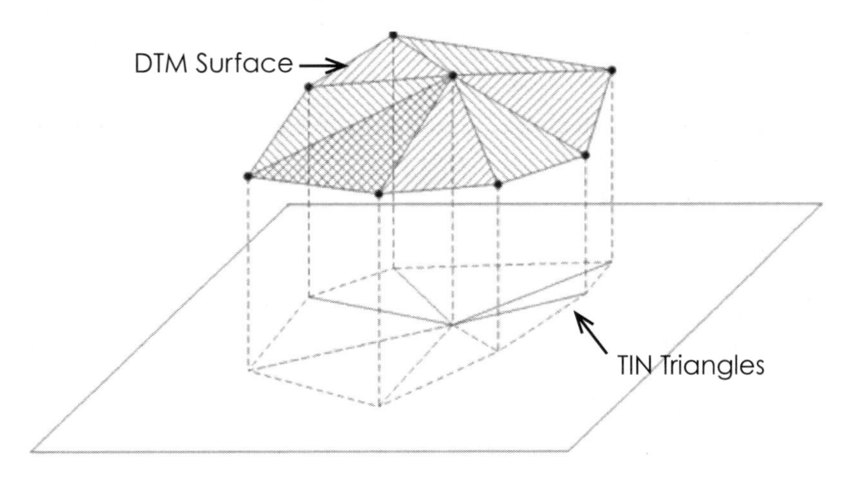

Figure 6-1 Graphical Representation of DTM and TIN

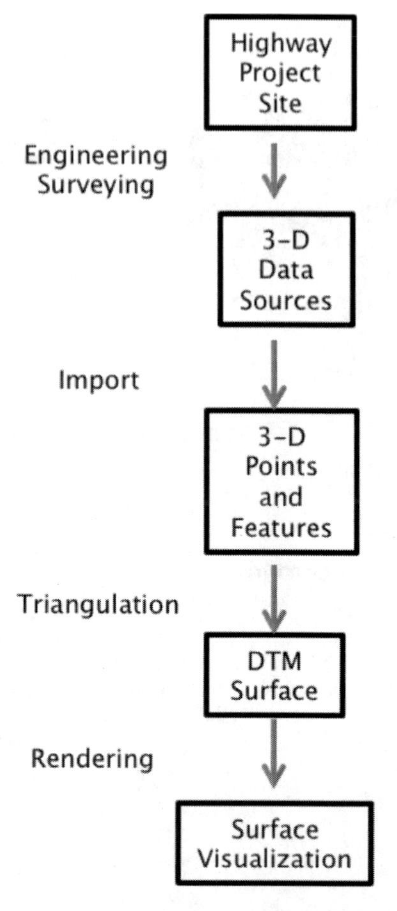

Figure 6-2 DTM Surface Generation Process

The DTM surface shown in Figure 6-1 consists of a set of contiguous, non-overlapping triangles. Each triangle has its own plane or facet. The TIN triangles represent the orthogonal projection of the DTM surface onto a reference plane.

Existing ground surfaces are constructed by connecting 3-D points to create a network of adjoining triangles. Highway engineers often obtain 3-D points through survey points and features, topographical maps, and existing design graphics.

Survey points are discrete points with X, Y, and Z coordinates. Z coordinates represent the elevation information of points. Survey features such as breaklines represent terrain information. Breaklines indicate discontinuities of a ground surface. A breakline is typically made of a series of survey points.

Topographical maps such as contours are often generated from aerial photographs using photogrammetric technologies or from existing quadrant maps using digitizers. A *contour map* is a 2-D representation of 3-D reality. A contour map uses contour lines to represents terrain elevation and morphological information.

X, Y, and Z coordinates inherited in existing design graphics or existing DTM models can also be used as inputs to create existing ground surfaces.

A typical process of creating DTM surfaces is shown in Figure 6-2.

Engineering Survey

The process involves the use of surveying technologies—Global Positioning Systems (GPS), Mobile Terrestrial Laser Scanning (MTLS), photogrammetric technologies, and conventional survey technologies (such as total stations)—to measure and determine the geospatial locations of points or features on or near the existing surface of highway project sites.

Accurately measured points or features establish a basic "network" of horizontal and vertical controls. These points or features form different 3-D data features or sources in different formats (X, Y, and Z coordinates, topographical maps, and design graphics).

Import

The DTM surface generation process includes loading or importing 3-D points and/or features from different data sources into a software application such as InRoads, GeoPAK, and AutoCAD Civil 3D.

Triangulation

Imported 3-D points and/or features are triangulated using the TIN technologies. DTM surfaces are then created.

Rendering

DTM surfaces can be visualized and applied as the foundation for geometric design (horizontal and vertical alignment design, cross section design, roadway design, and earthwork calculation).

This chapter describes the procedures for importing 3-D points and features to create the existing ground surface of the tutorial project site. The chapter further discusses how to visualize the DTM surface.

6.1 Import 3-D Features into DTM Surface

Before we start creating the project DTM surface, it is assumed that you have MicroStation and InRoads installed on your computer, and that you also have the MicroStation file **tutorial.dgn** available in the **C:/Temp/Book/Chp6/Master Files/** folder. (You can obtain the file from www.csupomona.edu/~xjia/GeometricDesign/Chp6/Master Files.)

6.1.1 Getting Started with tutorial.dgn

Copy the **Tutorial.dgn** file to **C:/Temp/Tutorial/Chp6/Master Files/** folder and do the following to check the **tutorial.dgn** design file:

1. Start MicroStation and InRoads.
2. When MicroStation prompts you to open a MicroStation file, select **tutorial.dgn** from the **C:/Temp/Tutorial/Chp6/Master Files/** folder.
3. Keep **View 1** window open and close other view windows if you see multiple view windows open. You will see the InRoads design environment similar to Figure 6-3.

   If you do not see the contours in **View 1** window, select the ***Fit View*** tool on the **View 1** Window toolbar in the MicroStation environment.
4. Go to **MicroStation → Settings → Design Files ... → Working Units.** Note that the **Master Unit** for the **tutorial.dgn** file is **feet**. Close the **Design File Settings** window when working units are verified.

   Note that the **tutorial.dgn** is a 3-D design file (see Figure 6-4).

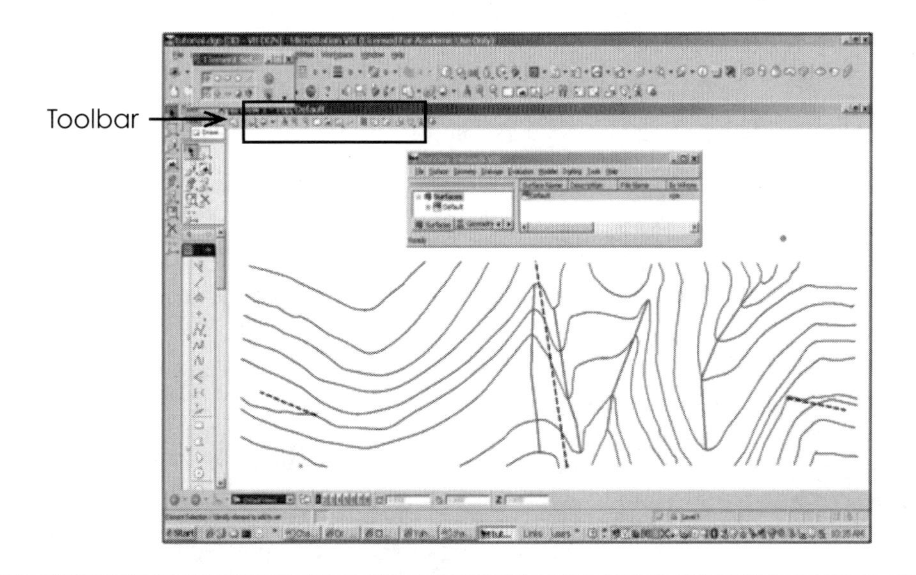

Figure 6-3 InRoads Design Environment

Tutorial.dgn (3D-V8 DGN)

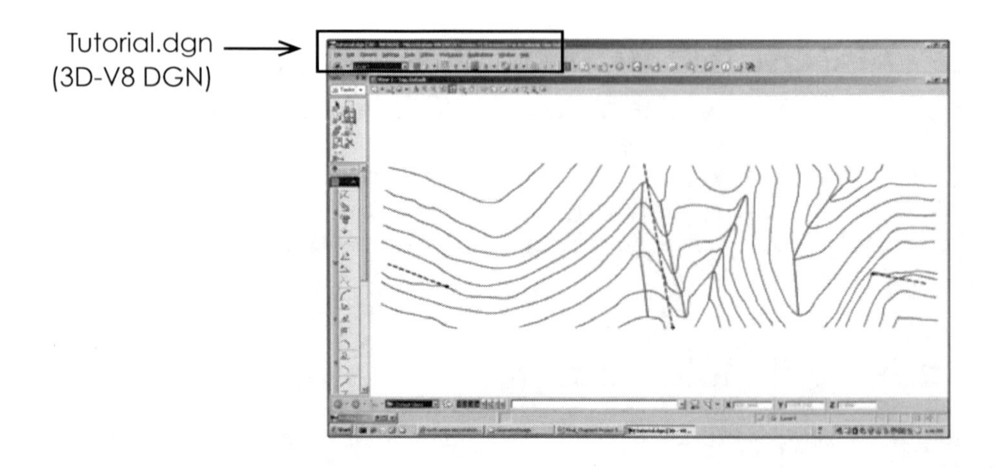

Figure 6-4 Tutorial.dgn

5. Go to **InRoads → Files → Project Options ...→ Units and Format**. Make sure **Imperial** is selected in the **Units → Linear** box. Close the **Project Options** window after verification.

   The unit in the InRoads working environment should match the unit of the **tutorial.dgn** file.

6. Go to **MicroStation → Settings → Levels → Level Manager.** You will see four levels (**Level 1, Level 2, Level 3, Default** and **North Arrow** level) in the **tutorial.dgn** file.

   Verify that contours are on **Level 1**, breaklines are on **Level 2**, and other features are on **Level 3**. **Default** and **North Arrow** levels are the default levels of MicroStation.

6.1.2 Create OG Empty DTM Surface

Now we create a new empty DTM surface following these steps:

1. Choose **InRoads → File → New.**
2. Click the ***Surface*** tab. The **New** dialog box appears (see Figure 6-5).

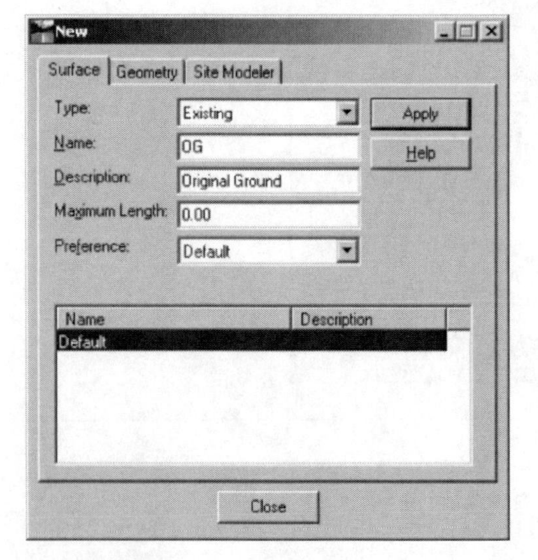

Figure 6-5 New Dialog Box

Enter the following in the dialog box:

Name:	***OG***
Description:	***Original Ground***
Maximum Length:	***0.00***
Preference:	***Default***

Confirm that your entered values are the same as those shown in Figure 6-5.

The **Maximum Length** field in the dialog box defines the maximum length of triangle sides. The 0.00 value indicates that the length can be as long as necessary. This parameter is important in controlling the shapes of triangles that will be generated later using the **Triangulate Surface** command in InRoads.

3. Click ***Apply*** and then ***Close***. InRoads creates an empty DTM surface called **OG** (see Figure 6-6).

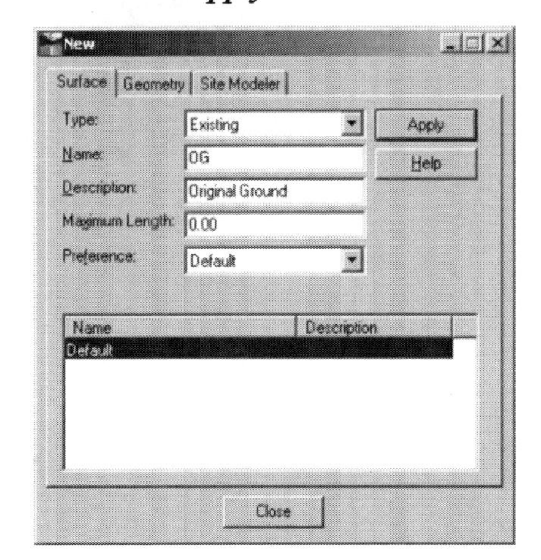

Figure 6-6 Created the OG Empty DTM Surface

There are two surfaces, **Default** and **OG**, under the tree of **Surfaces** (see Figure 6-7). **Default** surface is the default surface created by InRoads when you start InRoads. You can ignore it. The red block enclosing the **OG** surface indicates it is currently the active surface.

Note that the **OG** surface has 0 values for all the feature elements (Breakline Features, Contour Features, Exterior Features, Inferred Breaklines, Interior Features, Random Features, Range Points, and Triangles.)

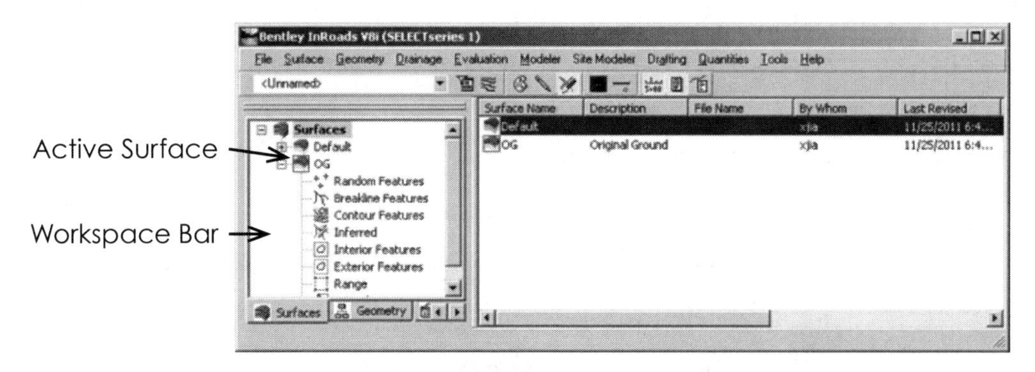

Figure 6-7 OG Empty DTM Surface

6.1.3 Load 3-D Features into the OG DTM Surface

After the empty project DTM surface (**og.dtm**) is created, the next step is to import or load the 3-D features stored in the project file, **tutorial.dgn**, into the **OG** surface.

Follow these steps to import the **contour** features into the **OG** surface:

1. Make sure the **OG** surface is the active surface. If not, pick the tab of **Surfaces** from the InRoads **Workspace Bar** (see Figure 6-7).
2. Go to **InRoads → Surface → Active Surface.** Select the **OG** surface.
3. Select ***Apply*** and ***Close***. OR, right click the **OG** surface → **Set Active.**
4. Select **InRoads → File → Import → Surface.**

   The **Import Surface** window appears (see Figure 6-8).
5. Select the ***From Graphics*** tab from the **Import Surface** window.
6. Enter the following information to import the contours into the **OG** surface:

   Surface: ***OG***
   Load From: ***Level***
   Level: ***Level 1***

Elevations: ***Use Element Elevations***

Thin Surface: ***Unchecked***

Use Tagged Graphics Name: ***Unchecked***

Seed Name: ***Tutorial***

Feature Style: ***Contour***

Point Type: ***Contour***

Maximum Segment Length: ***Unchecked***

Point density Interval: ***Unchecked***

Exclude from Triangulation: ***Unchecked***

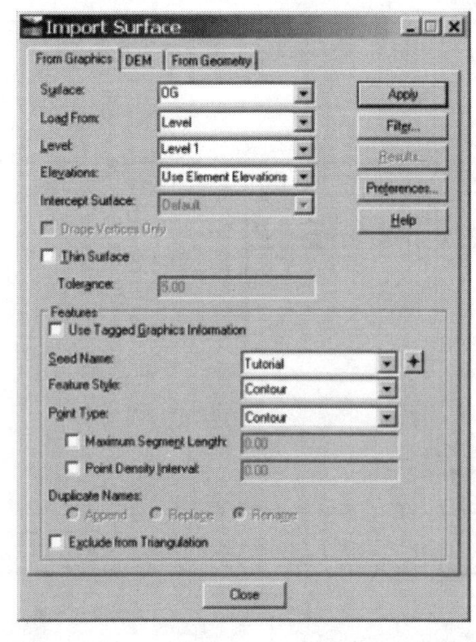

Figure 6-8 Import Contours into OG Surface

7. Select ***Apply***.

   The **Results** button is activated.

8. Click the ***Results*** button from the **Import Surface** window. The following information appears (see Figure 6-9):

   Number of points: 21579

   Number of lines: 30

9. Click ***Close*** on the **Results** window.

Now we import the breakline features into the **OG** surface:

1. Select **InRoads → File → Import → Surface**.
2. Select the ***From Graphics*** tab from the **Import Surface** window.

   Enter the following to import the breaklines into the **OG** surface (see Figure 6-10):

Surface: ***OG***

Load From: ***Level***

Level: ***Level 2***

Elevations: ***Use Element Elevations***

Thin Surface: ***Unchecked***

Use Tagged Graphics Name: ***Unchecked***

Seed Name: ***Tutorial***

Feature Style: ***Breakline***

Point Type: ***Breakline***

Maximum Segment Length: ***Unchecked***

Point density Interval: ***Unchecked***

Exclude from Triangulation: ***Unchecked***

3. Select ***Apply***.

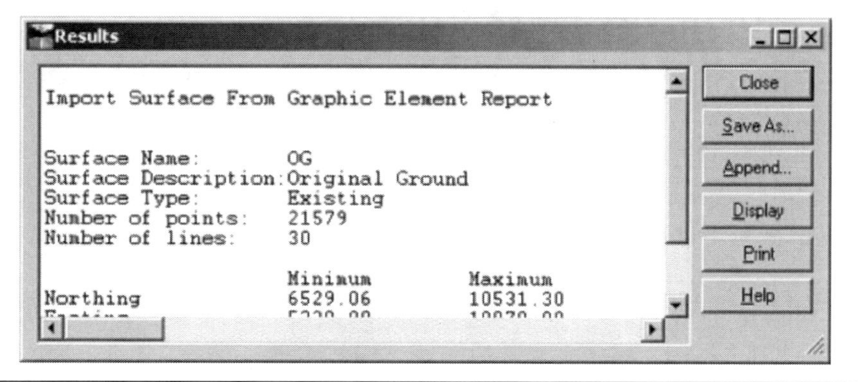

Figure 6-9 Results of Importing Contours in OG Surface

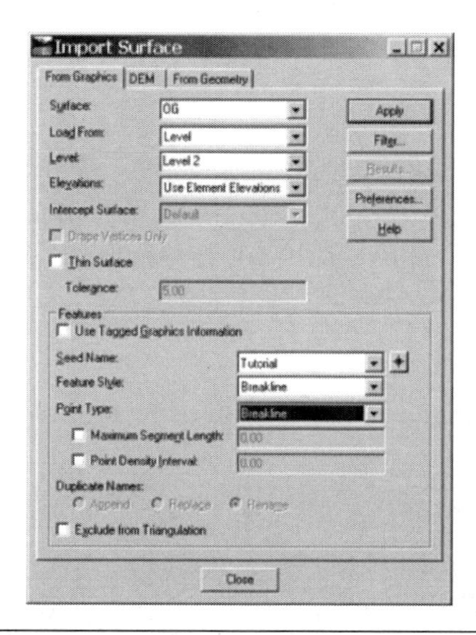

Figure 6-10 Import Breaklines into DTM Surface

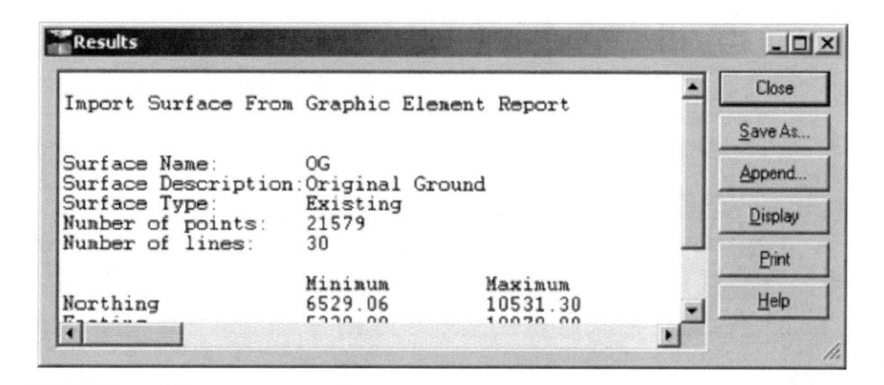

Figure 6-11 Results of Importing Breaklines in OG Surface

4. Click the ***Results*** button from the **Import Surface** window. The following information appears (see Figure 6-11):

   Number of points: 21

   Number of lines: 4

5. Click ***Close*** on the **Results** Window.
6. Close the **Import Surface** Window.

6.2 Create Project DTM Surface

When the contours and breaklines are imported into the **OG** surface, the surface has 3-D points available. The next step is to use InRoads utilities to triangulate these points and create the surface.

To triangulate the imported 3-D points, follow these steps:

1. Make sure the **OG** surface is the active surface. If not, set the **OG** surface as the active surface.
2. Select **Surface → Triangulate Surface.**
3. In the **Triangulate Surface** window, select the following (see Figure 6-12):

Surface:	***OG***
Maximum Length:	***0.00***
All Other Options:	***Check off***

   The Maximum Length setting specifies a distance that is used to define the maximum allowable triangle legs. After TIN triangles are formed, InRoads discards those triangles with a leg longer than the specified distance. For this tutorial project, the Maximum Length is set to 0.00. It means that InRoads does not discard any triangles once they are formed.
4. Select ***Apply***.

   When the surface has been triangulated, you will find that 21,600 points are used for the triangulation and 42,842 triangles are created.
5. To close the **Triangulate Surface** window, select ***Close***.

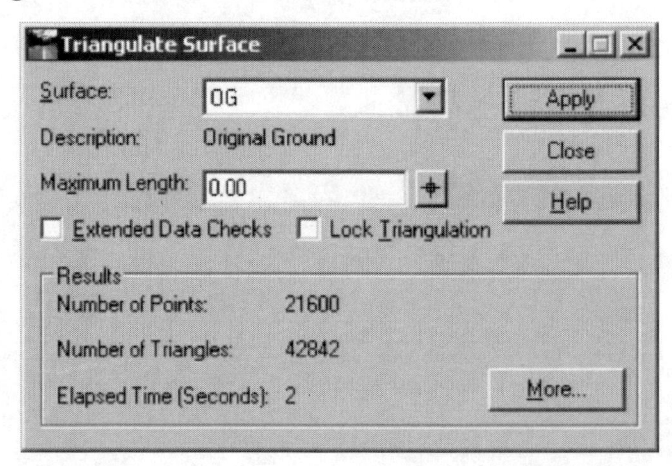

Figure 6-12 Triangulate Surface Window

6. Save the OG Surface. After the OG surface is triangulated, save it. After saving the surface, you can reload or open it later.
7. Select **InRoads → File → Save → Surface.**

   Since this is the first time you are saving the OG DTM surface onto hard disk, InRoads prompts you with the **Save As** window (see Figure 6-13).

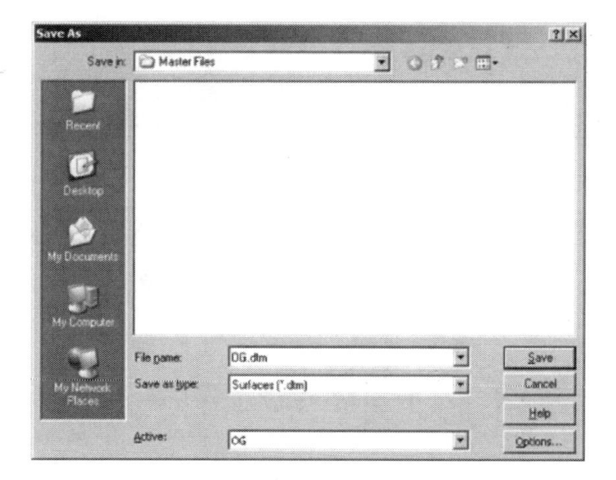

Figure 6-13 Save the OG DTM Surface

8. Enter this information in the **Save As** window:

Save in:	**C:/Temp/Tutorial/Chp6/Master Files/**
File Name:	**OG.dtm**
Save As Type:	**Surfaces (*.dtm)**
Active:	**OG**

The DTM surface file, **og.dtm**, is saved in the **C:/Temp/Tutorial/Chp6/Master Files/** folder.

9. Click the ***Save*** button. The DTM surface is saved.
10. To close the window, click ***Cancel***.

6.3 Explore the OG DTM Surface

The OG DTM surface has been created. Now we explore or view the surface using rendering technology. *Rendering* is the process of illustrating a 3-D model through the display of shaded surfaces. In the process, the mathematical data used to describe a solid model is translated into pixels and represented on a computer screen. MicroStation provides a comprehensive range of rendering choices, from the simple modes (such as hidden line display, smooth shading, and Phong shading) to the sophisticated rendering modes (such as Ray Tracing, Radiosity solving, and Particle Tracing).

Since the DTM surface for our project site is relatively flat, we need to vertically exaggerate the elevation of the displayed points within the surface. Follow these steps:

1. Go to **InRoads → Surface → View Surfaces → Options**. The **Surface Options** window appears (see Figure 6-14).
2. Enter the following information into the window:

Scale:	***3.0***
Scale Relative to Fixed Elevation:	***Unchecked***
Planarize	***Unchecked***

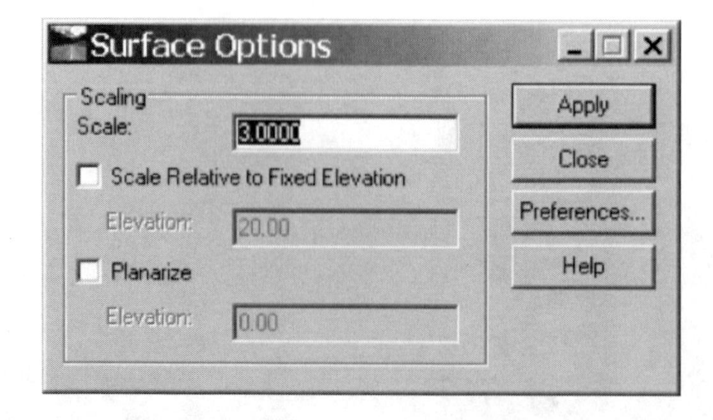

Figure 6-14 Surface Option Window

The scale factor is set to 3.0. That means the elevations of displayed points are vertically exaggerated 3 times the actual elevations. This factor scales only the displayed points; it does not modify the internal digital terrain model data.

3. To close the **Surface Options** window, select ***Close***.
4. Make sure the **OG** surface is loaded into InRoads. When you start InRoads, if you find the **OG** surface is not in the **Surface** list, you need to load or open it into InRoads and view it. Follow these steps to load the **OG** surface into InRoads:
   a. Select **InRoads → File → Open**. The **Open** window appears.
   b. Navigate to the folder where the **OG** surface file (**og.dtm**) is stored. (It should be located in the **C:/Temp/Tutorial/Chp6/Master Files/** folder).
   c. Select **Surface** (***.dtm**) in **Files of Type** field.
   d. When a list of surface files including **og.dtm** shows in the window, select **og.dtm → Open.**
   e. To close the window, select ***Cancel.***

To view the contours of the DTM surface, follow these steps:

1. Make sure the **OG** surface is the active surface. If not, set the **OG** surface as the active surface.
2. Select **InRoads → Surface → View Surface → Contours** .... The **View Contours** window appears (see Figure 6-15).
3. Enter the following information into the window:

Surface:	***OG***
Interval:	***2.00***
Minors per Major:	***4***
Major Contours:	***Check on***
Minor Contours:	***Check on***
Major Labels:	***Check on***

4. Check off all other options.
5. Select **Apply**.

The interval of 2.00 indicates the minor contour interval is 2 ft. The Minors per Major option indicates there are four minor contours between two major contours (see Figure 6-16).

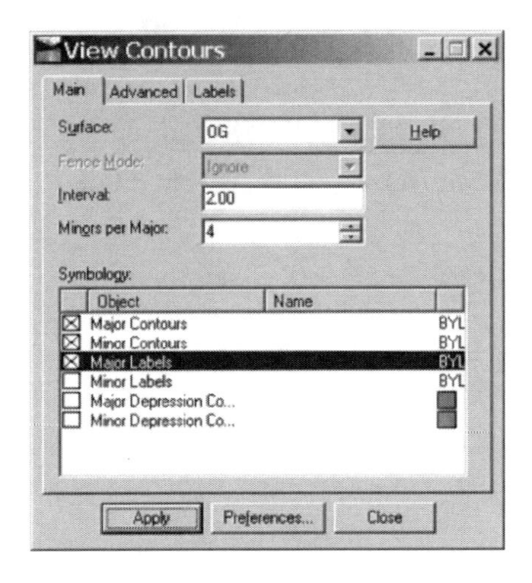

Figure 6-15 View Contours Window

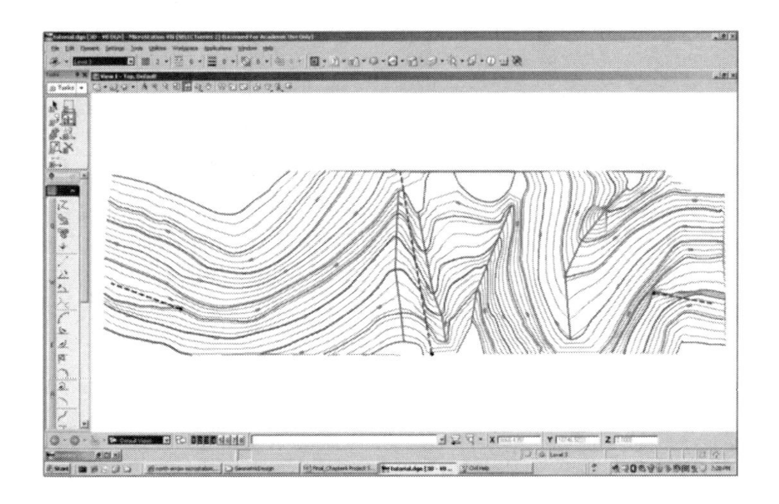

Figure 6-16 Contours Generated from the OG DTM Surface

6. To close the **View Contours** window, select ***Close***.
7. To delete the newly generated contours, use the MicroStation **Delete Element** command. The **Delete Element** can be found in **MicroStation → Main** tools.

To view the TIN triangles of the DTM surface, follow these steps:

1. Make sure the **OG** surface is the active surface. If not, set the **OG** surface as the active surface.
2. Select **InRoads → Surface → View Surface → Triangles** .... The **View Triangles** window appears (see Figure 6-17).
3. Enter the following information into the window:

Surface:	***OG***
Colored Model	***Check off***

4. Select ***Apply***. The TIN triangles for the OG surface are displayed over the original contours (see Figure 6-18).
5. To close the **View Triangles** window, select ***Close***.

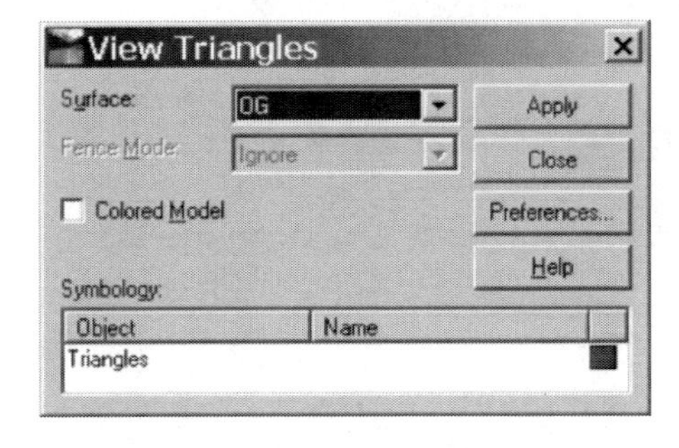

Figure 6-17 View Triangles Window

Figure 6-18 TIN Triangle

To render the OG DTM surface, follow these steps:

1. Adjust the global lighting to get a better rendering from the surface. Follow these steps:
   a. Select **MicroStation → Tools → Visualization → Lights → Light Manager.** The **Light Manager** window appears (see 6-19).
   b. Select ***Ambient*** and check the **On** option. Set **Lux** to ***30***. Select ***Flashbulb*** and check the **On** option. Set the **Lux** to ***30***.
   c. Select ***Solar*** and do the following:

On:	***Check on***
Intensity:	***100***
Other Options:	***see Figure 6-20***

   d. Close the **Light Manager** window.
2. Close the **View 1** window. Open the **View 2** window. The isometric view of the OG surface appears.
3. Go to **MicroStation → Settings → View Attributes.** The **View Attributes** window appears (see Figure 6-21).

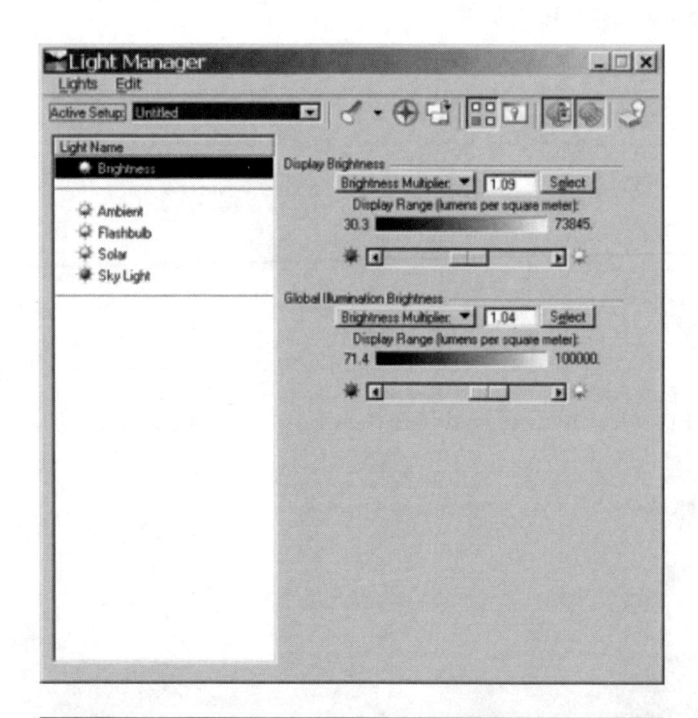

Figure 6-19 Light Manager Window

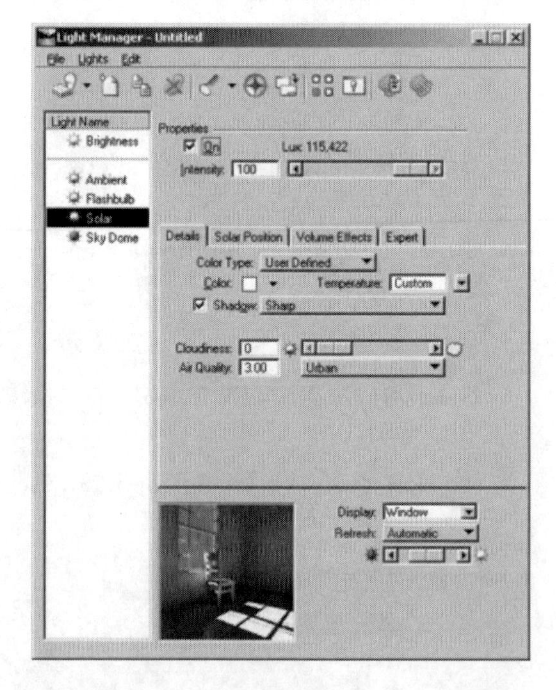

Figure 6-20 Solar Option in the Light Manager Window

4. Select ***Smooth with Shadows*** from the **Display Style** field. Note that the OG surface is changed to one similar to Figure 6-22.

Figure 6-21 View Attribute Window

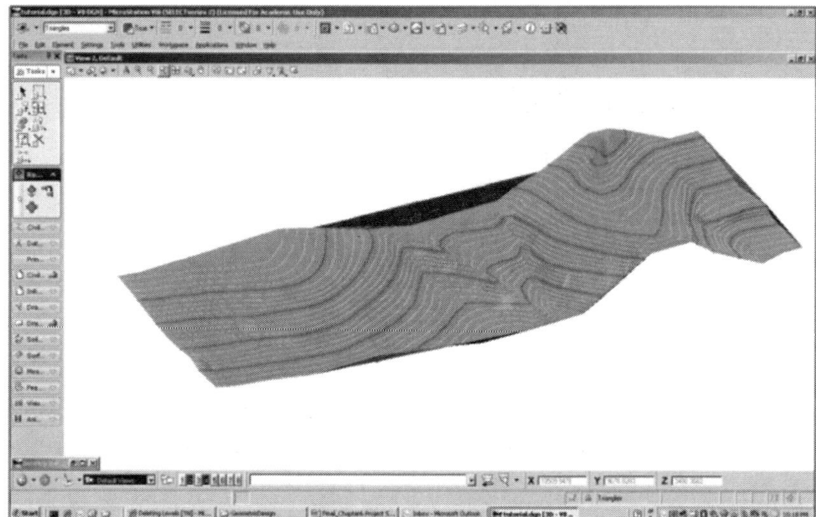

Figure 6-22 OG DTM Surface

In summary, the OG DTM surface is created by first importing the contours and breaklines (stored in **tutorial.dgn** design file) into InRoads. The vertices of contours and breaklines are then used as 3-D points for triangulation. The OD DTM surface is formed after the triangulation is successfully completed. The surface is saved into the **OG.dtm** file. The surface can be visualized using the MicroStation rendering technologies. Figure 6-23 summarizes the procedures for the OG DTM surface.

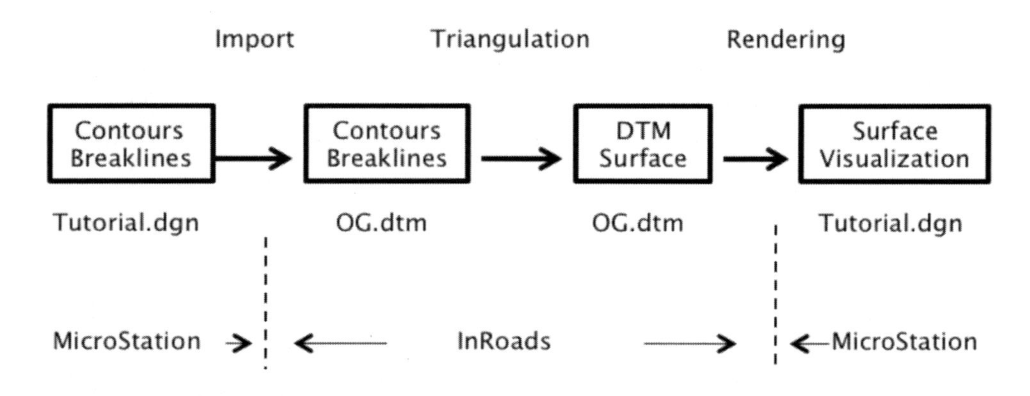

Figure 6-23 DTM Process for OG Surface

6.4 Further Understanding of DTM

A digital terrain model is a topographic model on computers that depicts the terrain relief or the existing ground surface of any site on earth. The model's accuracy largely depends on 3-D points and/or features imported from different data sources. A highly accurate DTM typically requires a significant amount of precisely measured 3-D points and/or features that are obtained from the surveying practice.

When 3-D points and/or features from different data sources are imported into a DTM, they should follow the requirements listed below (modified from the New York DOT and Caltrans Surveying Manuals and Guidelines):

1. 3-D points and features shall be representative of the existing ground at the time of the survey/mapping and shall have been field-edited. Roadway designers should not modify these points and features.
2. All survey points and features shall have the same coordinate system, horizontal and vertical datum.
3. The point density interval for breakline features shall be set to a maximum of 10 ft to allow for a sufficiently dense DTM surface.
4. Areas with several features, such as curbed areas, intersections, and areas of tight vertical curvature may require a smaller point density interval to accurately represent a DTM surface.
5. The point density interval for breakline features shall be set such that the chord height generated does not exceed 0.03 ft.
6. Superelevation transition points, the beginning and ending of vertical curves, and the beginning and ending of pavement width transitions should all be included in the DTM.
7. The minimum triangle length should be set so that triangles do not make erroneous connections.
8. Surfaces shall include an exterior boundary. Inaccurate or extraneous triangles around the perimeter of a proposed ground shall be deleted.
9. Problems encountered when developing a surface shall be corrected by fixing the features in the DTM, and not by modifying or deleting a point used to create a triangle vertex.
10. All existing ground nontriangulated data shall be combined and provided as one comprehensive surface. Similarly, all proposed nontriangulated data shall be combined and provided as one comprehensive surface.

DTM surfaces, once generated, need to be evaluated. Guidelines for such evaluation may include the following:

1. Display triangle features to look for inconsistencies in the DTM surfaces.
2. Generate DTM contours and look for inconsistencies, such as a large group of close contours or pavement areas that do not form Vs with the contours.
3. Review the DTM surface along the approximate centerline of all major roads and streams. The customer may review the profiles for unusual slopes or breaks.
4. Review DTM surfaces and check cross sections from the approximate centerline of all major roads and streams.

   Cross sections will be done at 10-ft intervals in most areas and 3-ft intervals in the areas of bridges or culverts. The limit of each section will be the approximate limit of survey. These sections may be reviewed for unusual slopes or breaks.
5. Display the triangles in a new MicroStation file to ensure that erroneous triangles have been removed.
6. Verify surface elevations using survey methods consistent with the accuracy attainable by the original method of collection.

Name________________________________ Date______________

6.5 Questions

Q6.1 What is a DTM?

Q6.2 When you create a new DTM surface, what is inside the surface?

Q6.3 Why do we use features on Level 1 to import contours into the empty DTM surface for the tutorial project?

Q6.4 Why do we use features on Level 2 to import breaklines into the DTM surface for the tutorial project? Are the existing points within the DTM surface deleted when the break lines are loaded into the surface?

Q6.5 Suppose you have a MicroStation file called **SurveyPoints.dgn** that contains a set of survey points within a project area. List the steps to get this set of points loaded and triangulated into the **OG** DTM surface.

Q6.6 Are all the imported points for the tutorial project used for triangulation?

Q6.7 Is it true that each imported point in the **OG** surface (for the tutorial project) has only X and Y (Easting and Northing) coordinates?

Q6.8 Suppose you have created a new DTM surface, but you forget where you saved this surface. How can you locate this surface? (Hint: Use InRoads to figure it out.)

Q6.9 Review the Caltrans Surveying Manual and write a summary related to DTM (www.dot.ca.gov/hq/row/landsurveys/SurveysManual/Manual_TOC.html).

Q6.10 Review New York DOT's Land Surveying Standards and Procedures Manual and write a summary related to DTM (https://www.nysdot.gov/divisions/engineering/design/design-services/land-survey/repository/LSSPM09.pdf).

Q6.11 Review Chapter 20 CADD Standards and Procedures of New York DOT's Highway Design Manual and write a summary related to DTM (https://www.nysdot.gov/divisions/engineering/design/dqab/hdm/hdm-repository/chapt_20.pdf).

Chapter 7

Horizontal Alignments

LEARNING OBJECTIVES

After completing this chapter you should know:

1. Design controls related to the tutorial project;
2. Caltrans standards related to the tutorial project;
3. How to create a new geometry project;
4. How to create six horizontal alignments;
5. How to check horizontal alignments for conformance to Caltrans design standards.

SECTIONS

Horizontal alignments, vertical alignments, and cross sections are the three key components in designing a roadway. They work together to form a facility and establish the character of a roadway (see Figure 7-1).

Figure 7-1a Smooth Roadway

Figure 7-1b Serpentine Roadway

A *horizontal alignment* typically consists of tangent segments (segments of straight lines), circular curves, and in some cases, spiral transition curves. A *vertical alignment* includes grades and parabolic curves. A *cross section* is defined by a series of cross slopes.

Roadway designers make a spatial arrangement of a horizontal alignment, a vertical alignment, and cross sections sequentially when they start the geometric design of a proposed roadway (see Figure 7-2). Designers first determine the number of lanes or the typical cross section for the roadway to be built. They then work on the horizontal alignment design to meet requirements of the design standards, controls, and guidelines as well as constraints associated with the roadway. Using the horizontal alignment and the existing ground DTM surface, roadway designers create vertical alignments.

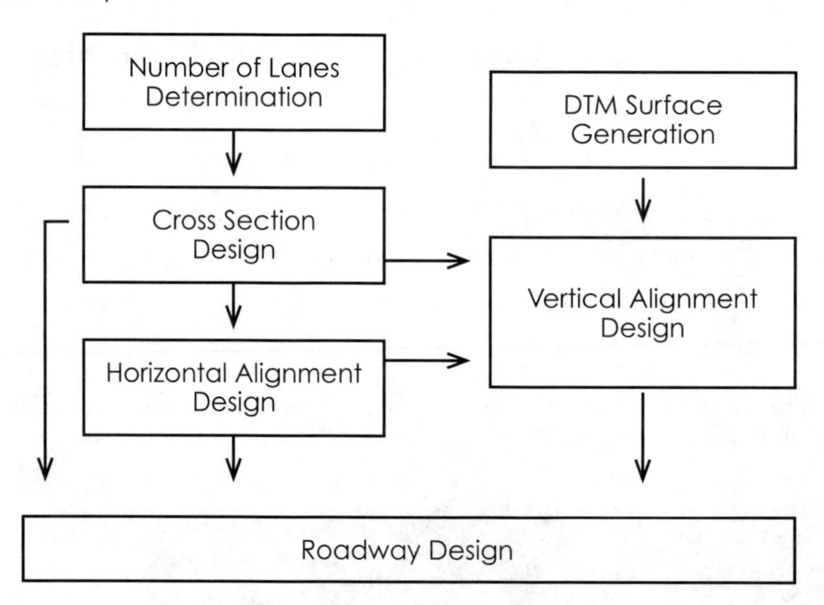

Figure 7-2 Geometric Design Procedures

Roadway designers then combine the DTM surface, the horizontal and vertical alignments, and the cross sections to form a roadway facility. The way by which these components are assembled significantly affects highway safety, operational performance, and aesthetics.

The design of a horizontal alignment is a trade-off process that determines a series of tangent segments and circular curves to ensure safe, smooth, and continuous vehicle operation on the proposed roadway. Factors that influence the location and configuration of tangent segments and circular curves include:

1. Highway functional classification, design designation, level of service, design speed, and design standards.
2. Safety considerations such as sight distance, consistency of alignment, facility-human interactions, and human factor considerations.
3. Physical controls including topography, hydrological features, geophysical conditions, land use, utilities, and natural and man-made features.

4. Environmental considerations that may affect adjacent land use, communities, and ecologically sensitive areas.

5. Economic considerations such as construction costs, right-of-way acquisition costs, utility impacts, and operating and maintenance costs.

Among the above factors, safety is always paramount and must be considered first, either directly or indirectly. Highway designers use their "artful" skills to create horizontal alignments with respect to these factors.

7.2.1 Design Standards for Horizontal Alignment

This chapter discusses the procedures for creating six horizontal alignments required for the interchange of Route 8 and Laguna Canyon Road. We will apply the following criteria and standards for the tutorial project:

1. Stopping Sight Distance (SSD)

Stopping sight distance is the continuous length of highway ahead visible to drivers. Drivers can safely stop before an object, if the object is located beyond the sight distance of the drivers.

A well-designed horizontal alignment shall provide at least the minimum stopping sight distance (as defined in Table 201.1, Caltrans *HDM*[1]) for the chosen design speed at all points on a highway. Given the design speed of Route 8 is 60 mph, the minimum SSD on Route 8 is 580 ft. The SSDs on the ramps vary because the operating speeds change along the ramps.

Where an object off the pavement such as a bridge pier, building, cut slope, or natural feature restricts sight distance, the minimum radius of curvature is determined by the following formula (see Figure 7-3).

$$SSD = \frac{R}{28.65}\left[\cos^{-1}\left(\frac{R-m}{R}\right)\right]$$

Where: R Radius of horizontal curve

M Clear distance from centerline of the lane nearest the obstruction (feet). In AASHTO, the clear distance is referred to as horizontal sightline offset (HSO).

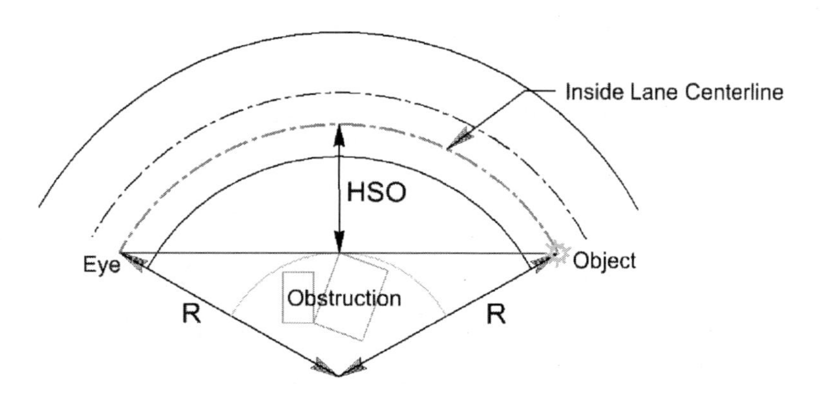

Figure 7-3 Stopping Sight Distance on Horizontal Curves

[1] *Caltrans HDM can be downloaded from http://www.dot.ca.gov/hq/oppd/hdm/hdmtoc.htm.*

Example 7-1:

Assume the design speed of an expressway is 60 mph. Bushes along the roadside block the drivers' view. The lateral clearance, *m*, is observed to be 20 ft from the centerline of the inside lane to the bushes. What is the minimum radius of the curve allowed for the geometric design?

Solution:

The minimum SSD is 580 ft for the design speed of 60 mph. The relationship among SSD, *m*, and *R* is given as:

$$SSD = \frac{R}{28.65}\left[\cos^{-1}\left(\frac{R-m}{R}\right)\right]$$

After trial-and-error efforts (or check Figure 201.6 of Caltrans *HDM*), *R* is estimated to be 2,150 ft.

2. Simple Curves

Table 203.2 of Caltrans *HDM* illustrates the minimum radius of curve for specific design speeds. The minimum radius of curve is 130 ft, 300 ft, 550 ft, 850 ft, 1,150 ft, 2,100 ft, and 3,900 ft for the design of speed of 20 mph, 30 mph, 40 mph, 50 mph, 60 mph, 70 mph, and 80 mph, respectively. This table is based on speed alone; it ignores the sight distance factor. If the minimum radius indicated in the table does not provide the desired lateral clearance to an obstruction, SSD standards on horizontal curves (or Figure 201.6 of Caltrans *HDM*) shall govern.

Example 7-2:

Route 8 is an undivided 4-lane freeway and has a median width of 62 ft (from inside edge of traveled way (ETW) to inside ETW). The lane width is 12 ft. Figure 7-4 shows the lane layout of Route 8. What is the minimum radius of the centerline curve?

Solution:

Given the design speed of 60 mph for Route 8, the minimum radius of the most inner curve, *R'*, is 1,150 ft. However, the minimum radius of the centerline curve (or the station line curve) is 1,150 ft + 12 ft + 12 ft + 62/2 ft = 1,205 ft.

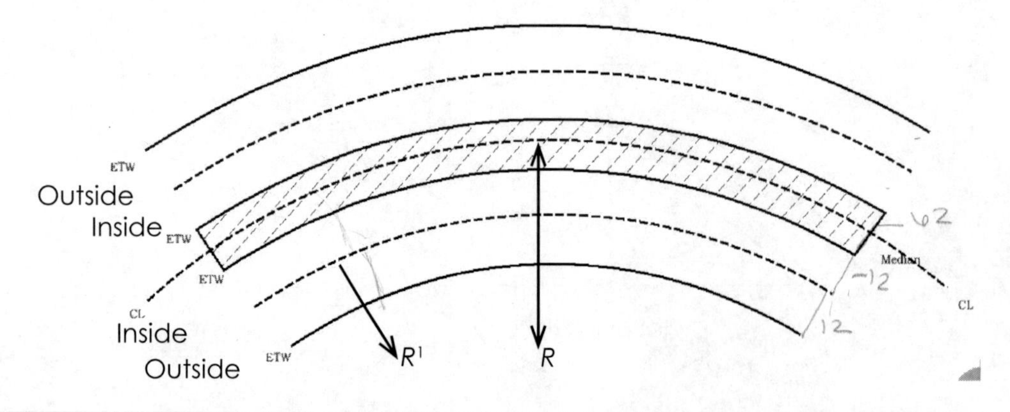

Figure 7-4 Route 8

3. Compound Curves

Compound curves should be avoided because drivers who have adjusted to the first curve could overdrive the second curve if the second curve has a smaller radius than the first. Exceptions can occur in mountainous terrain or other situations where use of a simple curve would result in excessive cost. Where a compound curve is necessary, the shorter radius should be at least two-thirds the longer radius when the shorter radius is 1,000 feet or less. On one-way roads, the larger radius should follow the smaller radius.

The total arc length of a compound curve should be not less than 500 feet.

Example 7-3:

Assume compound curves are considered in a roadway design for a ramp in California. The longer radius curve follows the shorter radius curve. The radius of the shorter curve is 1,000 ft. What is the minimum radius for the longer curve?

Solution:

According to the Caltrans standards on compound curves, the radius of the longer curve should be at least $3/2 \times 1{,}000$ ft = 1500 ft.

4. Reversing Curves

When horizontal curves reverse direction, the connecting tangents in between should be long enough to accommodate the standard superelevation runoffs (see Figure 202.5 of Caltrans *HDM*).

If this is impossible to have a sufficient tangent to link the reversing curves, the 6% per 100 feet rate of change should govern. When feasible, a minimum of 400 feet of tangent should be considered.

Superelevation is typically required when a vehicle moves in a circular lane. The vehicle experiences a centrifugal acceleration that acts toward the center of curvature. The superelevation, in working with the side friction developed between the vehicle's tires and the pavement surface, balances the centrifugal force.

Superelevation is measured in highway geometric design by superelevation rate (e) or cross slope. Since cross slopes on tangent segments and curves are different, a superelevation transition is required to gradually change the cross slope from tangent segments to the curves and from the curves to the tangent segments (see Figure 7-5; the superelevation rate is assumed to be 9%).

The superelevation transition consists of crown runoff and superelevation runoff. The crown runoff is the distance required for the cross slope change from –2% to 0%. The superelevation runoff is the distance to allow the cross slope to change from 0% to full superelevation rate (or e).

Example 7-4:

Suppose a rural 4-lane, undivided freeway is designed with the following elements (see Figure 7-6). Design speed V = 60 mph. Width of median along the curve is 62 ft. Lane width is 12 ft. Determine the following using Caltrans standards:

1. What are e_{max}, R_{min} at the station line, superelevation runoff (SE_{runoff}), and crown runoff (SE_{Crown})?
2. What are the superelevation rates at the beginning, middle, and end points of the SE_{runoff}?
3. What is the relative gradient used in Caltrans *HDM* for SE_{runoff} if the design superelevation rate (e) is 6% and 2 lanes are rotated.

Solution:

Since the highway facility is freeway, e_{max} = 10% according to Table 202.2 of Caltrans *HDM.*

Given the design speed is 60 mph, the minimum radius of the most inner ETW is 1,150 ft. The minimum curve radius (R_{min}) at the station line or the centerline of the median is 1,150 ft + 12 ft + 12 ft + 31 ft = 1,205 ft.

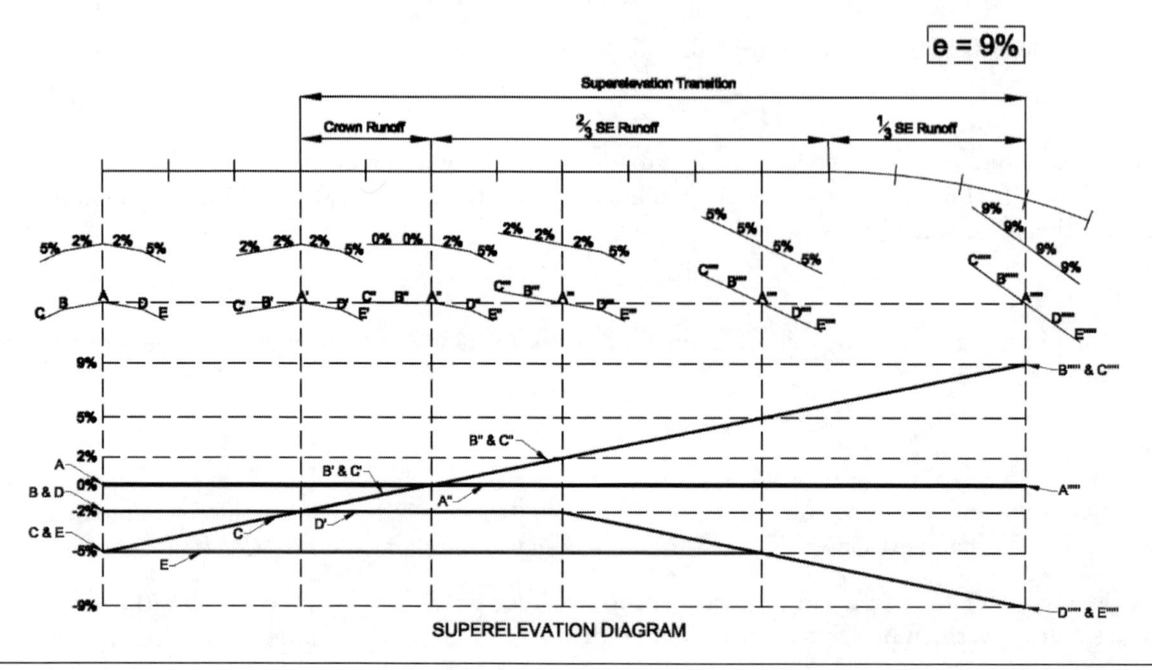

Figure 7-5 Superelevation Transition

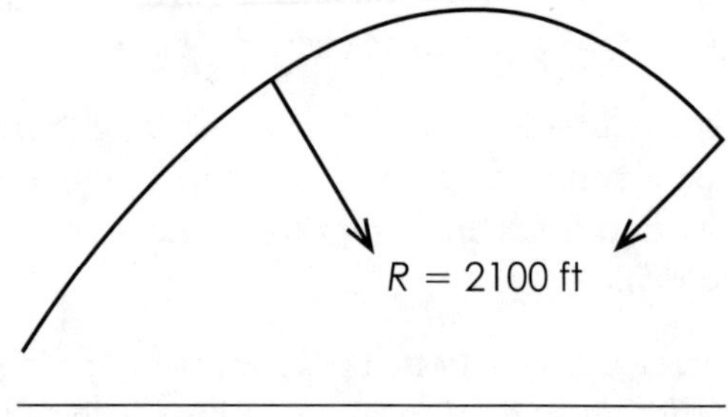

Figure 7-6 Curve for Example 7-4

The design radius of 2,100 ft for the curve is OK since it is greater than the radius of 1,205 ft.

Given e_{max} = 10% and R = 2,100 ft, design superelevation rate (e) is 6% according to Table 202.2 of Caltrans *HDM.*

Given two lanes are rotated and e = 6%, superelevation runoff (SE_{runoff}) = 210 ft, according to Figure 202.5A of Caltrans *HDM.*

Crown runoff (SEcrown) follows the same cross slope change rate as superelevation runoff and can be calculated as follows:

From 0% to 6%   From −2% to 0%

210 ft. SE_{crown}

$SE_{crown} = 210 \times 2/6 = 70$ ft

The superelevation rates at the beginning, middle, and end of the superelevation runoff are 0%, 3%, and 6%, respectively.

The relative gradient is defined to the change of cross slope over superelevation runoff. The total cross slope change from the tangent segment to the curve is 6% and the superelevation runoff is 210 ft. The relative gradient is

$= 6{:}\ 210 = 1{:}\ 35$

This indicates that in order to change 1% cross slope, the longitudinal distance of 35 ft is required for the transition.

Example 7-5:

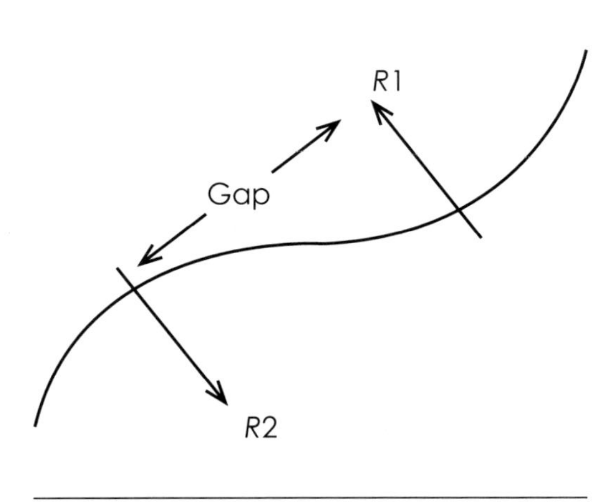

Figure 7-7 Reverse Curves

Suppose a rural 4-lane, undivided freeway in California is designed with the following reverse curves (see Figure 7-7):

Assume: $R1 = 1{,}250$ ft

$R2 = 1{,}500$ ft

Design Speed V = 60 mph

Two lanes are rotated

What is the minimum gap required for the transition between two reverse curves?

Solution:

Since the highway facility is freeway, emax = 10% according to Table 202.2 of Caltrans HDM.

Given the design speed is 60 mph, the minimum curve radius of the most inner ETW is 1,150 ft. The minimum curve radius (R_{min}) at the station line or the centerline of the median is 1,150 ft + 12 ft + 12 ft + 31 ft = 1,205 ft.

*R*1 and *R*2 are OK since they both are greater than the minimum radius of 1,205 ft.

Given $e_{max} = 10\%$ and $R1 = 1{,}250$ ft, design superelevation rate (e) is 9% according to Table 202.2 of Caltrans HDM.

Given $e_{max} = 10\%$ and $R2 = 1{,}500$ ft, design superelevation rate (e) is 8% according to Table 202.2 of Caltrans *HDM.*

Given two lanes are rotated and e = 9%, superelevation runoff ($SE_{runoff1}$) for the *R*1 curve = 330 ft, according to Figure 202.5A of Caltrans *HDM.*

Given two lanes are rotated and e = 8%, superelevation runoff ($SE_{runoff2}$) for the *R*2 curve = 300 ft, according to Figure 202.5A of Caltrans *HDM.*

The gap between the two reverse curves is the sum of 2/3 of $SE_{runoff1}$ and 2/3 of $SE_{runoff2}$, that is,

$$\text{Gap} = \frac{2}{3} \times 330 + \frac{2}{3} \times 300 = 420 \text{ ft}$$

The length of the design segment that links to the reverse curves shall be greater than 420 ft to meet Caltrans advisory standards on reverse curves.

5. Broken Back Curves

A broken back curve consists of two curves in the same direction joined by a short tangent. If possible, avoid designing highway horizontal alignments with broken back curves, as they are unsightly and against drivers' expectation when driving.

6. Alignments on Bridges

Due to the difficulty in constructing bridges with superelevation rates greater than 10%, the curve radii on bridges should be designed to accommodate superelevation rates of 10% or less.

Superelevation transitions on a bridge are difficult to construct and almost always result in an unsightly appearance of the bridge and the bridge railing. Therefore, if possible, horizontal curves should begin and end a sufficient distance from the bridge so that no part of the superelevation transition extends onto the bridge.

All the above criteria and standards shall be met and all the design considerations should be balanced to produce an alignment.

7.2.2 Horizontal Alignment Design Using InRoads

Horizontal alignment design in InRoads involves the use of a geometry project to store horizontal alignments, vertical alignments, and other design components (see Figure 7-8). A *geometric project* is a container that puts horizontal alignments and their associated elements (such as vertical alignments and superelevation definitions) into a file with the extension .alg. The *COGO point buffer* is a place where coordinates of geometric points are stored. In this book, we do not use the COGO features for the tutorial project.

To create a geometry project and horizontal alignments, InRoads must work with MicroStation. MicroStation provides a drawing environment within which designers can draw and display graphical elements (tangent segments and curves) for alignments. When certain graphical elements are finalized for an alignment, they can be imported directly into a geometry project as an alignment. (The original graphical elements remain unchanged in MicroStation.) Designers can also retrieve key information, such as coordinates of points of intersection (PIs) and radii of curves, from graphical elements and build horizontal alignments in InRoads.

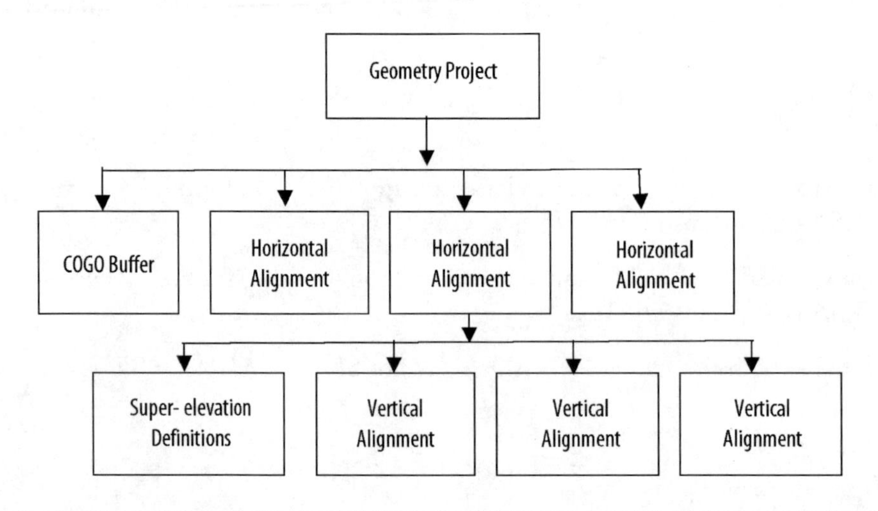

Figure 7-8 Structure of a Geometry Project in InRoads

InRoads provides a set of utilities and functions for managing entities (DTM surfaces, geometric projects, etc.) and displaying the entities in MicroStation. Once the entities are displayed in MicroStation as graphical elements, they belong to MicroStation. Deleting these graphical elements from MicroStation does not affect the entities stored in InRoads.

7.2.3 Create the Geometry Project for the Tutorial Project

Before we work on the horizontal alignments for Route 8, ramps, and Laguna Canyon Road, we have to create a geometry project in which the horizontal alignments can be stored. Follow these steps to create a new geometry project:

1. Start MicroStation and InRoads (if they are not open).
2. Copy the **tutorial_2D.dgn** file from the **C:/Temp/Book/Chp7/Master Files**[2] folder to the **C:/Temp/Turorial/Chp7/Master Files** folder.
3. Load the **tutorial_2D.dgn** file from **C:/Temp/Tutorial/Chp7/Master Files** folder or from the book website: www.csupomona.edu/~xjia/GeometricDesign/Chp7/Master Files/.

   The **tutorial_2D.dgn** file is derived from **tutorial.dgn** file (a 3-D file). It does not have z value. You can use either the **tutorial_2D.dgn** file or the **tutorial.dgn** file for horizontal alignment design. The **tutorial_2D.dgn** file is preferred since it can help you snap to graphical elements easily.
4. Create a geometry project by choosing ***InRoads → File → New***. The **New** window appears as shown in Figure 7-9.

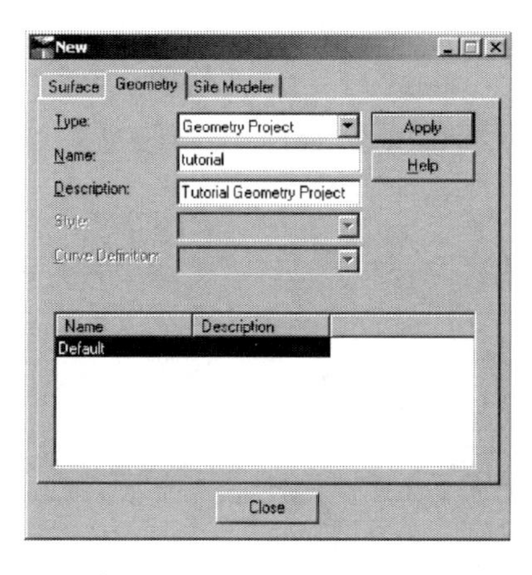

Figure 7-9 New Window for Creating Geometry Project

5. Select ***Apply*** and ***Close***.

   The geometry project **tutorial** is created. It is empty. It does not yet have any horizontal or vertical alignments.
6. Make sure the geometry project **tutorial** is active (see Figure 7-10).

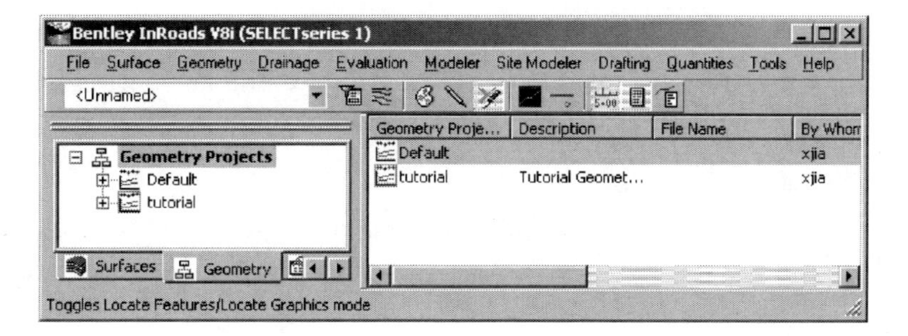

Figure 7-10 Active Geometry Project – Tutorial

[2] *C:/Temp/Book/Chp7/Master Files and C:\Temp\Book\Chp7\Master Files are interchangeable.*

7. If the tutorial project is not active:
   a. Go to ***InRoads* → *Geometry* → *Active Geometry*.**
   b. Select ***Tutorial*** and ***Apply***.
   c. Select ***Close***. OR, you can right click the **tutorial** geometry project and select **Set Active**.

7.2.4 Horizontal Alignment for Route 8

Now we design the Route 8 horizontal alignment. Follow these steps:

1. Go to ***MicroStation* → *Settings* → *Levels* → *Manager*.**

   The **Level Manager** window appears (see Figure 7-11).

2. In the **Level Manager** window, go to ***Levels* → *New*.**
   a. Name the new level **CL** and set the **Color, Line Style,** and **Line Weight** to your preference.
   b. Set the ***CL*** level as the active level.
   c. Close the **Level Manager** window.

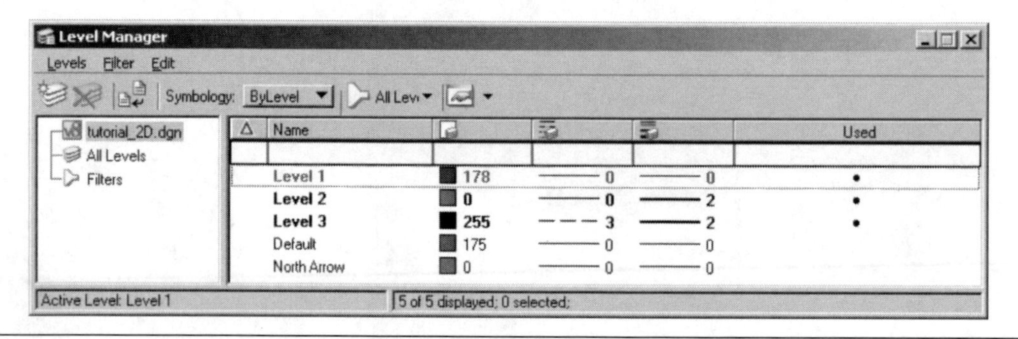

Figure 7-11 Level Manager

3. Create graphical elements for Route 8 by following these steps:
   a. Select **MicroStation → Place Line**. Snap to Point A and Key in the following in the Key-in box: xy = 8197.5641, 7069.9315

      Note that x and y values are the easting and northing coordinates of the first PI:

Easting:	8197.5641
Northing:	7069.9315

      This step creates a tangent segment to connect to Point A. The tangent segment shall have the same bearing as S70°E and follow the required entering direction.

   b. Select **MicroStation → Place Line.**

      Key in the following in the Key-in box: xy = 15032.9951, 8289.3462

   c. Snap to Point B.

Note that x and y values form the easting and northing coordinates of the second PI:

Easting:	15032.9951
Northing:	8289.3462

This step creates a tangent segment to connect to Point B. The tangent segment shall have the same bearing as S80°E.

d. Select **MicroStation → Place Line** and snap to the first and second PIs.

   This step creates a tangent segment perpendicular to the Laguna Canyon Road centerline that passes through Point C (see Figure 7-12). This keeps the horizontal alignment of a local road perpendicular to an undercrossing bridge, a good practice in design.

e. Select the radius of the first curve (close to Point A) to be 1,250 ft, which is greater than 1,205 ft (the minimum radius of curve on Route 8).

   Note that the design radius meets Caltrans standards on minimum curve radius.

f. Use the ***MicroStation → Construct Circular Fillet*** to place the first curve for the first PI Point (close to Point A). Type ***1250*** into **Radius** field and select ***Both*** for the **Truncate** field.

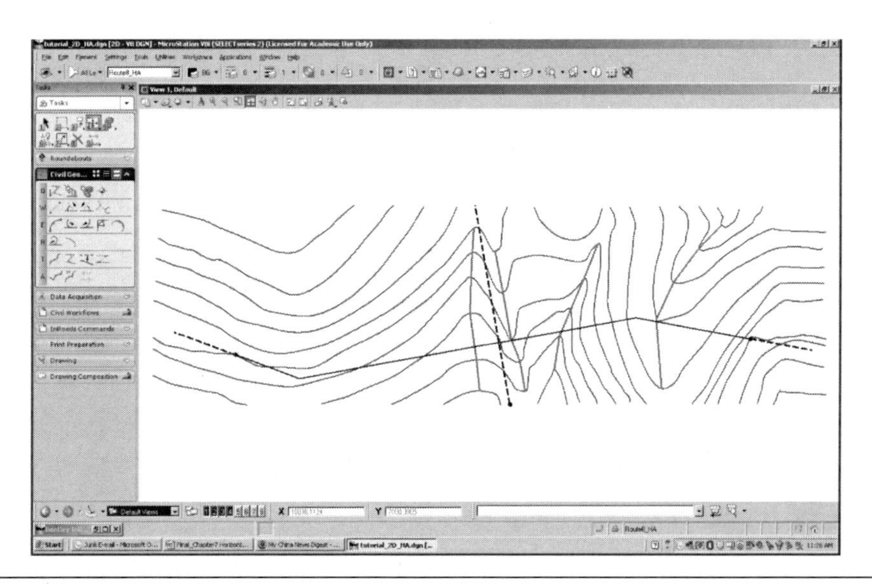

Figure 7-12 Route 8 Design Process

g. Select the radius of the second curve (close to Point B) to be 1,500 ft, which is greater than 1,205 ft (the minimum radius of curve on Route 8).

h. Use the ***MicroStation → Construct Circular Fillet*** to place the second curve for the second PI Point (close to Point B). Type ***1500*** into **Radius** field and select ***Both*** for the **Truncate** field.

   Figure 7-13 shows the graphical elements created for the Route 8 horizontal alignment.

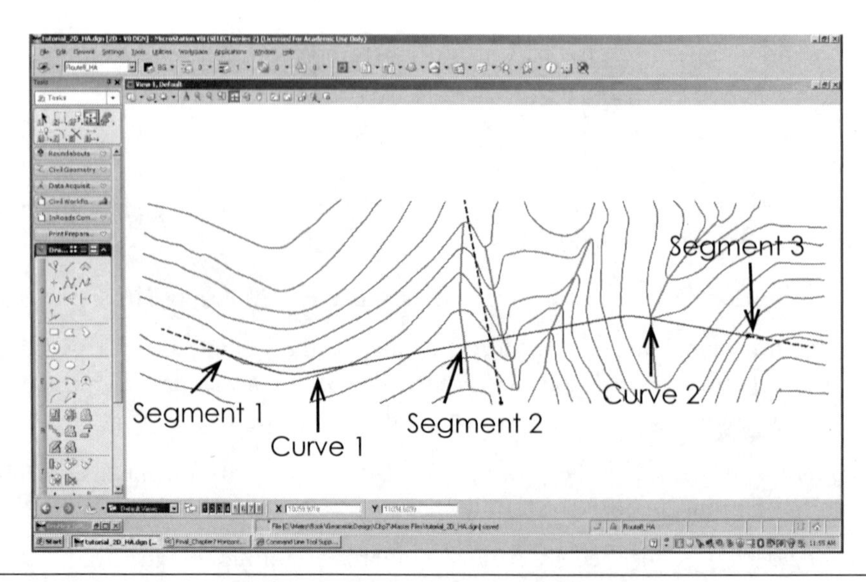

Figure 7-13 Graphical Elements for Route 8 Horizontal Alignment

Before we put the graphical elements into InRoads to form the Route 8 horizontal alignment, we need to further check their conformance to Caltrans standards.

There are five elements created for Route 8 (see Table 7-1). The conformance check shows that these graphical elements altogether meet Caltrans standards.

4. Import the graphical elements into InRoads to create Route 8 horizontal alignment.
    a. Make sure the **CL** level is active in MicroStation. Make sure the **tutorial** geometry project is active. If it is not active, set it to be active.
    b. Within MicroStation, create a complex chain using **Create Complex Chain** command (see Figure 7-14).
    c. Select the five graphical elements in the order of segment 1, curve 1, segment 2, curve 2, and segment 3 for Route 8 and create a chain for Route 8.

Table 7-1 Conformance Check of Caltrans Standards

Element Type	Length (L) /Radius (R)	Conformance Check
Segment 1	1055.30 (L)	$e_{max} = 10\%$ e = 8%, given $R1$ = 1,250 ft SE_{runoff} = 300 ft 2/3 × SE_{runoff} = 200 ft 1,055.30 ft > 200 ft. OK
Curve 1	657.29 (L) 1250.00 (R)	Deflection angle > 10°. OK

Segment 2	6332.34 (L)	2/3 × SE_{runoff} of Curve 1 = 200 ft 2/3 × SE_{runoff} of Curve 2 = 200 ft 6,332.34 ft > 200 ft + 200 ft. OK
Curve 2	525.97 (L) 1500.00 (R)	Deflection angle > 10°. OK
Segment 3	2134.94 (L)	emax = 10% e = 8%, given R2 = 1,500 ft SE_{runoff} = 300 ft 2/3 × SE_{runoff} = 200 ft 2,134.94 ft > 200 ft. OK

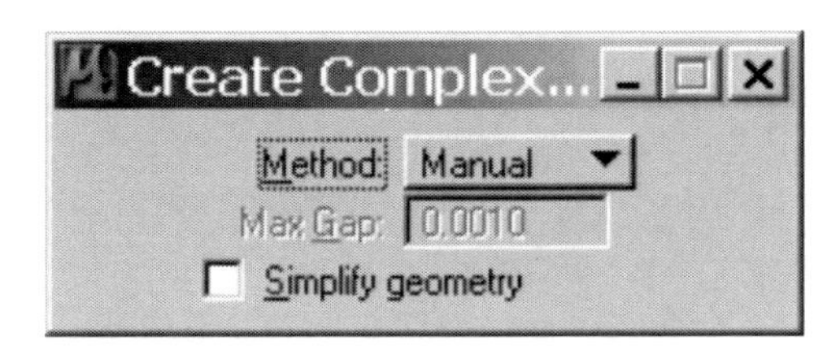

Figure 7-14 Create Complex Chain Command

d. Go to ***InRoads → Import → Geometry ....***

The **Import Geometry** window appears (see Figure 7-15).

e. Select the **From Graphics** tab and type the following:

Type:	**Horizontal Alignment**
Name:	**Route 8**
Description:	**Horizontal Alignment**
Style:	**Default**
Horizontal Curve Definition:	**Arc**
Geometry Project:	**Tutorial**
Other fields:	Same as shown in Figure 7-15

f. Choose ***Apply***.

InRoads prompts you to select the graphical elements.

g. Select the chain you just created.

h. Choose ***Close***. Check in the **Workspace Bar** window to see if Route 8 is created.

Your **Workspace Bar** window will look similar to Figure 7–16.

i. Select ***InRoads → Geometry → Horizontal Curve Set → Stationing ....***

j. Type **10+00** in the Starting Station field. Select **Apply** and **Close**.

Note: This step sets the starting station of Route 8 to be 10+00. In practice, we do not use 0+00 as the starting station for a new alignment.

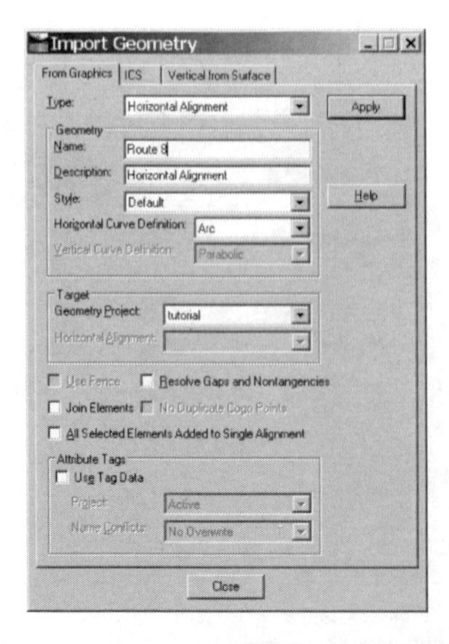

Figure 7-15 Import Geometry Window

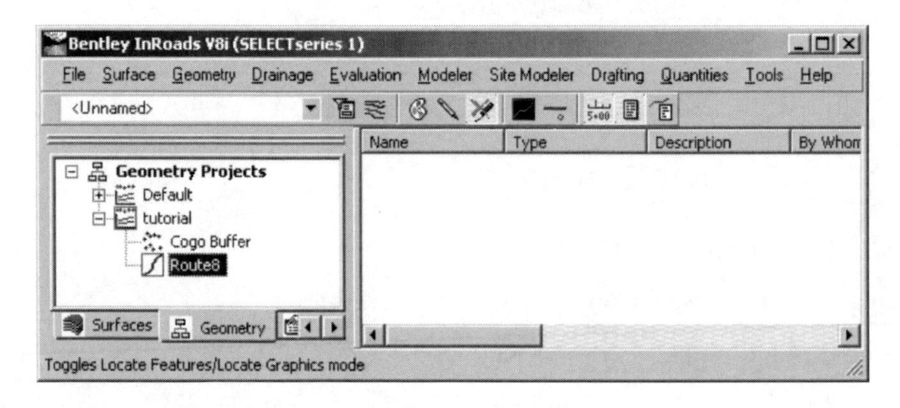

Figure 7-16 Route 8 Created

5. Save the **tutorial** geometry project.
   a. Make sure the **tutorial** geometry project is set active.
   b. Go to ***InRoads → File → Save → Geometry Project.*** Save the **tutorial** geometry project as ***tutorial.alg*** in the **C:/Temp/Tutorial/Chp7/Master Files** folder.
6. The Route 8 horizontal alignment is created and the **tutorial** geometry project is saved. Now we can view Route 8 using InRoads tools.
   a. Make sure the **tutorial** geometry project is set as active.
   b. Make sure Route 8 is set as active.
   c. Select ***InRoads → Geometry → View Geometry → Active Horizontal.***
   d. Select ***InRoads → Geometry → View Geometry → Stationing ….***

   The **View Stationing** window appears (see Figure 7-17).

e. In the **View Stationing** window, do the following:

» Choose Regular Stations (see Figure 7-18).
» Enter these fields with the following options or values:

Major Station: **Check on**
Major Ticks: **Check on**
Minor Ticks: **Check on**
Set the format on Major Station: **SS+SS.SS**

» Click thc Text Symbology icon of the **Major Station** (see Figure 7-18).

The Text Symbology window appears (see Figure 7-19).

» In the **Text Symbology** window, type ***0.20*** in the **Horizontal** field.
» Select ***OK***.
» Choose ***Apply*** and ***Close*** in the **View Stationing** window.

Stations are displayed along the horizontal alignment of Route 8 (see Figure 7-20).

Sometimes the stations are displayed with other symbols overlapping each other, which make it difficult to see the stations clearly.

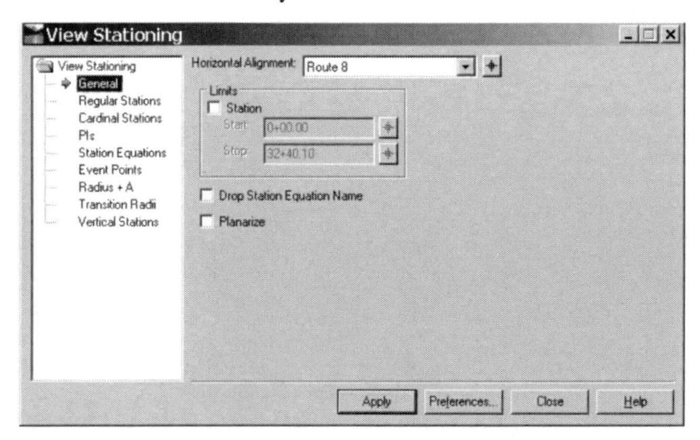

Figure 7-17 View Stationing Window

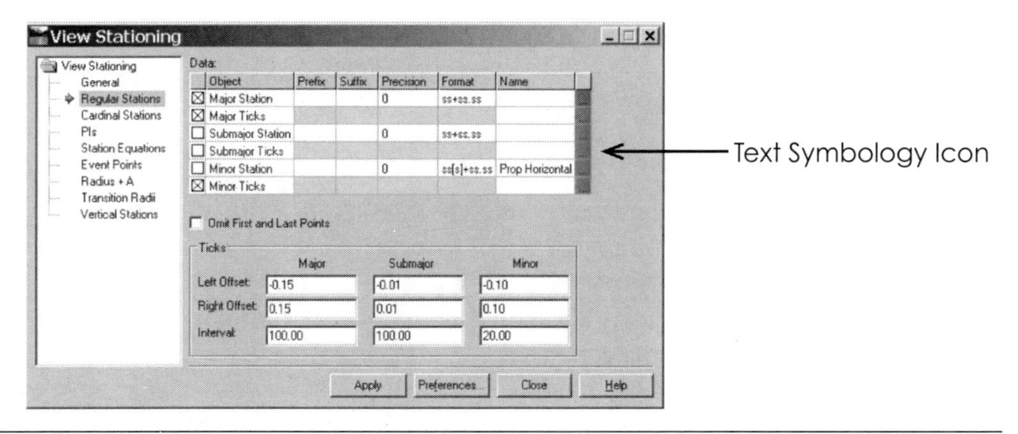

Figure 7-18 View Stationing Settings

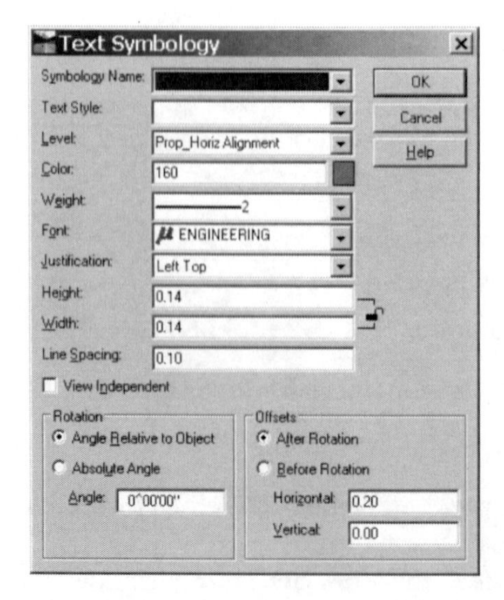

Figure 7-19 Text Symbology Window

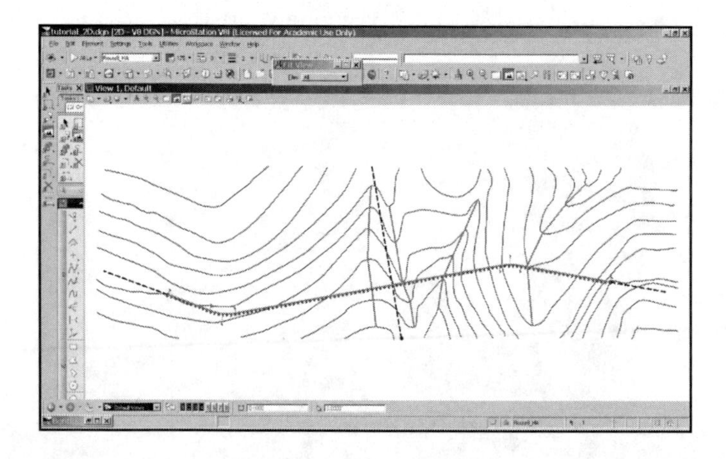

Figure 7-20 Route 8 Stations

Follow these steps to clean the noisy symbols:

a. Choose ***MicroStation → Settings → View Attributes.***

b. Turn the **Text Nodes** off.

c. Close the **View Attributes** window.

Note that the stations on Route 8 increase from left to right and meet Caltrans standards.

7. Create lanes for Route 8.

The lane configuration of Route 8 is shown in Figure 7-21. Create a new level called **ETW** in MicroStation using **Level Manager** command.

a. Set the below settings as:

Color:	***Your choice***
Line Style:	***0***
Line Weight:	***1***

b. Use the MicroStation **Move/Copy Parallel** command to create ETW lines for Route 8. Taking two ETWs (close to the CL) as an example, you need to do the following:

c. Go to ***MicroStation → Move/Copy Parallel.*** The **Move/Copy Parallel** command window appears (see Figure 7-22). Set the following within the **Move/Copy Parallel** window:

Mode:	***Original***
Distance:	***31***
	Check on
Keep original:	***Check on***
Use Active Attributes:	***Check on***

Distance of 31 ft is measured from the Centerline to its closest ETW.

d. Select the Route 8 Centerline and create the two ETWs from the centerline (see Figure 7-23).

e. Set the following within the **Move/Copy Parallel** window:

Mode:	***Original***
Distance:	***55***
	Check on
Keep original:	***Check on***
Use Active Attributes:	***Check on***

Distance of 55 ft is measured from the Centerline to the two left side ETWs.

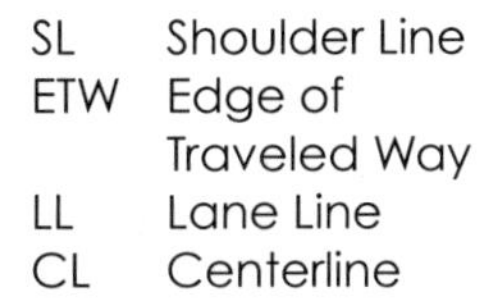

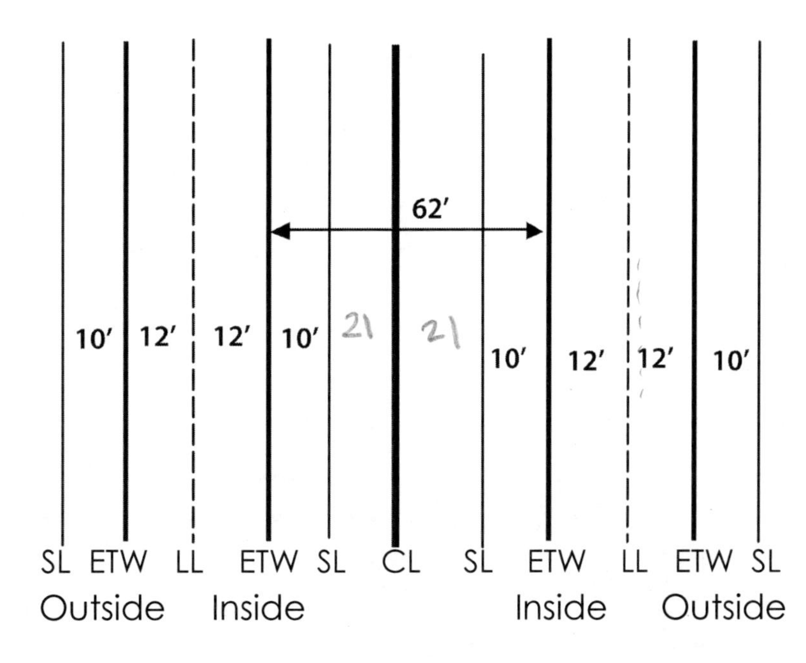

Figure 7-21 Lane Configurations for Route 8

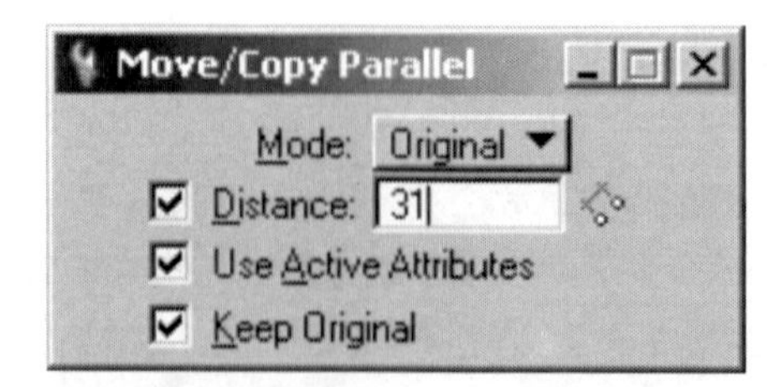

Figure 7-22 Move/Copy Parallel Window

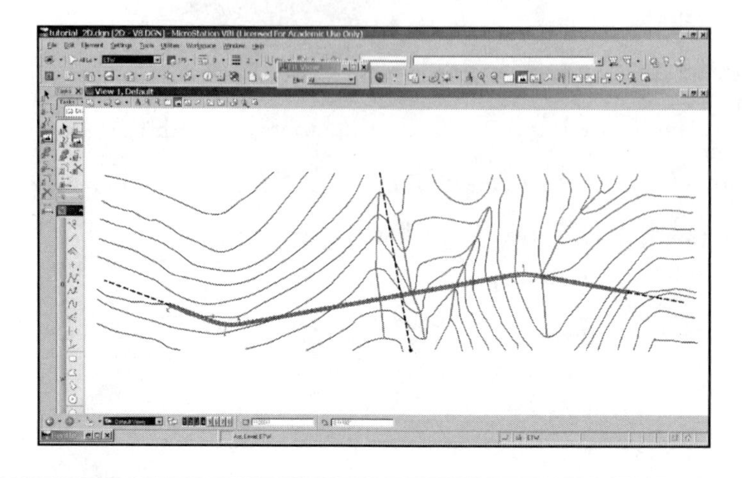

Figure 7-23 Edges of Traveled Way (ETWs)

f. Create the new **SL** level. The **line style** is **solid line**, the **line weight** is **0** and the **color** is your choice.

g. Select **SL** level as the active level and create SLs using the **Move/Copy Parallel** command. The distances from the centerline to the right and left SLs (along the driving direction of Route 8) are **21** and **65**, respectively.

h. Create the new **LL** level. The **line style** is **dashed line**, the **line weight** is **0** and the **color** is your choice.

i. Select **LL** level as the active level and create LLs using the **Move/Copy Parallel** command. The distance from the Centerline to the LLs is **43**.

Figure 7-24 shows all the lanes configured for Route 8.

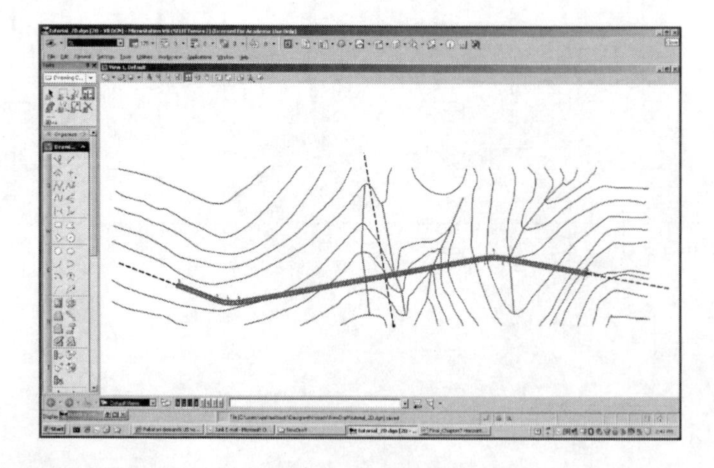

Figure 7-24 Created Lane Configuration for Route 8

7.2.5 Horizontal Alignment for Laguna Canyon Road

Now we start working on the design of horizontal alignment for Laguna Canyon Road. Steps involved in the design are as follows:

1. Go to ***InRoads → File → New.***

   The **New** window appears.

   a. Select the **Geometry** tab and type the following:

Type:	**Horizontal Alignment**
Name:	**Laguna Canyon Road**
Description:	**Laguna Canyon Road**
Style:	**Default**
Curve Definition:	**Arc**

   b. Select ***Apply*** and ***Close.***

   The empty horizontal alignment **Laguna Canyon Road** is created.

2. Make sure the **Laguna Canyon Road** horizontal alignment is active. If it is not, right click the horizontal alignment in the **Workspace Bar** window under the **Tutorial** geometry project (see Figure 7-25).

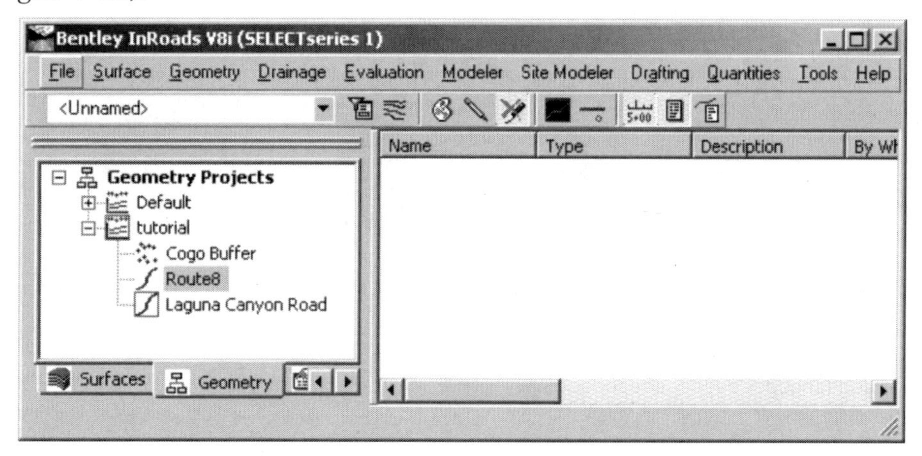

Figure 7-25 Set Laguna Canyon Road Active

3. Create the horizontal alignment for Laguna Canyon Road.

   a. Set **CL** level as the active one. If **CL** level is not available, go to ***MicroStation → Level Manager*** to create the **CL** level. Select the color, line style, and line weight of your choice.

   b. Select ***InRoads → Geometry → Horizontal Curve Set → Add PI.***

   The InRoads prompts you to identify first point.

   c. Select **Point C** or key in ***NE = 6561.6798, 12467.1916.***

   d. Select the North end point of the Laguna Canyon Road or key in ***NE = 9921.8132, 11874.7094.***

   e. If the MicroStation **Key In** window is not available, select ***MicroStation → Utilities → Key in.***

   f. Select ***InRoads → Geometry → Horizontal Curve Set → Stationing ….***

g. Type **10+00** in the Starting Station field. Select **Apply** and **Close**.

   This step sets the starting station of Laguna Canyon Road to be 10+00. The Laguna Canyon Road horizontal alignment is created.

h. Make sure the **tutorial** geometry project is set active.

   » Go to ***InRoads → File → Save → Geometry Project.***

   » Save the **tutorial** geometry project as ***tutorial.alg*** in the **C:/Temp/Tutorial/Chp7/Master Files** folder.

4. Now we view the Laguna Canyon Road horizontal alignment.

   a. Go to ***InRoads → Geometry → View Geometry → Stationing.***

   b. Select all the settings similar to those for Route 8.

   The stations are displayed along the Laguna Canyon Road (see Figure 7-26). Note that the stations increase in the direction from South to North, which is consistent with Caltrans standards.

5. Create lanes for Laguna Canyon Road

   The lane configuration for Laguna Canyon Road is shown in Figure 7-27.

   » Use the MicroStation **Move/Copy Parallel** Command to create the SLs, LLs, and ETWs.

   Figure 7-28 shows the lanes created from the centerline of Laguna Canyon Road.

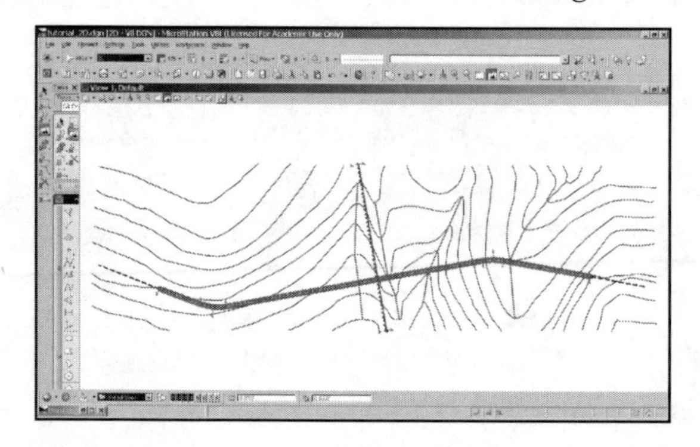

Figure 7-26 Laguna Canyon Road Horizontal Alignment

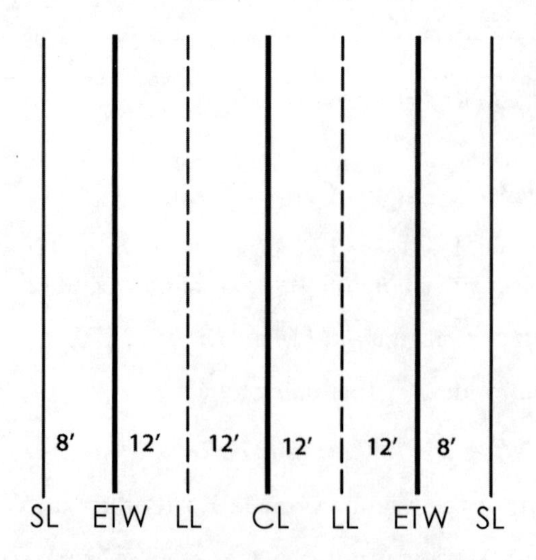

Figure 7-27 Lane Configuration of Laguna Canyon Road

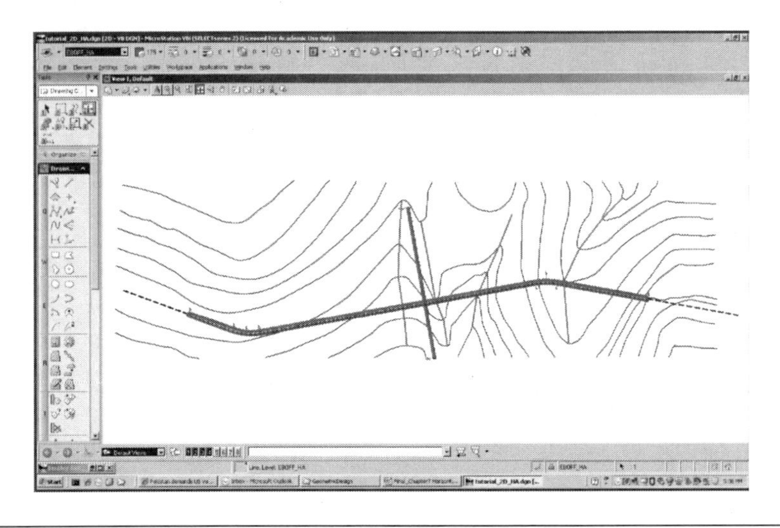

Figure 7-28 Lanes Created for Laguna Canyon Road

7.2.6 Horizontal Alignment for Eastbound Off Ramp

Horizontal alignment design for the Eastbound Off Ramp is a process that lays out the station line of the ramp (or the left ETW of the ramp[3]). It consists of three components: freeway exit, ramp intersection (or ramp terminal), and the ramp itself.

1. Location and Design of Freeway Exit

The freeway exit design involves 1) locating a point along the *right side* ETW[4] of Route 8 Eastbound (or the intersection point between the *right side* ETW of Route 8 Eastbound and the ramp station line) and 2) superimposing the standard design of freeway exit for the ramp.

The intersection point should be located on a tangent segment wherever possible to provide maximum sight distance and optimum traffic operation. If the location of the intersection point on a curve is necessary, the ramp exit taper should also be curved. The radius of the exit taper should be about the same as the freeway edge of traveled way in order to develop the same degree of divergence as the standard design.

Caltrans standard design of freeway exit is the advisory standard. We should design the ramp exit exactly same as shown in Figure 7-29 or Figure 504.2B of Caltrans *HDM.*

The minimum deceleration length (DL) shown on Figure 7-29 shall be provided prior to the first curve beyond the exit nose to assure adequate distance for vehicles to decelerate before entering the curve. The range of minimum DL (distance) is 570 ft, 470 ft, 420 ft, and 270 ft for the curve radius of less than 300 ft, 300–499 ft, 500–999 ft, and 1,000 ft and over, respectively.

[3] *Left ETW means the ETW on the left side of drivers in the driving direction along the ramp.*
[4] *Right Side ETW means the ETW on the right side of drivers in the driving direction along the ramp.*

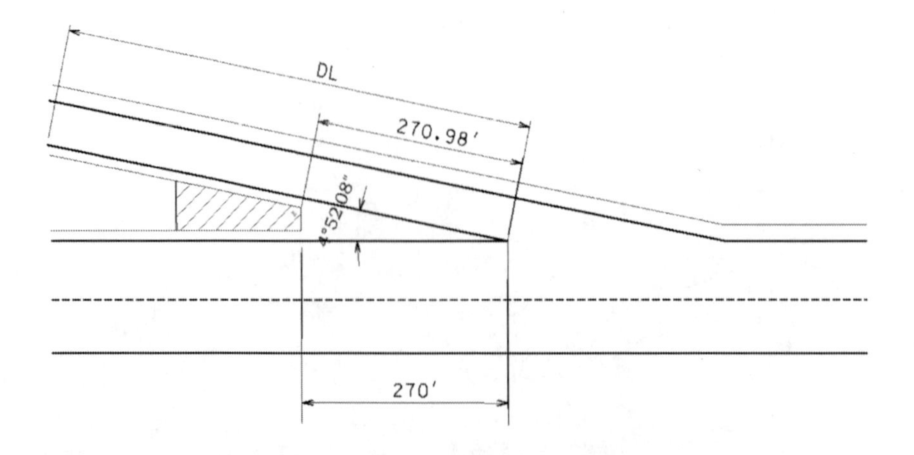

Figure 7-29 Standard Design of Freeway Exit

Note that the minimum length between the exit node and the end of ramp is 525 ft for a full stop at the end of ramp. The deceleration length is from the exit node to the beginning of curve (BC) of the curve. The exit nose shown in Figure 7-29 is located downstream of the 23-foot dimension.

Example 7-6:

Given the triangle below, what is the angle β?

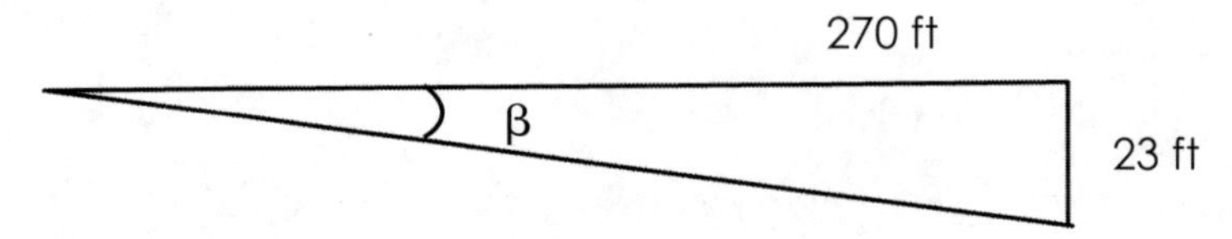

Solution:

$\beta = \tan^{-1}(23/270) = 4.869° = 4°52'08"$

Example 7-7:

Given the freeway exit layout below, what is the DL?

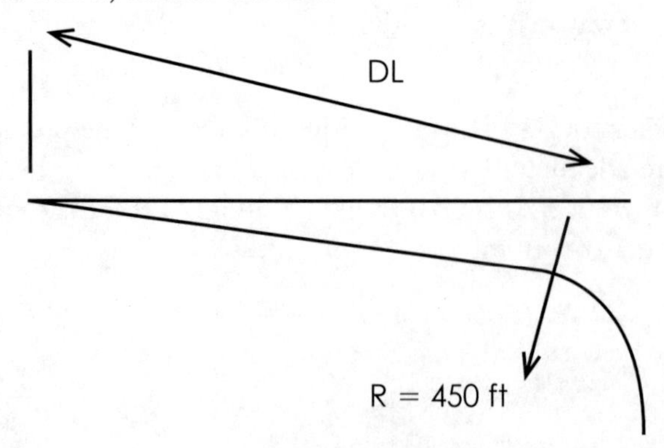

Solution:

According to Figure 7-29 or Figure 504.2B of Caltrans *HDM*, DL = 470 ft from the exit nose to BC of the curve.

Use the following steps to create the freeway exit:

a. Set the **ETW** level to be active. If **ETW** level is not available, go to ***MicroStation → Level Manager*** to create the **ETW** level. Select the color, line style, and line weight of your choice.

b. Find a point (later called the start point) on the outside ETW of Route 8 in the eastbound direction (between Station 46+00 and Station 47+00). The point coordinates are:

Northing:	7423.1405
Easting:	10487.6413

c. Select ***MicroStation → Place Line.*** Key in **ne=7423.1405, 10487.6413.**

d. Set the snap mode to be the ***Perp*** mode and snap to the nearest SL line (see Figure 7-30).

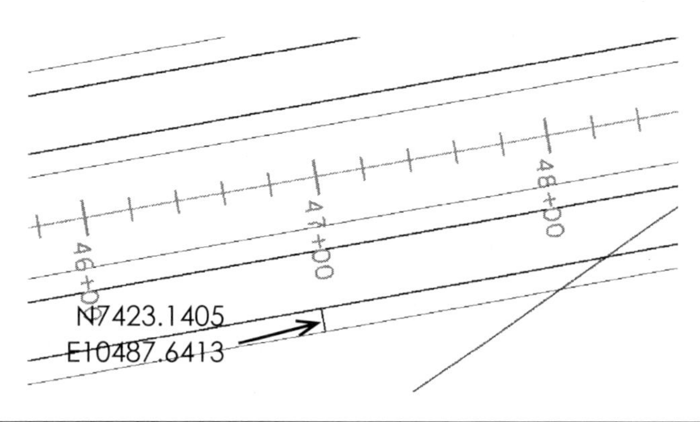

Figure 7-30 Draw the 23-ft Line Segment

e. Use ***MicroStation → Move/Copy Parallel*** command to parallel copy the 23-ft line segment 270 ft to the right.

f. Use ***MicroStation → Move/Copy Parallel*** command to parallel copy the Route 8 outside ETW (in the eastbound direction) 23 ft down or to the South (see Figure 7-31).

The two parallel-copied segments intersect at the point with the following coordinates:

Northing:	7447.9778
Easting:	10757.4785

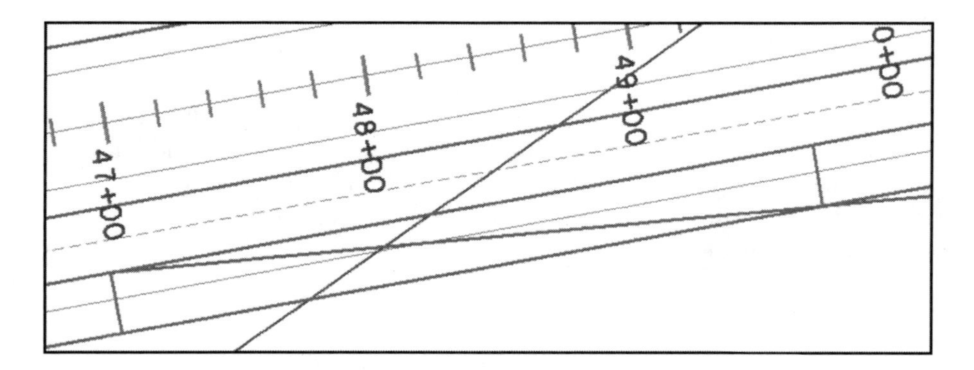

Figure 7-31 Draw the 270-ft Exit Line Segment

g. Draw a line segment passing through the two points.

h. Use ***MicroStation → Measure Angle*** command to measure the angle between the Route 8 outside ETW and the Eastbound Off Ramp alignment.

The result should be 4.869° or 4°52'08".

2. Location and Design of Ramp Intersection and Terminal

The design of the ramp intersection is to balance the factors such as sight distance, construction and right-of-way costs, storage requirements for left-turn movements off the Laguna Canyon Road, and the proximity of other local road intersections. In this project, we concentrate only on sight distance requirements since data for other factors is not available.

Horizontal sight restrictions to the Eastbound Off Ramp drivers are caused by side slopes of the Laguna Canyon Road in the area of the Route 8 bridge.

It is assumed that the ramp intersection is controlled by Stop signs. The sight distance is then measured between the center of the outside lane (close to the Eastbound Off Ramp terminal) and the eye of the driver of the ramp vehicle. The vehicle is assumed to be 10 feet back from the edge of shoulder at Laguna Canyon Road (see Figure 7-32).

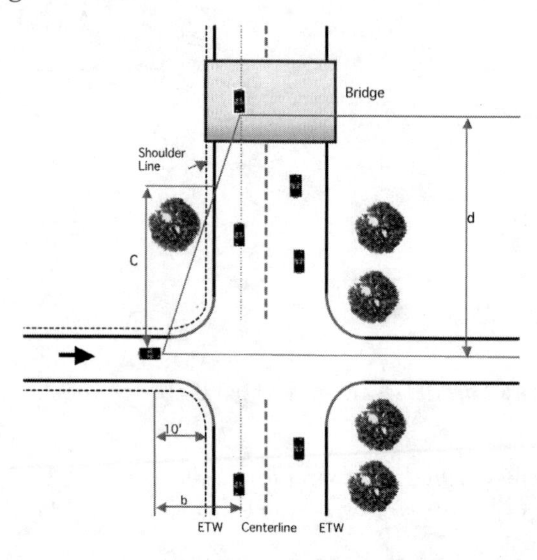

Figure 7-32 Sight Distance Requirements for Ramp Intersection

Ramp setback from the bridge structure can be calculated on the basis of sight distance controlled by the slopes. The ramp setback indicates the station line or the left ETW of the Eastbound Off Ramp shall be at least 343 ft away from the Route 8 shoulder line in the eastbound direction.

In addition to the setback requirement, the length of the crossroad (or the Laguna Canyon Road) should be set open to view left-turn maneuvers coming from the off ramp. Caltrans standards indicate that the length of the crossroad should be greater than the product of the prevailing speed of vehicles on the crossroad, and the time required for a stopped vehicle on the ramp to execute a left-turn maneuver. This time is estimated to be 7½ seconds. The design speed of the Laguna Canyon Road is 40 mph. The length of the crossroad open to view therefore is 495 ft.

Given the above analysis, the ramp intersection for the Eastbound Off terminal is set 500 ft away from the Route 8 bridge.

$$c = d \times \left[\frac{b - a - 6}{b} \right]$$

$$= 495 \times \left[\frac{(10 + 10 + 6) - 2 - 6}{10 + 10 + 6} \right]$$

$$= 343 \text{ ft}$$

Where: *a* Distance from ETW to bridge railing (not shown in Figure 7-32). It is assumed to be 2 ft.

b Distance from the centerline of the near lane to the eye of a ramp vehicle driver. The ramp driver's eye is assumed to be located 10 ft from the edge of shoulder, but not less than 15 ft from the ETW. The shoulder width is assumed to be 10 ft according to Caltrans standards. The lane width is 12 ft.

c Ramp set back from the end of bridge abutment.

d Corner sight distance. It is measured from the 3.5 ft eye height of drivers on the ramp to the 4.5 ft height of an object on the crossroad. Corner sight distance is 275 ft, 330 ft, 385 ft, 440 ft, 495 ft, 550 ft, 605 ft, 660 ft, 715 ft, and 770 ft for design speed of 25 mph, 30 mph, 35 mph, 40 mph, 45 mph, 50 mph, 55 mph, 60 mph, 65 mph, and 70 mph, respectively.

V Anticipated prevailing speed on crossroad.

Use the following steps to create the ramp terminal:

a. Set the **ETW** level to be active. If **ETW** level is not available, go to ***MicroStation → Level Manager*** to create the **ETW** level. Select the color, line style, and line weight of your choice.

b. Find a point on the right side ETW of Laguna Canyon Road (within the area of Station 17 + 00). Coordinates of this point are:

Northing:	7242.1034
Easting:	12322.8443

Since this point is more than 500 ft away from the Route 8 bridge, it meets Caltrans sight distance requirements of the ramp intersection.

c. Draw a line segment perpendicular to the right side ETW of Laguna Canyon Road (see Figure 7-33).

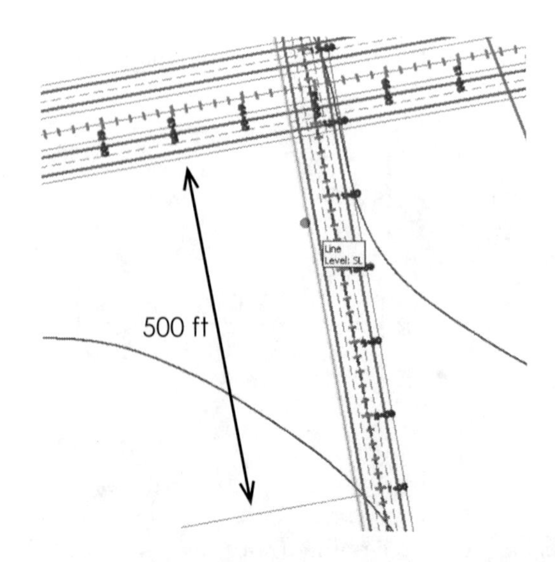

Figure 7-33 Ramp Terminal Design for Eastbound Off Ramp

3. Ramp Design

The ramp design is to fill the gap between the freeway exit and the ramp terminal.

a. Set the **ETW** level to be active. If **ETW** level is not available, go to ***MicroStation → Level Manager*** to create the **ETW** level. Select the color, line style, and line weight of your choice.

b. Draw a line segment that goes through the two PI points (see Figure 7-34):

First Point	Northing:	7485.5465
	Easting:	11165.6310
Second Point	Northing:	7155.6886
	Easting:	11833.7809

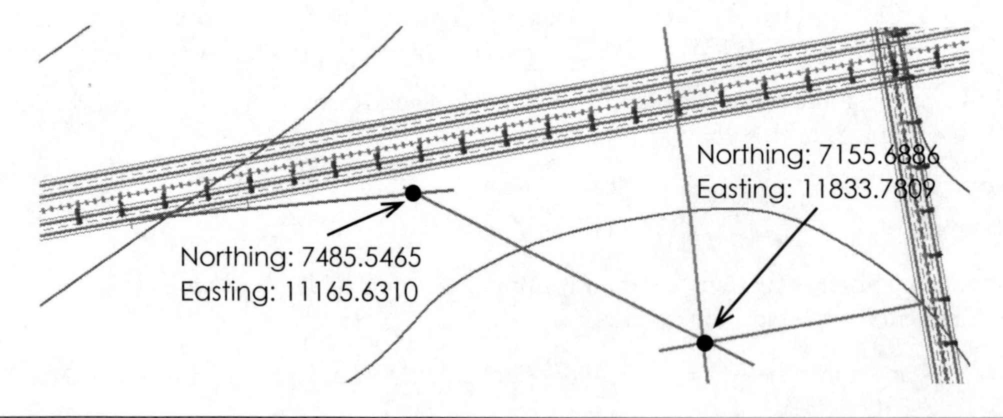

Figure 7-34 Filling the Gap

For the two PIs, use ***MicroStation → Construct Circular Fillet*** command.

c. Type the following in the **Construct Circular Fillet** window for the first PI:

Radius:	***850***
Truncate:	***Both***

d. Type the following in the **Construct Circular Fillet** window for the second PI:

Radius:	***400***
Truncate:	***Both***

e. Delete the parallel-copied line segments and the graphical elements for the Eastbound Off Ramp shown in Figure 7-35.

There are five graphical elements for this ramp: three line segments and two circular curves.

f. Select ***Close***.

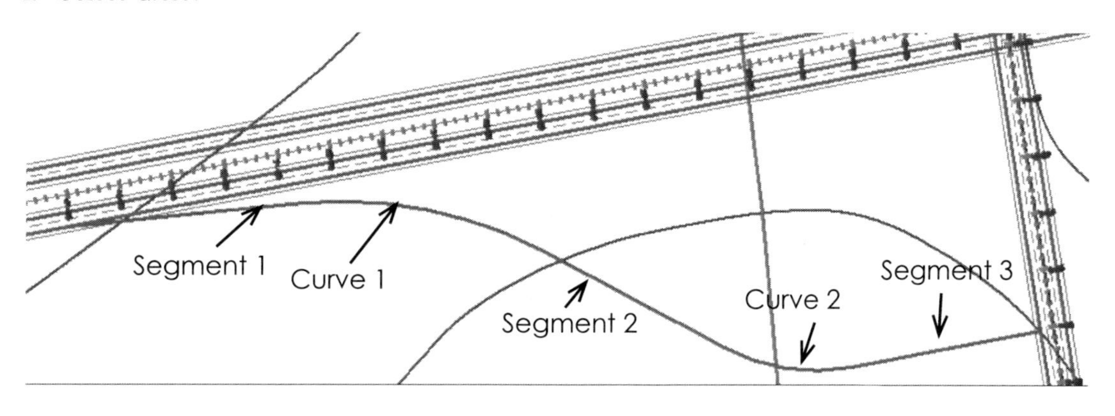

Figure 7-35 Graphical Elements Generated for Eastbound Off Ramp

4. Caltrans Standards Conformance Check

   The results of the conformance checks are shown in Table 7-2.

5. Import the Graphical Elements into InRoads to Create the Eastbound Off Ramp

   a. Make sure the **ETW** level is active in MicroStation.

   b. Make sure the **tutorial** geometry project is active. If it is not active, set it to be active.

   c. In MicroStation, create a complex chain (Segment 1, Curve 1, Segment 2, Curve 2, and Segment 3) using **Create Complex Chain** command.

   d. Select the five graphical elements and create a chain for Eastbound Off Ramp.

   e. Go to ***InRoads → Import → Geometry ...***

   f. In the **Import Geometry** window, select the **From Graphics** tab and type the following:

Type:	**Horizontal Alignment**
Name:	**EBOFF**
Description:	**Horizontal Alignment**
Style:	**Default**
Horizontal	
Curve Definition:	**Arc**
Geometry Project:	**Tutorial**
Other fields:	Same as those for Route 8

Table 7-2 Standards Check for Eastbound Off Ramp

Element Type	Length (L)/ Radius (R)	Conformance Check
Segment 1	440.86 (L)	According to Figure 7-29 (or 504.2B of Caltrans *HDM*), the deceleration length (DL) for Curve 1 is 420 ft since the curve radius is 850 ft. 440.86 ft > 420 ft. OK
Curve 1	850.00 (R) 467.82 (L)	The curve radius is 850 ft, same as 850 ft for the design speed of 50 mph. OK
Segment 2	373.62 (L)	Segment 2 acts as the superelevation (SE) transition for two reverse curves. e_{max} = 12% 2/3 × SE runoff of Curve 1 = 2/3 × 240 = 160 ft. 2/3 × SE runoff of Curve 2 = 2/3 × 300 = 200 ft. 373.62 ft > 360 ft. OK
Curve 2	400.00 (R) 254.14 (L)	The curve radius is 400.00 ft, greater than 300 ft for the design speed of 30 mph. Curve 2 is close to the ramp terminal. It is reasonable to assume the design speed for Curve 2 is 30 mph. OK
Segment 3	365.12 (L)	This segment is used for SE transition from e of −12% to 2% on Laguna Canyon Road. 2/3 × SE runoff = 200 ft. 2% transition = 300/12 × 2 = 50 ft. 365.12 ft > 250 ft. OK

g. Choose ***Apply***. InRoads prompts you to select the graphical elements. Select the chain you just created.

h. Check in the **Workspace Bar** window to make sure the **EBOFF** horizontal alignment is created. You will see a **Workspace Bar** window similar to Figure 7-36.

i. Select ***InRoads → Geometry → Horizontal Curve Set → Stationing ...***.

j. In the Stating Station field, type **10+00**. Select **Apply** and **Close**.
This step sets the starting station of Eastbound Off Ramp to be 10+00.

6. Save the **tutorial** geometry project.

   a. Make sure the **tutorial** geometry project is set active.

   b. Go to ***InRoads → File → Save → Geometry Project.***

   c. Save the **tutorial** geometry project as ***tutorial.alg*** in the **C:/Temp/Tutorial/Chp7/Master Files** folder.

7. View EBOFF Horizontal Alignment

   a. Make sure the **tutorial** geometry project is set active.

b. Make sure **EBOFF** is set active.

c. Go to ***InRoads → Geometry → View Geometry → Active Horizontal.***

d. Go to ***InRoads → Geometry → View Geometry → Stationing***

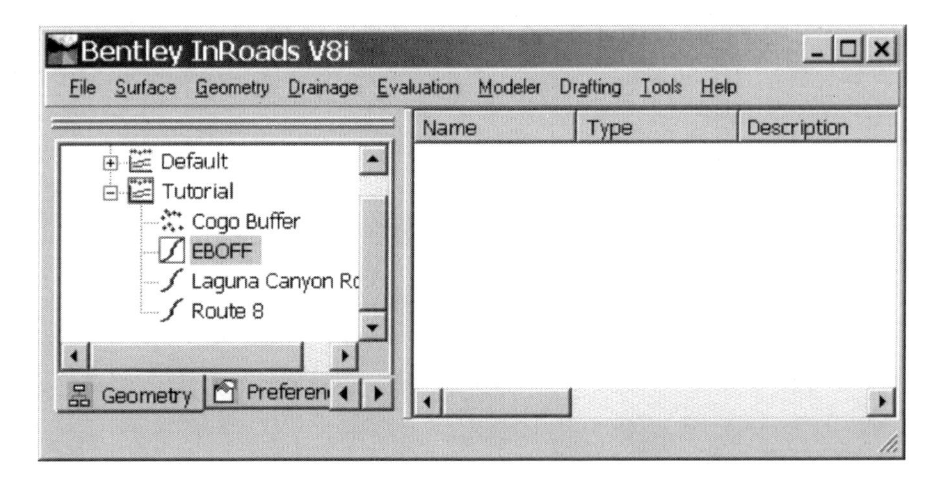

Figure 7-36 EBOFF Created

e. In the **View Stationing** window, select Regular Stations and do the following:

Major Station:	**Check on**
Major Ticks:	**Check on**
Minor Ticks:	**Check on**
Set the format on Major Station:	**SS+SS.SS**

f. Click the **Text Symbology** icon of the **Major Station.**
The **Text Symbology** window appears.

g. In the **Text Symbology** window, type ***0.20*** in the **Horizontal** field. Select ***OK.***

h. Select ***Apply*** and ***Close*** the **View Stationing** window.

Stations (increasing from left to right) are displayed along the EBOFF horizontal alignment (see Figure 7-37).

i. When the stations are displayed with other symbols overlapping each other, to clean the noisy symbols, follow these steps:

» Choose ***MicroStation → Settings → View Attributes.***
» Turn the **Text Nodes** off.
» Close the **View Attributes** window.

8. Create Lanes for Eastbound Off Ramp

The lane configuration of EBOFF horizontal alignment is shown in Figure 7-38.

a. Set the **ETW** level active. If **ETW** level is not available, using the **Level Manager** command, create a new level called **ETW** in MicroStation.

b. Set the settings below for the **ETW** level:

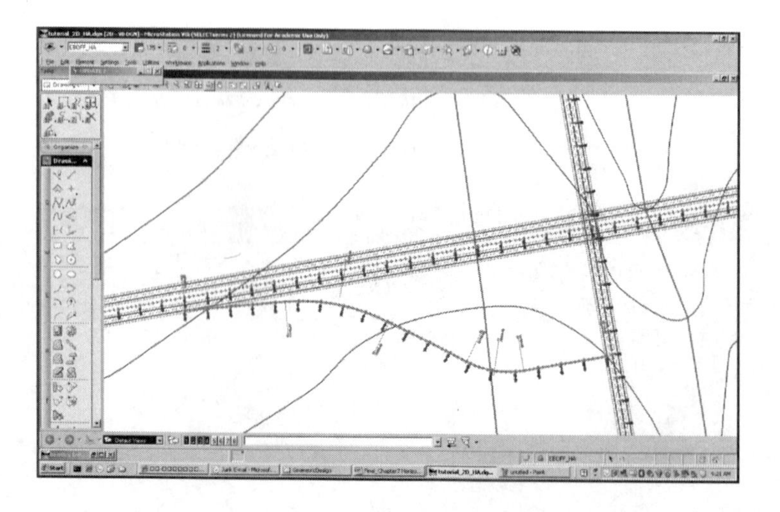

Figure 7-37 EBOFF Horizontal Alignment Stations

Color:	***Your choice***
Line Style:	***0***
Line Weight:	***1***

c. Use the MicroStation ***Move/Copy Parallel*** command to create ETW lines for EBOFF.

d. Set the **SL** level to be active.

e. Create a new **SL** level if the level does not exist.

f. Select the **SL** level to be active and set the settings below for the SL level:

Color:	***Your choice***
Line Style:	***0***
Line Weight:	***1***

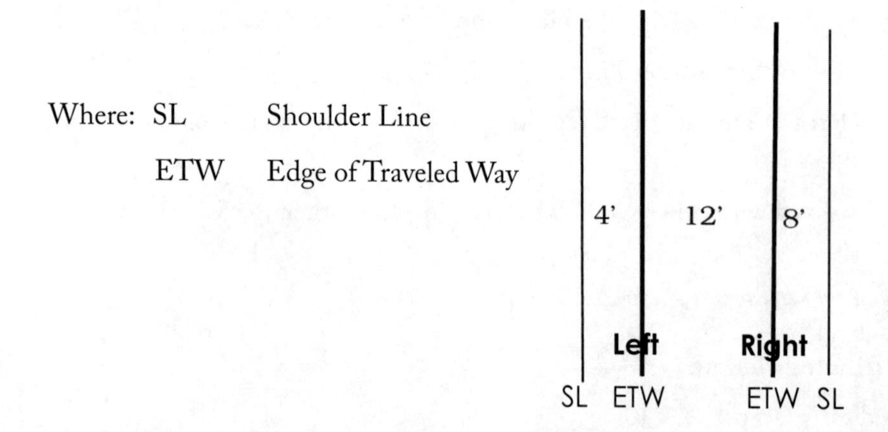

Figure 7-38 Lane Configuration for EBOFF

g. Create SLs using the **Move/Copy Parallel** command. The distances from the left ETW to the SLs are **4** and **20**, respectively.

h. Do the gore area treatment using MicroStation commands. Figure 7-39 shows the results of the treatment. You need to use Figure 504.2B of Caltrans *HDM* as a reference to polish the gore area.

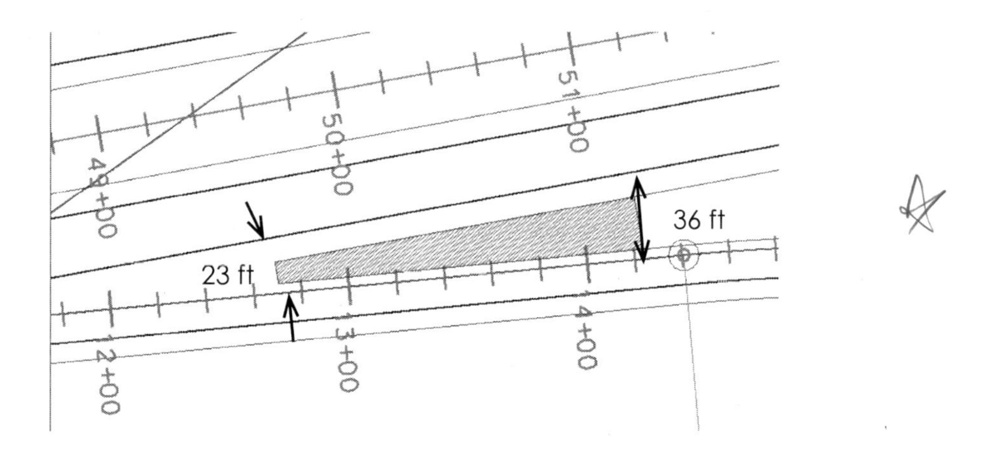

Figure 7-39 EBOFF Gore Area Treatment

The shaded area indicates the contrasting surface treatment beyond the gore pavement. This surface treatment enhances aesthetics and minimizes maintenance efforts. It helps drivers easily identify and differentiate the contrasting surface treatment from the pavement areas that are intended for regular vehicle use.

Figure 7-40 shows all the lanes configured for EBOFF horizontal alignment.

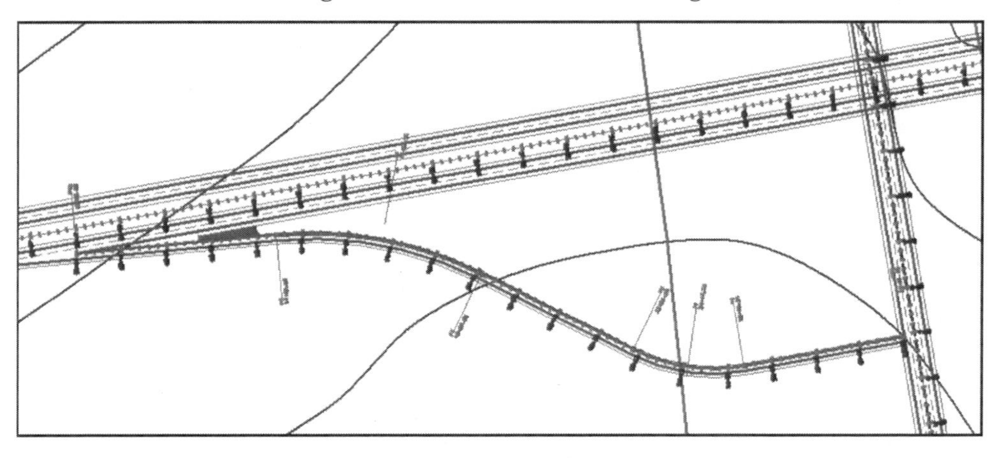

Figure 7-40 Created Lanes for EBOFF Horizontal Alignment

7.2.7 Horizontal Alignment for Eastbound On Ramp

Horizontal alignment design for Eastbound On Ramp is a process to determine the station line of the ramp (or the left ETW of the ramp). It consists of three components: freeway entrance, ramp intersection (or ramp terminal), and the ramp itself.

1. Location and Design of Freeway Entrance

The freeway entrance design involves 1) locating a point along the *right side* ETW of Route 8 Eastbound (or the intersection point between the *right side* ETW of Route 8 Eastbound and the ramp station line) and 2) superimposing the standard design of freeway entrance for the ramp.

The intersection point should be located on tangent sections of Route 8 Eastbound wherever possible to provide maximum sight distance and optimum traffic operation. In this project, we will find a point between Station 84+00 and Station 85+00.

The Caltrans standard design of freeway entrance used for this project is shown in Figure 7-41. Note that the distance from the inlet nose (14-foot point) to the end of the acceleration lane taper should equal the sum of the distances (300 ft + 167.11 ft) shown on Figure 7-41 (or Figure 504.2A of Caltrans *HDM*).

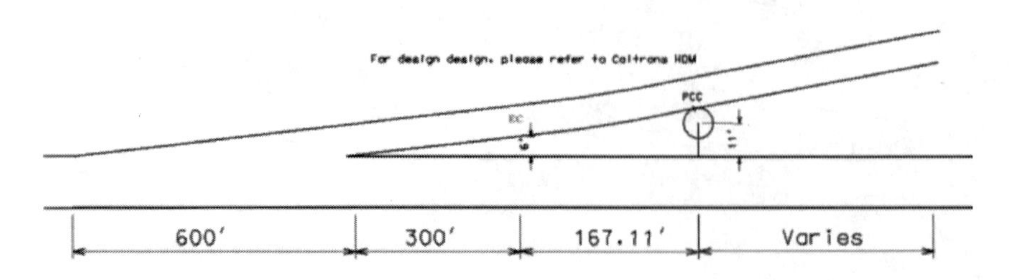

Figure 7-41 Freeway Entrance Design

The design speed at the inlet nose should be consistent with Caltrans standards. Since the approach is a diamond ramp with high alignment standards, the design speed should be at least 50 mph.

Use the following steps to create the freeway entrance for the Eastbound On ramp:

a. Set the ETW level to be active. If the level does not exist, use ***MicroStation → Level Manager*** to create the **ETW** level. Select the color, line style, and line weight of your choice.

b. Find a point on the outside (or the right side) ETW of Route 8 in the eastbound direction (between Station 84+00 and Station 85+00). The point coordinates are:

Northing:	8081.1087
Easting:	14171.0266

c. Draw a perpendicular line segment to the ETW line and parallel copy this line segment 300 ft and 467.11 ft to the west side of the line segment. The 300-ft offset line will be used to get the EC point of the 3,000-ft curve. The 467.11-ft offset will be used to get the inlet nose (or the BC of the 3,000-ft curve) within the 14-ft zone.

d. Use the ***MicroStation → Move/Copy Parallel*** command to parallel copy the ETW line with 6 ft and 11 ft offset to the South of the ETW line. The intersection between the 300-ft offset and 6-ft offset is the EC point of the 3,000-ft curve.

e. Create a line segment to pass through the two points (or the freeway entrance point and the EC point).

f. Zoom in to the 14-ft zone (see Figure 7-42).

g. Draw a 3-ft radius circle at the center point (or the intersection point of the 11-ft ETW offset line and the 467.11-ft offset; see Figure 7-43).

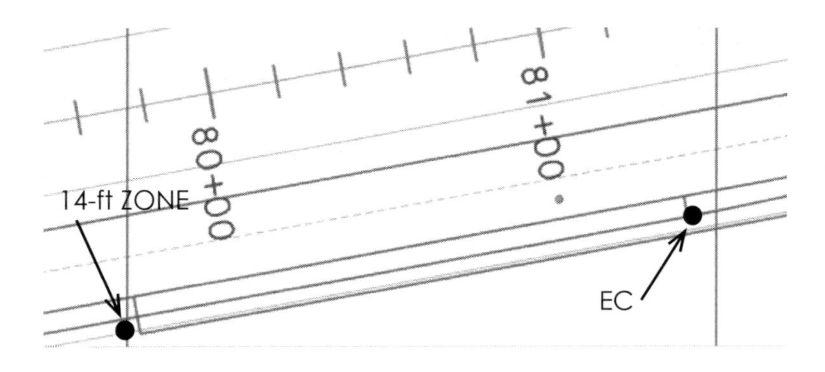

Figure 7-42 Freeway Entrance Point

h. Draw the inlet nose.

i. Go to MicroStation. In the **Key-in** window, type ***Rotate View Element.***

j. When MicroStation prompts you to identify an element, select the 467.11 ft offset.

The result is shown in Figure 7-44.

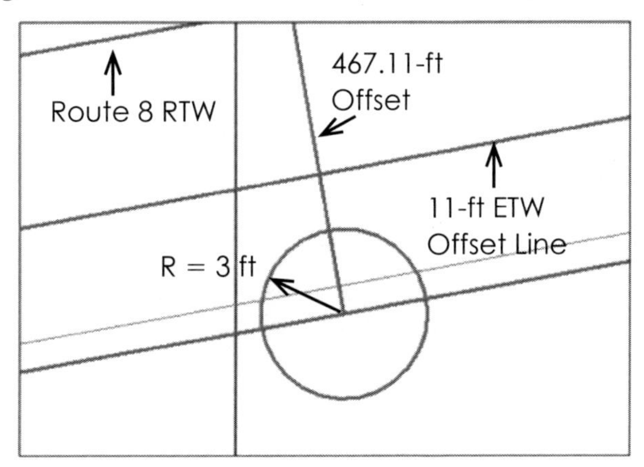

Figure 7-43 3-ft Circle

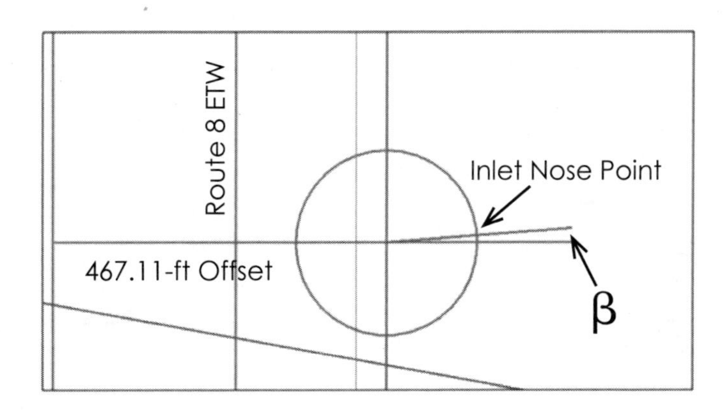

Figure 7-44 Rotated View

Note that the view is rotated and the 467.11-ft offset is the x-axis. Such rotation makes the drawing of a line 4°20'14" from the 467.11-ft offset much easier.

k. Draw a line segment from the center of the circle and at the angle of 4:20:14 (meaning 4°20'14" or 4.3372°; see Figure 7-45).

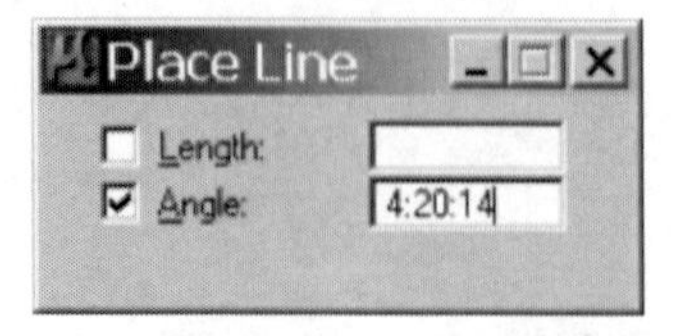

Figure 7–45 Place a line at Angle of 4°20'14"

l. Go to ***MicroStation → Tools → View → View Control → View Rotation → Rotate View.*** The Rotate View window appears.

m. Select the ***Unrotated*** method (see Figure 7-46).

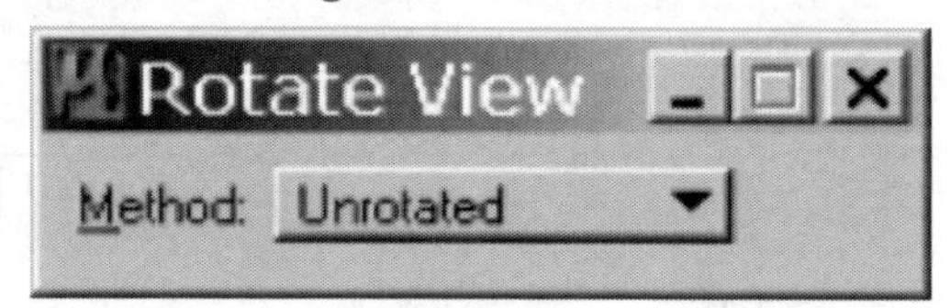

Figure 7–46 Unrotated View

n. Measure the angle β (see Figure 7-46) and ensure the angle is 4.3372°.

o. Draw an arc using ***MicroStation → Place Arc*** command. Make sure the radius is locked with ***3000.00*** ft. and the method is ***Start, End, Mid*** (see Figure 7-47).

p. Make sure the ***Tangent Point*** snap mode is selected and the line segment passing through the EC point and the freeway entrance point is snapped on.

q. Make sure the ***Intersection*** snap mode is selected for snapping the Inlet Nose point.

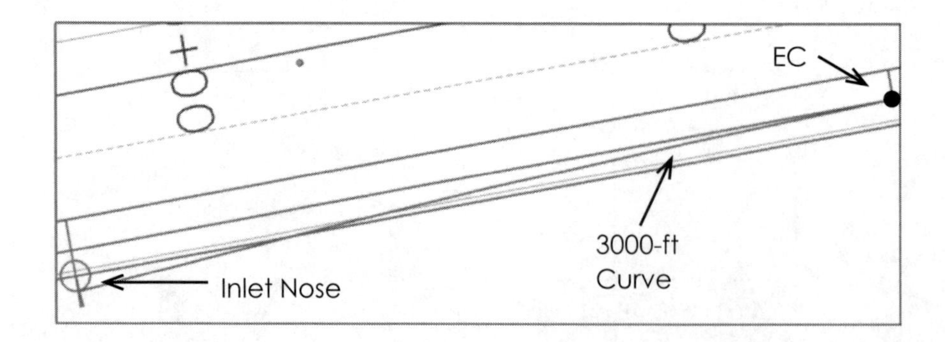

Figure 7–47 3,000-ft Curve Design for the EBON Ramp

2. Location and Design of EBON Ramp Terminal

The EBON Ramp terminal is a part of the ramp intersection that links to the EBOFF Ramp terminal. The design of the EBON Ramp terminal is the extension of the EBOFF terminal as shown in Figure 7-48.

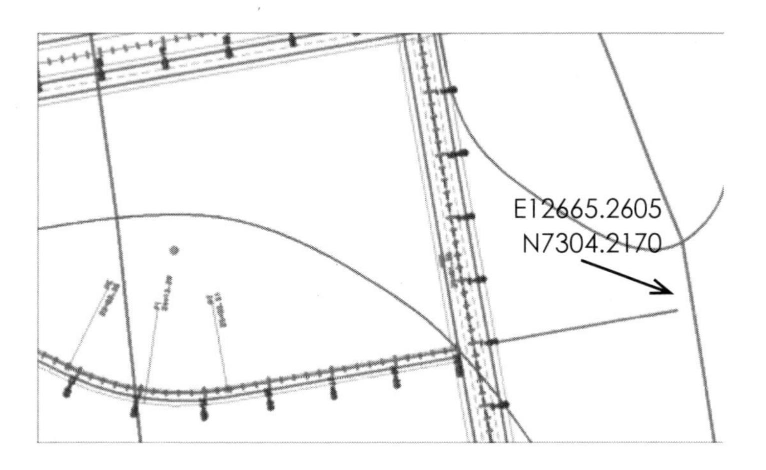

Figure 7-48 EBON Ramp Terminal Design

a. Use the ***MicroStation → Extend Line*** command to extend the station line of the EBOFF Ramp terminal 300 ft beyond the right ETW of the Laguna Canyon Road. The Easting and Northing coordinates of the end point are 12,665.2605 and 7,304.2170, respectively.

b. Use ***MicroStation → Partial Delete*** command to delete the extended segment of the Laguna Canyon Road.

3. Ramp Design

The design of the freeway entrance and the ramp terminal provides two ends for the EBON Ramp. The ramp design therefore is to fill the gap between them using these steps:.

a. Set the ETW level to be active. If the **ETW** level does not exist, select ***MicroStation → Level Manager*** to create it. Select the color, line style, and line weight of your choice.

b. Draw an arc using the ***MicroStation → Place Arc*** command.

c. Make sure the radius is locked with ***1000.00*** ft and the method is ***Start, End, Mid.***

d. Make sure the ***Tangent Point*** snap mode is selected and the 3,000-ft radius curve is snapped on (see Figure 7-49).

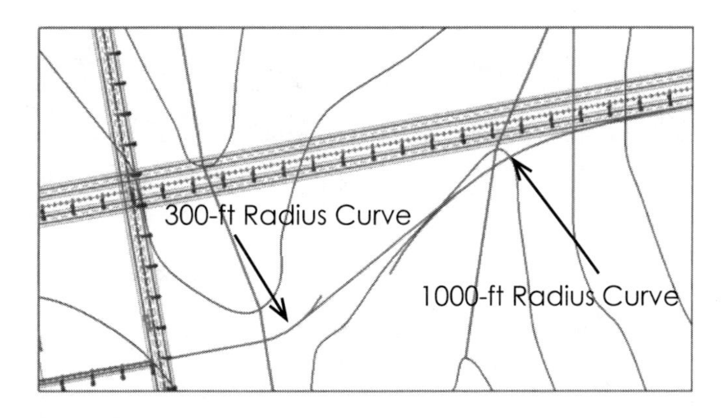

Figure 7-49 Construction Elements for EBON Ramp

a. Draw an Arc Using ***MicroStation → Place Arc Command.***
Make sure the radius is locked with ***300.00 ft***. and the method is ***Start, End, Mid.***

b. Make sure the ***Tangent Point*** snap mode is selected and the end of the EBON Ramp terminal is snapped on (see Figure 7-49).

c. Draw a line to be the tangent of the 300-ft and 1,000-ft radius curves.

d. Make sure the **Tangent** snap mode is selected and the two curves are snapped on one after another (see Figure 7-50).

e. Trim the entire EBON Ramp and delete all construct elements.

The final graphical elements for the EBON Ramp are shown in Figure 7-50.

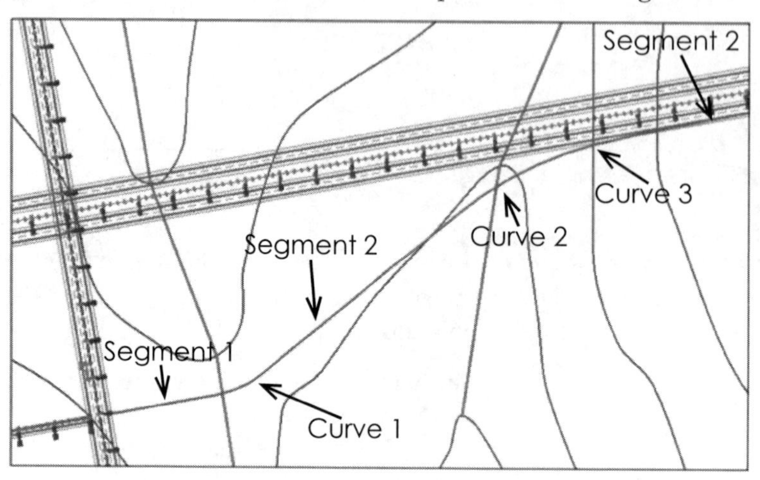

Figure 7-50 Graphical Elements for the EBON Ramp

4. Caltrans Standards Conformance Check

Table 7-3 shows the result of the conformance check.

Table 7-3 Check of Caltrans Standards for Eastbound On Ramp

Element Type	Length (L)/ Radius (R)	Conformance Check
Segment 1	300.00 (L)	Segment 1 acts as the SE transition from 2% (matching to the grade of Laguna Canyon Road) to e (=12%) for Curve 1. 2/3 × SE runoff of Curve 1 = 2/3 × 300 = 200 ft. 300 ft > 200 ft. OK
Curve 1	300.00 (R) 149.83 (L)	The curve radius is 300.00 ft. It is equal to the minimum radius of curve for the design speed of 30 mph. OK

Segment 2	689.25 (L)	Segment 2 acts as the SE transition for two reverse curves. e_{max} = 12% 2/3 × SE runoff of Curve 1 = 2/3 × 300 = 200 ft. 2/3 × SE runoff of Curve 2 = 2/3 × 240 = 160 ft. 689.25 ft > 360 ft. OK
Curve 2	1000.00 (R) 423.72 (L)	The curve radius is 1,000.00 ft, greater than 850 ft for the design speed of 50 mph. This curve is close to the freeway entrance point. It is reasonable to assume the design speed for this curve is 50 mph. OK
Curve 3	3000.00 (R) 167.10 (L)	The curve is a part of the Caltrans standard design of freeway entrance. This curve is a compound curve to Curve 2. It meets the compound curve requirements. OK
Segment 3	300.06 (L)	It is a part of the Caltrans standard design of freeway entrance.

5. To create the Eastbound On Ramp, import the Graphical Elements into InRoads
   a. Make sure the **ETW** level is active in MicroStation.
   b. Make sure the **tutorial** geometry project is active. If it is not active, set it to be active.
   c. Within MicroStation, create a complex chain using **Create Complex Chain** command.
   d. Select the six graphical elements (in the order of Segment 1, Curve 1, Segment 2, Curve 2, Curve 3, and Segment 3) and create a chain for the Eastbound On Ramp.
   e. Go to ***InRoads → Import → Geometry ….*** In the **Import Geometry** window, select the **From Graphics** tab and type the following:

Type:	**Horizontal Alignment**
Name:	**EBON**
Description:	**Horizontal Alignment**
Style:	**Default**
Horizontal	
Curve Definition:	**Arc**
Geometry Project:	**Tutorial**
Other fields:	Same as those for Route 8

   f. Select ***Apply***.

      InRoads prompts you to select the graphical elements.
   g. Select the chain you just created. Select ***Close***.
   h. Select ***InRoads → Geometry → Horizontal Curve Set → Stationing ….***
   i. Type **10+00** in the Starting Station field. Select **Apply** and **Close**.

      This step sets the starting station of Eastbound On Ramp to be 10+00.

j. Check in the **Workspace Bar** window to be sure the **EBON** horizontal alignment is created.

You will see a **Workspace Bar** window similar to Figure 7-51.

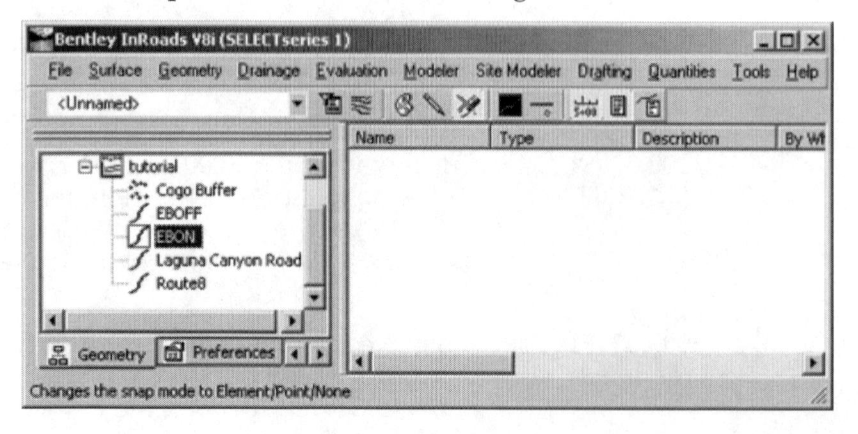

Figure 7–51 EBON Created

6. Save the **tutorial** project

   a. Make sure the **tutorial** geometry project is set to be active.

   b. Go to ***InRoads → File → Save → Geometry Project.***

   c. Save the **tutorial** geometry project as ***tutorial.alg*** in the **C:/Temp/Tutorial/Chp7/Master Files folder.**

7. View **EBON** Horizontal Alignment

   a. Make sure the **tutorial** geometry project is set to be active.

   b. Make sure **EBON** is set to be active.

   c. Go to ***InRoads → Geometry → View Geometry → Active Horizontal.***

   d. Go to ***InRoads → Geometry → View Geometry → Stationing****....*

   e. In the **View Stationing** window, do the following:

      Major Station: **Check on**
      Major Ticks: **Check on**
      Minor Ticks: **Check on**
      Set the format on Major Station: **SS+SS.SS**

   f. Click the **Text Symbology** icon of the **Major Station.**
   The **Text Symbology** window appears.

   g. In the **Text Symbology** window, type ***0.20*** in the **Horizontal field**. Select ***OK.***

   h. From the **View Stationing** window, select ***Apply*** and ***Close.***

   Stations are displayed along the EBON horizontal alignment (see Figure 7-52).

   i. When the stations are displayed with other symbols overlapping each other, to clean the noisy symbols, follow these steps:

   » Select ***MicroStation → Settings → View Attributes.***
   » Turn the **Text Nodes** off.
   » Close the **View Attributes** window.

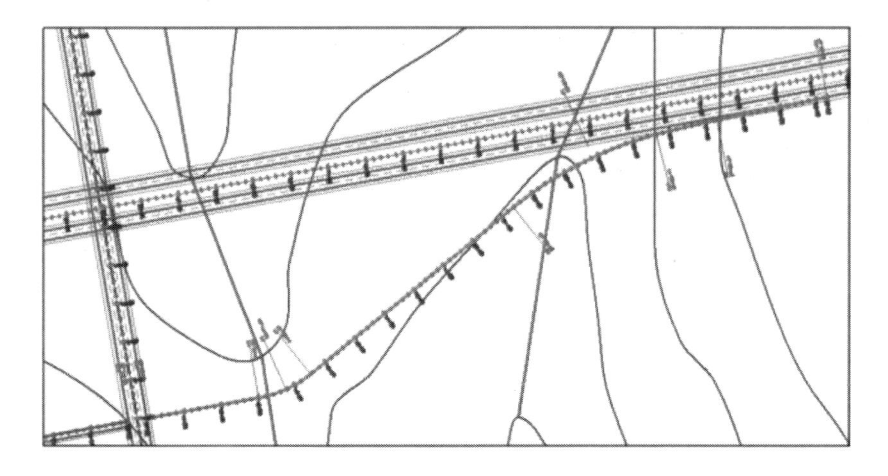

Figure 7–52 EBON Horizontal Alignment Stations

8. Create Lanes for the Eastbound ON Ramp

The lane configuration of EBON horizontal alignment is shown in Figure 7-53. Set the **ETW** level to be active.

a. Create the **ETW** level in MicroStation using the **Level Manager** command if the **ETW** level does not exist. Select the **ETW** level as the active level and set the settings below to:

Color:	***Your choice***
Line Style:	***0***
Line Weight:	***1***

Where: SL Shoulder Line

ETW Edge of Traveled Way

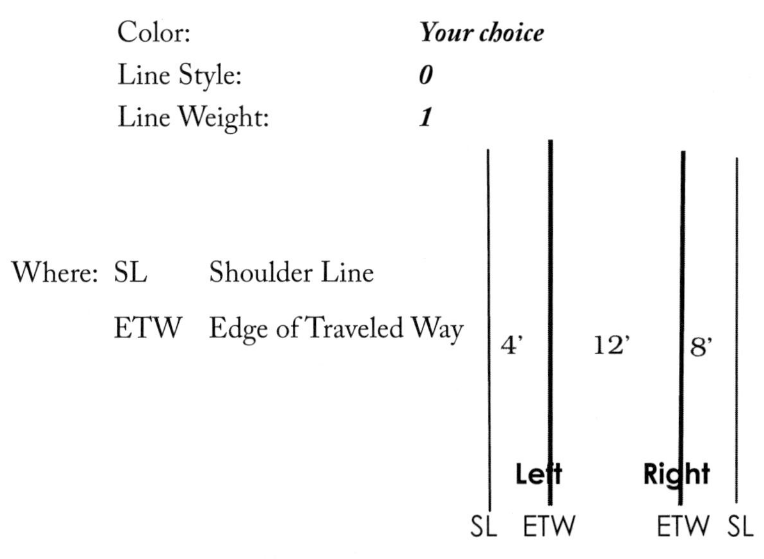

Figure 7–53 Lane Configuration for the EBON Ramp

b. Use the MicroStation **Move/Copy Parallel** command to create ETW lines for the EBON Ramp.

c. Create a new **SL** level if the level does not exist. Select the SL level to be active and set the settings below to:

Color:	***Your choice***
Line Style:	***0***
Line Weight:	***0***

d. Create SLs using the **Move/Copy Parallel** command. The distances from the left ETW to the SLs are **4** and **20**, respectively.

e. Do the gore area treatment using MicroStation commands.

Figure 7-54 shows the results of the treatment. Figure 7-55 shows all the lanes configured for the EBON horizontal alignment.

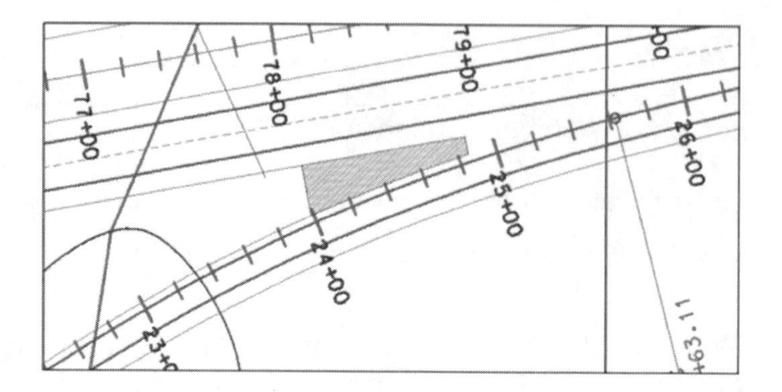

Figure 7-54 EBON Gore Area Treatment

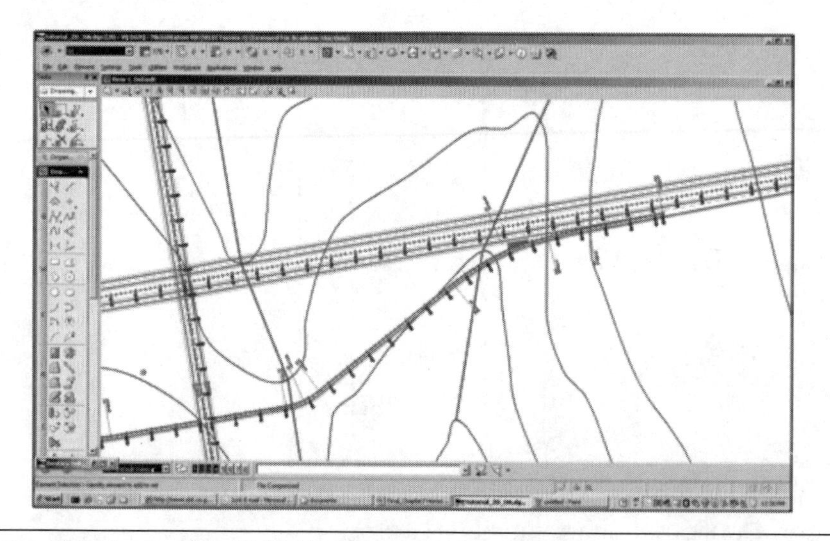

Figure 7-55 Created Lanes for EBON Horizontal Alignment

7.2.8 Horizontal Alignment for the Westbound Off Ramp

The horizontal alignment design for the Westbound Off Ramp consists of three components: freeway exit, ramp intersection (or ramp terminal), and the ramp itself.

1. Location and Design of Freeway Exit

The freeway exit design involves 1) locating a point along the *right side* ETW of Route 8 Westbound (or the intersection point of the *right side* ETW of Route 8 Westbound and the ramp station line) and 2) superimposing the standard design of freeway exit for the ramp.

Use these steps to create the freeway exit for the Westbound Off ramp:

a. Set the ETW level to be active. If the level does not exist, select ***MicroStation → Level Manager*** to create the ETW level. Select the color, line style, and line weight of your choice.

b. Find a point (later called the start point) on the outside ETW of Route 8 in the Westbound direction (within the area of Station 81+00; see Figure 7-56). The point coordinates are:

Northing:	8136.4582
Easting:	13855.3392

c. Draw a line segment *perpendicular* to the Route 8 ETW in the Westbound direction. The segment should be outside of the shoulder line. The length of segment is greater than 23 ft.

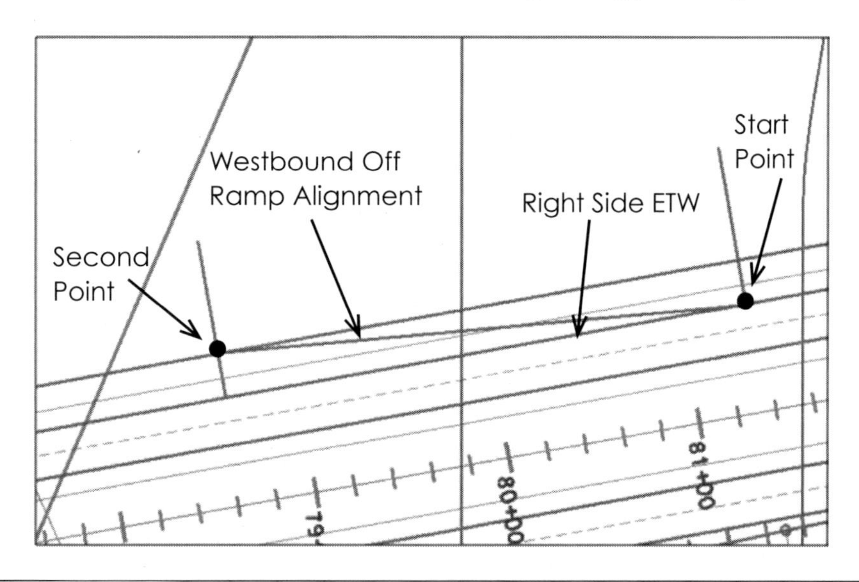

Figure 7-56 Find the Starting Point for the Westbound Off Ramp

d. Use ***MicroStation → Move/Copy Parallel*** command to parallel copy the line segment 270 ft to the left (see Figure 7-56).

e. Use the ***MicroStation → Move/Copy Parallel*** command to parallel copy the Route 8 outside ETW (in the westbound direction) 23 ft up or to the North (see Figure 7-56).

The two parallel-copied segments intersect at the point (later called the second point) with the following coordinates:

Northing:	8111.6209
Easting:	13585.5020

f. Draw a line segment passing through the two points.

g. Use the ***MicroStation → Measure Angle*** command to measure the angle between the Route 8 outside ETW and the Westbound Off Ramp alignment.

The result should be 4.869° or 45°2'08".

2. Location and Design of the Ramp Intersection and Terminal

Stop signs are assumed to be used to control traffic at the Westbound Off Ramp terminal.

The location of the ramp terminal is governed by the sight distance and visibility requirements for left-turn drivers coming out of the ramp terminal.

The sight distance is then measured between the center of the outside lane (close to the Westbound Off Ramp terminal) and the eye of the driver of the ramp vehicle. The vehicle is assumed to be 10 feet back from the edge of shoulder at Laguna Canyon Road.

The ramp setback from the bridge structure can be calculated on the basis of the sight distance controlled by the slopes.

$$c = d \times \left[\frac{b - a - 6}{b}\right]$$

$$= 495 \times \left[\frac{(10 + 10 + 6) - 2 - 6}{10 + 10 + 6}\right]$$

$$= 343 \text{ ft}$$

The ramp setback indicates the station line or the left ETW of the Westbound Off Ramp shall be at least 343 ft away from the Route 8 shoulder line in the Westbound Off direction.

Clear visibility requires the crossroad (or the Laguna Canyon Road) be set open to left-turn drivers coming out of the off ramp. Caltrans standards indicate that the length of the crossroad should be greater than the product of the prevailing speed of vehicles on the crossroad and the time required for a stopped vehicle on the ramp to execute a left-turn maneuver. This time is estimated to be 7.5 seconds. The design speed of the Laguna Canyon Road is 40 mph. The length of crossroad open to view therefore is 495 ft.

Given the above analysis, the ramp intersection for the Westbound Off terminal is set 500 ft away from right side ETW of the Route 8.

Use these steps to create the ramp terminal for the Westbound Off ramp:

a. Set the **ETW** level to be active. Select ***MicroStation → Level Manager*** to create the **ETW** level if this level does not exist. Select the color, line style, and line weight of your choice.

b. Find a point on the right side ETW of Laguna Canyon Road (between Station 28+00 and Station 29+00). Coordinates of this point are:

Northing:	12177.1846
Easting:	8344.6018

   Since this point is more than 500 ft away from the Route 8 bridge, it meets Caltrans sight distance and visibility requirements of the ramp intersection.

c. Draw a 300-ft line segment *perpendicular* to the right side ETW of the Laguna Canyon Road (see Figure 7-57). The Easting and Northing coordinates of the end point are 12,472.5105 and 8,397.3563, respectively.

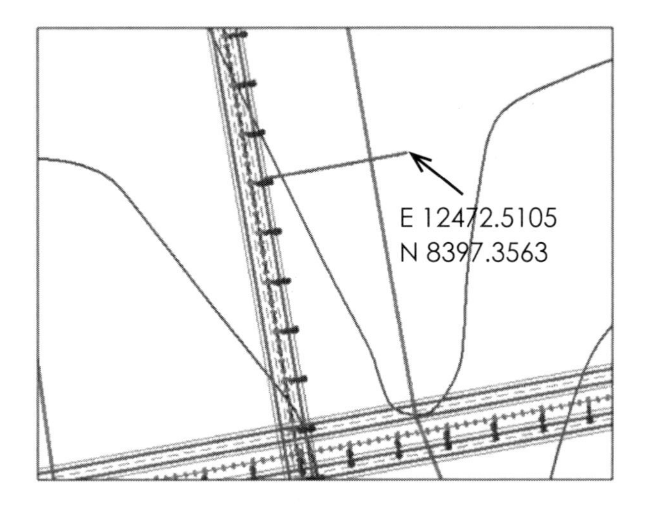

Figure 7-57 Ramp Terminal Design for the Westbound Off Ramp

3. Ramp Design

With the design of the freeway exit and the ramp terminal, the ramp design itself is to fill the gap between them using these steps:

a. Set the **ETW** level to be active.
b. Draw a line segment that goes through the two PI points below (see Figure 7-58):

First Point	Northing:	8397.3563
	Easting:	12472.5105
Second Point	Northing:	8075.6652
	Easting:	13194.8725

c. For the two PIs, use the ***MicroStation → Construct Circular Fillet*** command. Type the following in the **Construct Circular Fillet** window for the first PI:

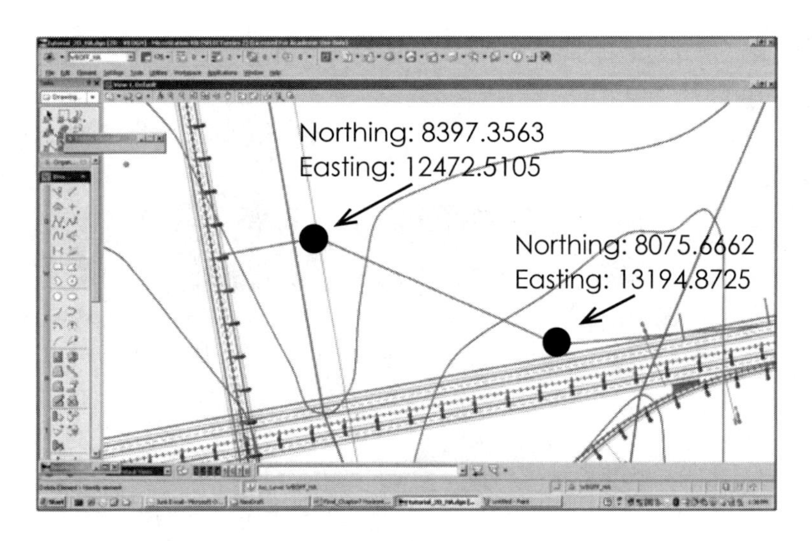

Figure 7-58 Filling the Gap

Radius: ***300***
Truncate: ***Both***

d. Type the following in the **Construct Circular Fillet** window for the second PI:

Radius: ***1000***
Truncate: ***Both***

e. Delete the parallel-copied line segments. The graphical elements for the Westbound Off Ramp are shown in Figure 7-59.

There are five graphical elements for this ramp: three line segments and two circular curves.

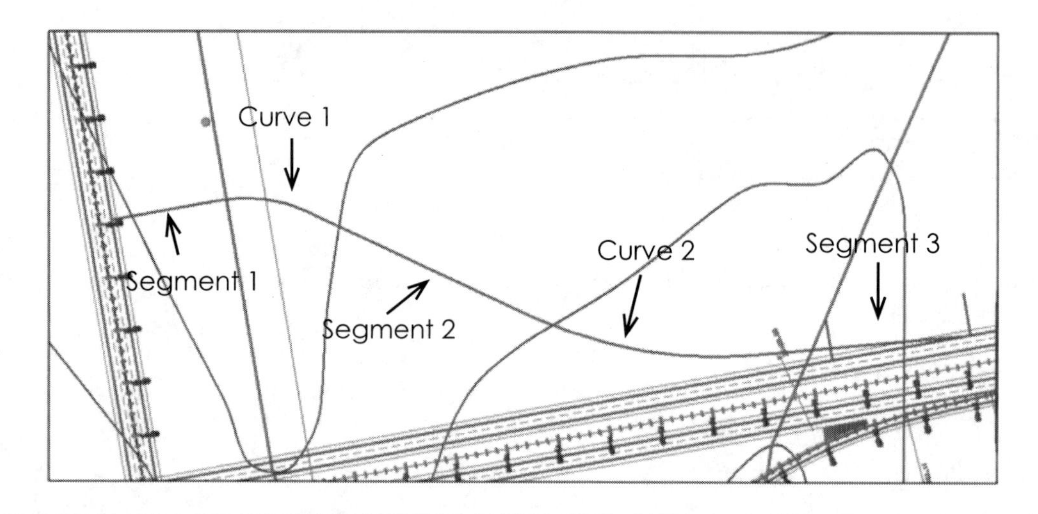

Figure 7-59 Graphical Elements Generated for the Westbound Off Ramp

4. Caltrans Standards Conformance Check

Table 7-4 shows the result of the conformance check.

Table 7-4 Check of Caltrans Standards for Westbound Off Ramp

Element Type	Length (L)/ Radius (R)	Conformance Check
Segment 1	207.90 (L)	This segment is used for SE transition from e of 12% to 2% on Laguna Canyon Road. 2/3 × SE runoff = 200 ft 207.901 ft > 200 ft. OK
Curve 1	300.00 (R) 178.72 (L)	The curve radius is 300.00 ft equal to the minimum radius of the curve for the design speed of 30 mph. Curve 1 is close to the ramp terminal. It is reasonable to assume the design speed for Curve 1 is 30 mph. OK

Segment 2	437.58 (L)	Segment 2 acts as the SE transition for two reverse curves. e_{max} = 12% 2/3 × SE runoff of Curve 1 = 2/3 × 300 = 200 ft. 2/3 × SE runoff of Curve 2 = 2/3 × 240 = 200 ft. 437.58 ft > 360 ft. OK
Curve 2	1000.00 (R) 510.75 (L)	The curve radius is 1,000.00 ft, greater than 850 ft. for the design speed of 50 mph. OK
Segment 3	402.18 (L)	According to Figure 7-29 (or 504.2B of Caltrans HDM), the Deceleration Length (DL) for Curve 2 is 270 ft. since the curve radius is 1000 ft. 402.18 ft > 270.0 ft. OK

5. Import the graphical elements into InRoads to create the Westbound Off Ramp
   a. Make sure the **ETW** level is active in MicroStation.
   b. Make sure the **tutorial** geometry project is active. If it is not active, set it to be active.
   c. Within MicroStation, create a complex chain using **Create Complex Chain** command.
   d. Select the five graphical elements (Segment 1, Curve 1, Segment 2, Curve 2, and Segment 3) and create a chain for the Westbound Off Ramp.
   e. Go to ***InRoads → Import → Geometry …***.
   f. In the **Import Geometry** window, select the **From Graphics** tab and type the following:

      Type: **Horizontal Alignment**
      Name: **WBOFF**
      Description: **Horizontal Alignment**
      Style: **Default**
      Horizontal
      Curve Definition: **Arc**
      Geometry Project: **Tutorial**
      Other fields: Same as those for Route 8

   g. Select ***Apply***.

      InRoads prompts you to select the graphical elements.
   h. Select the chain you just created. Select ***Close***.
   i. Check in the **Workspace Bar** window if the **WBOFF** horizontal alignment is created.
   j. Select ***InRoads → Geometry → Horizontal Curve Set → Stationing …***.
   k. Type **10+00** in the Starting Station field. Select **Apply** and **Close**.

      This step sets the starting station of Westbound Off Ramp to be 10+00.
6. Save the **tutorial** geometry project
   a. Make sure the **tutorial** geometry project is set to be active.
   b. Go to ***InRoads → File → Save → Geometry Project***.

c. Save the **tutorial** geometry project as ***tutorial.alg*** in the **C:/Temp/Tutorial/Chp7/Master Files** folder.

7. View **WBOFF** horizontal alignment

   a. Make sure the **tutorial** geometry project is set to be active.

   b. Make sure **WBOFF** is set to be active.

   c. Go to ***InRoads → Geometry → View Geometry → Active Horizontal.***

   d. Go to ***InRoads → Geometry → View Geometry → Stationing …***.

   e. Within the **View Stationing** window, do the following:

Major Station:	**Check on**
Major Ticks:	**Check on**
Minor Ticks:	**Check on**
Set format on Major Station:	**SS+SS.SS**

   f. Click the **Text Symbology** icon of the **Major Station.**

   The **Text Symbology** window appears.

   g. In the **Text Symbology** window, type ***0.20*** in the **Horizontal field**. Select ***OK***.

   h. Select ***Apply*** and ***Close*** in the **View Stationing** window.

   Stations are displayed along the WBOFF horizontal alignment (see Figure 7-60).

   i. When the stations are displayed with other symbols overlapping each other, do the following to clean the noisy symbols:

   » Select ***MicroStation → Settings → View Attributes.***
   » Turn the **Text Nodes** off.
   » Close the **View Attributes** window.

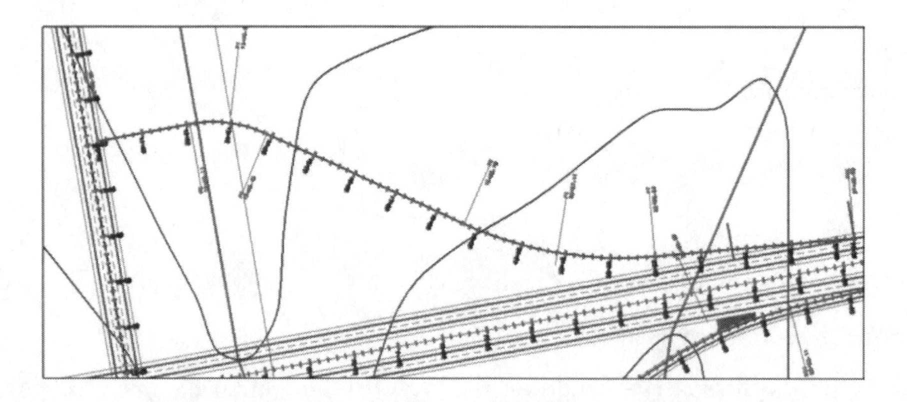

Figure 7-60 WBOFF Horizontal Alignment Stations

8. Create lanes for the Westbound Off Ramp

The lane configuration of WBOFF horizontal alignment is shown in Figure 7-61.

   a. Set the **ETW** level to be active. Select ***MicroStation → Level Manager*** command if the ETW level does not exist. Select the **ETW** level as the active level and set the below settings as:

Color: *Your choice*
Line Style: ***0***
Line Weight: ***1***

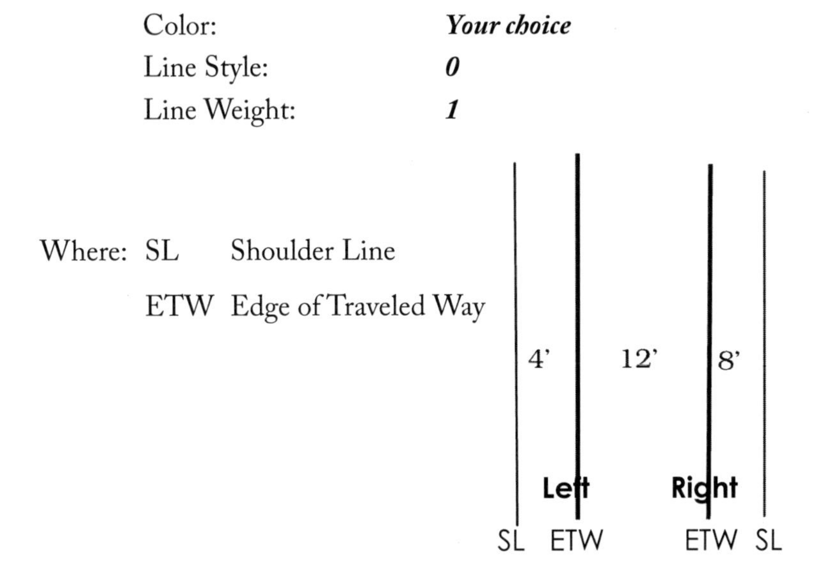

Figure 7–61 Lane Configuration for the WBOFF Ramp

b. Use the MicroStation **Move/Copy Parallel** command to create ETW lines for the WBOFF Ramp.

c. Create a new **SL** level if the level does not exist. Select the **SL** level as the active and set the below settings as:

Color: *Your choice*
Line Style: ***0***
Line Weight: ***0***

d. Create SLs using the **Move/Copy Parallel** command. The distances from the left ETW to the SLs are **4** and **20**, respectively.

e. Do the gore area treatment using MicroStation commands.

Figure 7-62 shows the results of the treatment. Figure 7-63 shows all the lanes configured for WBOFF horizontal alignment.

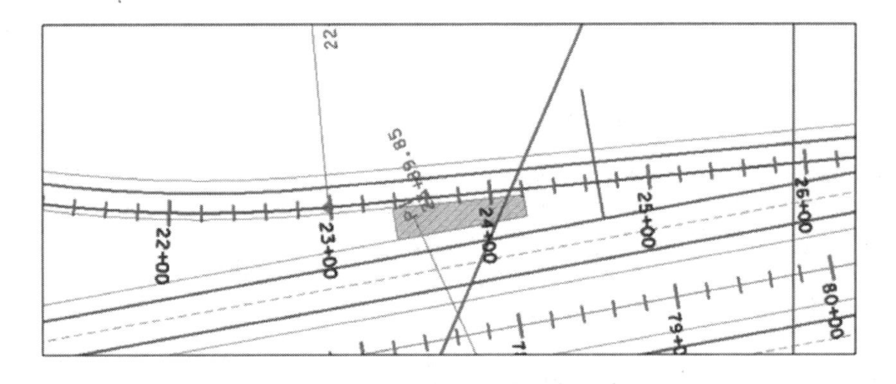

Figure 7-62 WBOFF Gore Area Treatment

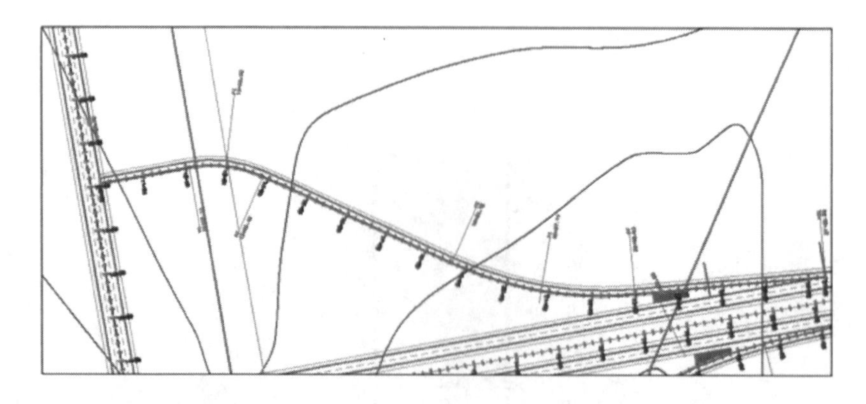

Figure 7-63 Created Lanes for WBOFF Horizontal Alignment

7.2.9 Horizontal Alignment for Westbound On Ramp

Horizontal alignment design for the Westbound On Ramp involves three components: freeway entrance, ramp intersection (or ramp terminal), and ramp itself.

1. Location and Design of Freeway Entrance

The freeway entrance design involves 1) locating a point along the *right side* ETW of Route 8 Westbound and 2) superimposing the standard design of freeway entrance for the ramp.

The intersection point should be located on tangent sections of Route 8 Westbound wherever possible to provide maximum sight distance and optimum traffic operation. In this project, we will find a point between Station 42+00 and Station 43+00.

The Caltrans standard design is used for the freeway entrance of the WBON ramp. The design speed at the inlet nose should be consistent with approach alignment standards. Since the approach is a diamond ramp with high alignment standards, the design speed should be at least 50 miles per hour.

Use these Steps to create the freeway entrance of the Westbound On ramp:

a. Set the **ETW** level to be active.

b. Find a point on the outside (or the right side) ETW of Route 8 in the Westbound direction (between Station 42+00 and Station 43+00; see Figure 7-64). The point coordinates are:

Northing:	7449.9812
Easting:	10012.3576

c. Draw a perpendicular line segment to ETW and parallel copy this line segments 300 ft and 467.11 ft to the eastside of the line segment.

d. The 300-ft offset line will be used to get the EC point of the 3,000-ft curve.

The 467.11-ft offset will be used to get the inlet nose (or the BC point of the 3,000-ft curve) within the 14-ft zone.

e. Use ***MicroStation → Move/Copy Parallel*** command to parallel copy the ETW line with 6 ft and 11 ft offset to the North of the ETW line.

The intersection between the 300-ft offset and 6-ft offset is the EC point of the 3000-ft curve.

f. Create a line segment to pass through the two points (or the freeway entrance point and the EC point). Zoom in to the 14-ft zone (see Figure 7-64).

g. Draw a 3-ft radius curve at the center point (or the intersection point between the 11-ft ETW offset line and the 467.11-ft offset; see Figure 7-65).

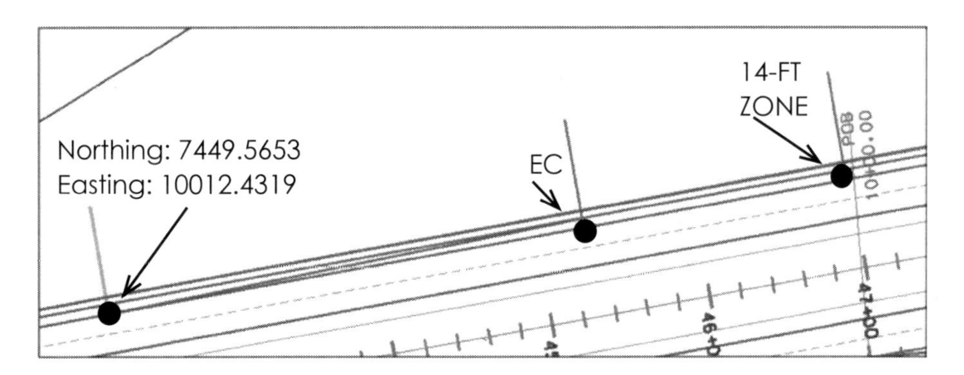

Figure 7-64 Freeway Entrance Point

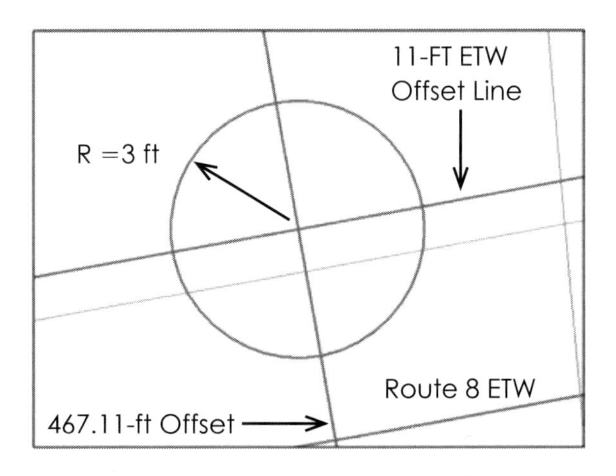

Figure 7-65 3-ft Circle for WBON Ramp

h. Draw the inlet nose.

i. Go to MicroStation and type ***Rotate View Element*** in the **key-in** window.

j. When MicroStation prompts you to identify an element, select the 467.11-ft offset.

The result of the selection is shown in Figure 7-66.

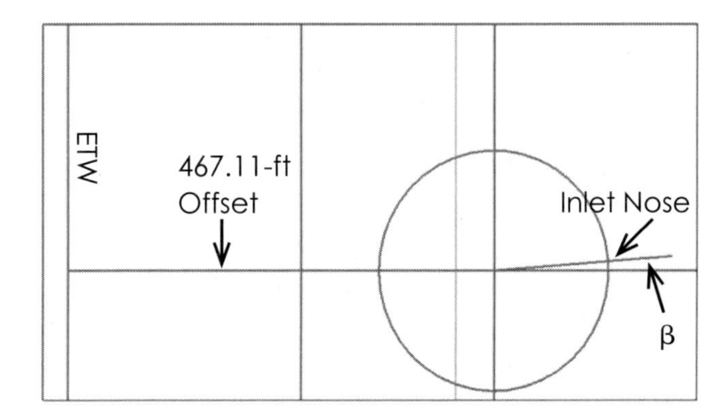

Figure 7-66 Rotated View

Note that the view is rotated and the 467.11-ft offset is the x-axis. Such rotation makes the drawing of a line 4°20'14" from the 467.11-ft offset easier.

k. Draw a line segment from the center of the circle and at the angle of 4:20:14 (meaning 4°20'14" or 4.3372°).

l. Go to ***MicroStation → Tools → View → View Control → View Rotation → Rotate View.*** The Rotate View window appears.

m. Select the ***Unrotated*** method.

n. Measure the angle β (see Figure 7-66) and ensure the angle is 4.3372°.

o. Draw an arc using the ***MicroStation → Place Arc*** command.

p. Make sure the radius is locked with ***3000.00*** ft. and the method is ***Start, End, Mid.***

q. Make sure the ***Tangent Point*** snap mode is selected and the line segment (passing through the EC point and the freeway entrance point) is snapped on.

r. Make sure the ***Intersection*** snap mode is selected for snapping the Inlet Nose point (see Figure 7-67).

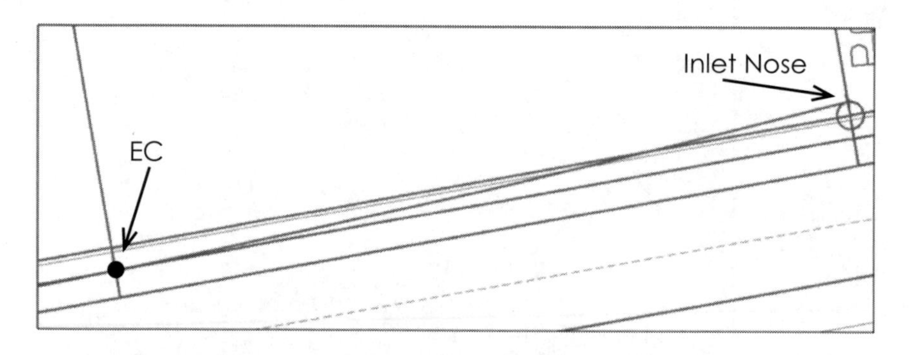

Figure 7-67 3,000-ft Curve Design for WBON Ramp

2. Location and Design of WBON Ramp Terminal

The WBON Ramp terminal is a part of the ramp intersection that links to the WBOFF Ramp terminal. The design of the WBON Ramp terminal is the extension of the WBOFF terminal.

Figure 7-68 WBON Ramp Terminal

a. Use ***MicroStation → Extend Line*** command to extend the station line of the WBOFF Ramp terminal 300 ft. beyond the ETW of the Laguna Canyon Road. The northing and easting coordinates of the end point are 8283.4066 and 11834.6065, respectively.

b. Use the ***MicroStation → Partial Delete*** command to delete the extended segment within the Laguna Canyon Road. Figure 7-68 shows the WBON Ramp terminal.

3. Ramp Design

The design of the freeway entrance and the ramp terminal provides two ends for the WBON Ramp. The ramp design therefore is to fill the gap between them.

a. Set the **ETW** level to be active.

b. If the **ETW** level does not exist, select ***MicroStation → Level Manage***r to create the level. Select the color, line style, and line weight of your choice.

c. Draw an arc using ***MicroStation → Place Arc*** command.

d. Make sure the radius is locked with ***1000.00 ft.*** and the method is ***Start, End, Mid.***

e. Make sure the ***Tangent Point*** snap mode is selected and the 3,000-ft. radius curve is snapped on (see Figure 7-69).

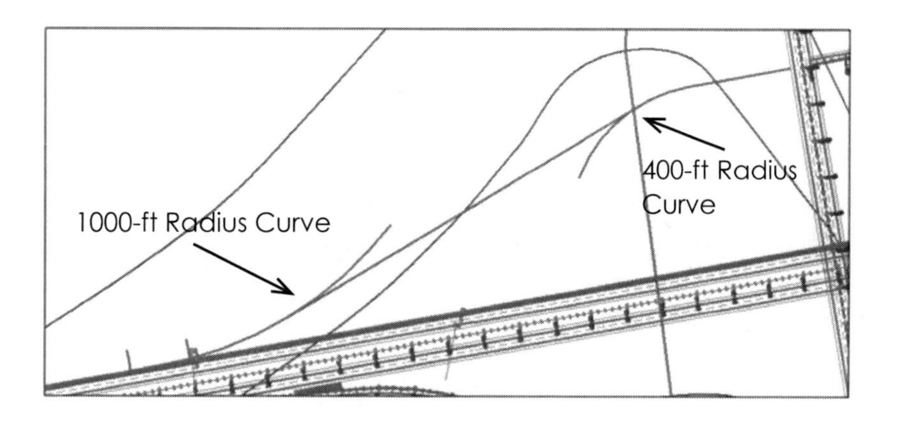

Figure 7-69 Construction Elements for WBON Ramp

f. Draw an arc using the **MicroStation → Place Arc** command.

g. Make sure the radius is locked with ***400.00*** ft and the method is ***Start, End, Mid.***

h. Make sure the ***Tangent Point*** snap mode is selected and the end of the WBON Ramp terminal is snapped on.

i. Draw a line to be the tangent of the 400-ft and 1,000-ft radius curves.

j. Make sure the ***Tangent*** snap mode is selected and the two curves are snapped on one after another (see Figure 7-70).

k. Trim the entire WBON Ramp and delete all construct elements.

The final graphical elements for the WBON Ramp are shown in Figure 7-70.

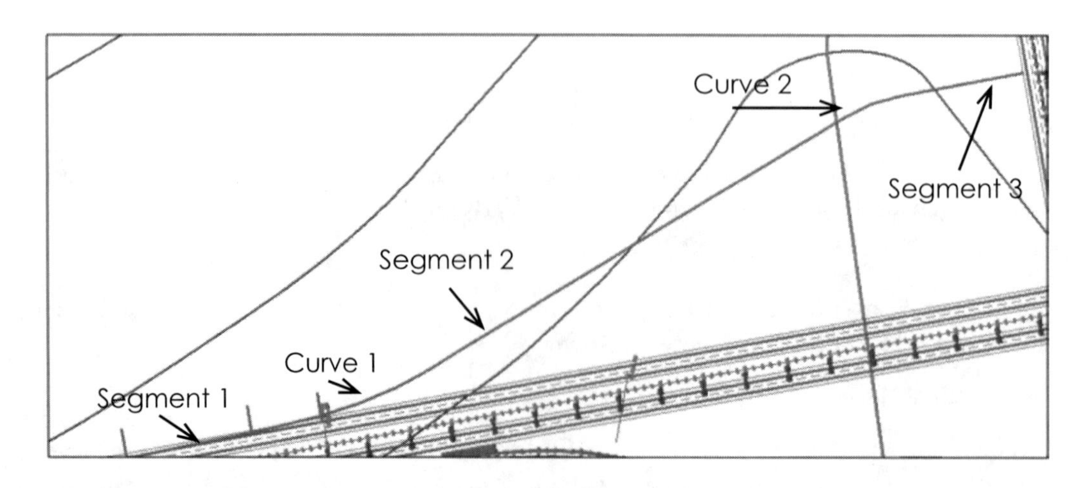

Figure 7-70 Graphical Elements for WBON Ramp

4. Conformance Check of Caltrans Standards

Table 7-5 shows the result of the conformance check.

Table 7-5 Check of Caltrans Standards for Westbound On Ramp

Element Type	Length (L)/ Radius (R)	Conformance Check
Segment 1	300.06 (L)	It is a part of the Caltrans standard design of freeway entrance.
Curve 1	3000.00 (R) 167.10 (L)	The curve is a part of the Caltrans standard design of freeway entrance. This curve is a compound curve to Curve 2. It meets the compound curve requirements. OK
Curve 2	1000.00 (R) 285.68 (L)	The curve radius is 1,000.00 ft, greater than 850 ft. for the design speed of 50 mph. This curve is close to the freeway entrance point. It is reasonable to assume the design speed for this curve is 50 mph. OK
Segment 2	1126.95 (L)	Segment 2 acts as the SE transition for two reverse curves. e_{max} = 12% 2/3 × SE runoff of Curve 1 = 2/3 × 300 = 200 ft. 2/3 × SE runoff of Curve 2 = 2/3 × 240 = 160 ft. 1126.95 ft > 360 ft. OK
Curve 3	400.00 (R) 144.55 (L)	The curve radius is 400.00 ft. It is greater than the minimum radius (300 ft) of curve for the design speed of 30 mph. OK
Segment 3	300.00 (L)	Segment 3 acts as the SE transition from 2% (matching to the grade of Laguna Canyon Road) to e (=12%) for Curve 3. 2/3 × SE runoff of Curve 1 = 2/3 × 300 = 200 ft. 300.00 ft > 270 ft. OK

5. Import the graphical elements into InRoads to create the Westbound ON Ramp
   a. Make sure the **ETW** level is active in MicroStation.
   b. Make sure the **tutorial** geometry project is active. If it is not active, set it to be active.
   c. Within MicroStation, create a complex chain (Segment 1, Curve 1, Curve 2, Segment 2, Curve 3, and Segment 3) using the **Create Complex Chain** command.
   d. Select the six graphical elements and create a chain for the Westbound On Ramp.
   e. Go to ***InRoads* → *Import* → *Geometry* ….**
   f. In the **Import Geometry** window, select the **From Graphics** tab and type the following:

Type:	**Horizontal Alignment**
Name:	**WBON**
Description:	**Horizontal Alignment**
Style:	**Default**
Horizontal	
Curve Definition:	**Arc**
Geometry Project:	**Tutorial**
Other fields:	Same as those for Route 8

   g. Select ***Apply***.

      InRoads prompts you to select the graphical elements.
   h. Select the chain you just created. Select ***Close***.
   i. Check in the **Workspace Bar** window to ensure the **WBON** horizontal alignment is created.
   j. Select ***InRoads* → *Geometry* → *Horizontal Curve Set* → *Stationing* ….**
   k. Type **10+00** in the Starting Station field. Select **Apply** and **Close**.

      This step sets the starting station of the Westbound On Ramp to be 10+00.
6. Save the **tutorial** geometry project
   a. Make sure the **tutorial** geometry project is set to be active.
   b. Go to ***InRoads* → *File* → *Save* → *Geometry Project*.**
   c. Save the **tutorial** geometry project as ***tutorial.alg*** in the **C:/Temp/Tutorial/Chp7/Master Files** folder.
7. View **WBON** horizontal alignment
   a. Make sure the **tutorial** geometry project is set to be active.
   b. Make sure **WBON** is set to be active.
   c. Go to ***InRoads* →*Geometry* → *View Geometry* → *Active Horizontal*.**
   d. Go to ***InRoads* →*Geometry* → *View Geometry* → *Stationing* ….**
   e. In the **View Stationing** window, do the following:

      Select Regular Stations

Major Station:	**Check on**
Major Ticks:	**Check on**

Minor Ticks: **Check on**

Set the format of Major Station: **SS+SS.SS**

f. Click the **Text Symbology** icon of the **Major Station.**

The **Text Symbology** window appears.

g. In the **Text Symbology** window, type ***0.20*** in the **Horizontal** field. Select ***OK.***

h. In the **View Stationing** window, select ***Apply*** and ***Close.***

Stations are displayed along the WBON horizontal alignment (see Figure 7-71).

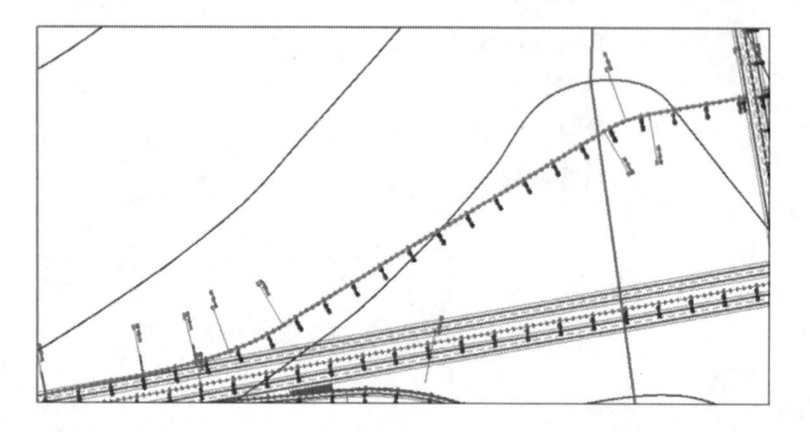

Figure 7-71 WBON Horizontal Alignment Stations

i. When the stations are displayed with other symbols overlapping each other, do the following to clean the noisy symbols:

» Select ***MicroStation → Settings → View Attributes.***
» Turn the **Text Nodes** off.
» Close the **View Attributes** window.

8. Create lanes for the Westbound ON Ramp

The lane configuration of WBON horizontal alignment is shown in Figure 7-72. Steps to create this lane configuration are as follows:

a. If ETW level does not exist, using the **Level Manager** command create a new level called **ETW** in MicroStation. Select the **ETW** level as the active level and set the settings below as:

Color: ***Your choice***

Line Style: ***0***

Line Weight: ***1***

b. Use the MicroStation **Move/Copy Parallel** command to create ETW lines for WBON.

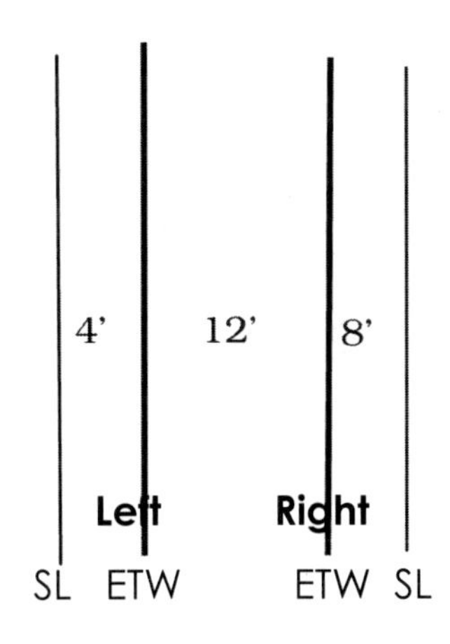

Figure 7–72 Lane Configuration for WBON Ramp

c. Create a new **SL** level if the level does not exist. Select the **SL** level as active and set the settings below as:

Color:	***Your choice***
Line Style:	***0***
Line Weight:	***0***

d. Create SLs using the **Move/Copy Parallel** command. The distances from the left ETW to the SLs are **4** and **20**, respectively.

e. Do the gore area treatment using MicroStation commands.

Figure 7-73 shows all the lanes configured for WBON horizontal alignment. Figure 7-74 shows all the lanes configured for WBON horizontal alignment.

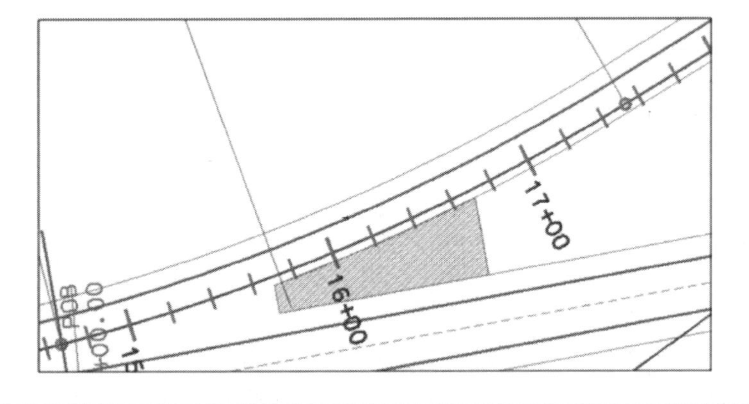

Figure 7-73 WBON Gore Area Treatment

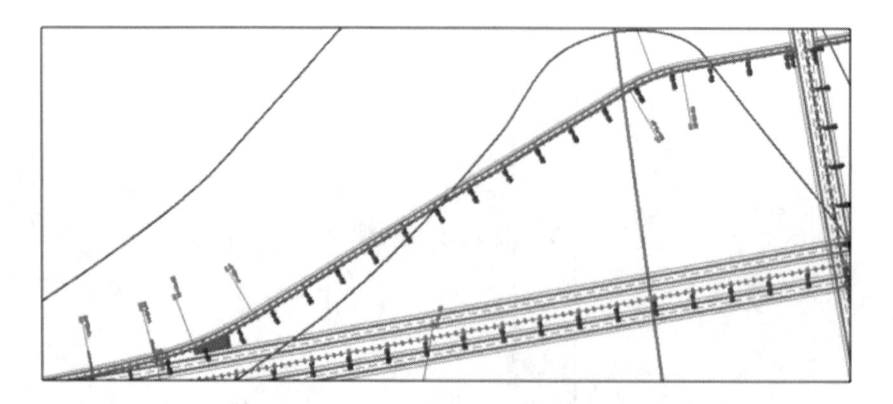

Figure 7-74 Created Lanes for WBON Horizontal Alignment

f. Save the working **tutorial_2D.dgn** as **C:/Temp/Tutorial/Chp7/Master Files/tutorial_2D_HA.dgn.**

The difference between the two files is that the **tutorial_2D_HA.dgn** has horizontal alignment and gore treatments, while the **tutorial_2D.dgn** does not. In later chapters, the **tutorial_2D_HA.dgn** will be used to create profiles and vertical alignments.

7.3 Summary of Horizontal Alignment Design

This chapter describes the process of designing six horizontal alignments for the tutorial project. The design process applies Caltrans standards (design speed, minimum radius of curve, standard design of freeway entrance and exit, superelevation transition of reverse curves, spacing of ramp intersection away from freeway bridge, etc.) to the tutorial project.

Through this design process, you should begin to understand that horizontal alignments are the results of your "artful" work that balances all the constraints: standards, project control agreements (A, B, and C points), and driver behaviors (variance of design speed along ramps). There are many alternatives possible when you design a horizontal alignment. The horizontal alignment we create is one of the alternatives that meets the Caltrans standards.

The design process emphasizes the importance of safety (or Caltrans standards) in creating horizontal alignments. Given all the factors that govern the shape of horizontal alignment, the safety factors (such as stopping sight distance and superelevation transition) should be considered directly or indirectly.

This chapter takes advantage of your solid knowledge of MicroStation. We first create graphical elements for each horizontal alignment. We then check the conformance of these elements to Caltrans standards. After all the graphical elements are checked to meet Caltrans standards, we import them into InRoads to form a horizontal alignment. InRoads acts as a container to store horizontal alignments. It communicates with MicroStation to display stations and other attributes (such as BCs, ECs, and PIs) of horizontal alignments.

Each horizontal alignment provides station line for a roadway (Route 8, Laguna Canyon Road, and ramps). To produce PS & E[5] documents, other roadway features such as lane lines, shoulder lines, and ETW lines should be added to form lane configuration of the roadway. Entrance and exit gore areas should be designed to meet Caltrans standards.

[5] *PS & E: stands for Plans, Specifications, and Cost Estimates.*

Name____ Date________________

7.4 Questions

Q7.1 Design speed is a critical design consideration in horizontal alignment. List the Caltrans standards related to design speed.

Q7.2 Why do we consider Points A, B, and C in the interchange design of the Route 8/Laguna Canyon Road?

Q7.3 Caltrans has these types of standards: Mandatory, Advisory, and Permissive. Describe these three types of standards. When a mandatory standard cannot be met in the design of a horizontal alignment, designers must go through the design exception approval process. Identify this approval process from Caltrans websites and review this process.

Q7.4 Given the design speed of a freeway to be built in California is 70 mph, what are the stopping sight distance and the minimum radius of curve on this freeway?

Q7.5 Assume the design speed of an expressway is 70 mph. Three bushes along the roadside block the view of drivers. The lateral clearance, m, is observed to be 20 ft. from the centerline of inside required for the geometric design. What is the minimum radius of the circle allowed for the geometric design?

Q7.6 An undivided 4-lane freeway has a median width of 24 ft. (from inside edge of traveled way (ETW) to inside ETW). The lane width is 12 ft. What is the minimum radius of the centerline curve?

Q7.7 Suppose a roadway has a curve with the following information

Curve Length 567.50 ft

Central angle 8°

Does this curve meet the Caltrans permissive standards?

Q7.8 Assume that compound curves are considered in a roadway design for a ramp in California. The larger curve follows the smaller curve. The radius of the larger curve is 2,000 ft. What is the minimum radius required for the shorter curve?

Q7.9 Crown runoff is an element to be considered in superelevation transition. What is the similar term used in AASHTO's "A Policy of Geometric Design of Highways and Streets"?

Q7.10 Redraw Figure 7-5. Is the superelevation runoff (L) the length of the curve?

Q7.11 A rural 6-lane, undivided freeway is designed with the following elements:

Design Speed V	60 mph
Width of median along the curve	62 ft
Lane width	12 ft

Determine the following using Caltrans standards:

e_{max}, R_{min} at the station line,

Superelevation $(SE)_{runoff}$, $SE_{crown\ runoff}$

What are the superelevation rates at the beginning, the middle, and the end points of the SE_{runoff}?

What is the relative gradient used in Caltrans HDM for SE_{runoff} if the design superelevation rate (e) is 6% and two lanes are rotated.

Q7.12 A rural 4-lane, undivided freeway (SR44) in California (median width = 24 ft) has a curve (with the radius of 1,500 ft) followed by a reverse curve (with the radius of 1,200 ft), what is the minimum gap required between the reverse curves?

Q7.13 What are broken back curves?

Q7.14 What is a Geometry Project in InRoads? Can a geometry project have more than one horizontal alignment?

Q7.15 Assume a horizontal alignment is set to be active. When you click ***Geometry → View Geometry → Active Horizontal*** to view an active horizontal alignment, you cannot see the horizontal alignment. What could be wrong?

Q7.16 When an alignment is created using InRoads, at which level is the horizontal alignment located?

Q7.17 When you use ***Geometry → View Geometry → Stationing*** to display stations for a horizontal alignment, you cannot display the stations in the MicroStation view window. What is the possible reason for this problem?

Q7.18 Can we use commands in InRoads to create the lane configuration of Route 8, instead of using the MicroStation **Move/Copy Parallel** command?

Q7.19 Do we consider minimum radius of the curve and supereleveation transition for the Laguna Canyon Road? Why?

Q7.20 List the steps used to delete an existing horizontal alignment in InRoads.

Q7.21 List the steps used to empty an existing horizontal alignment in InRoads.

Q7.22 List the steps used to create an empty horizontal alignment, assuming a project geometry test is created in InRoads.

Q7.23 What is the file extension for a geometry project? Does a horizontal alignment have its own file name?

Q7.24 Draw Caltrans standard gore area treatment for a freeway entrance.

Q7.25 Why is the "tutorial_2D.dgn", rather than "tutorial.dgn", preferred when we design horizontal alignments?

Chapter 8
Profiles

LEARNING OBJECTIVES

After completing this chapter you should know:

1. The concepts of profiles;
2. How to create profiles for the tutorial project.

SECTIONS

8.1 Load DTM Surface and Horizontal Alignment into InRoads
8.2 Create the Profile for Route 8
8.3 Create the Profile for the Laguna Canyon Road
8.4 Create the Profile for the Eastbound Off Ramp
8.5 Create the Profile for the Eastbound On Ramp
8.6 Create the Profile for the Westbound Off Ramp
8.7 Create the Profile for the Westbound On Ramp
8.8 Questions

Profiles are 2-D graphs that show elevations extracted from DTM surfaces along horizontal alignments. They provide a drafting environment on which vertical alignments can be developed.

A *profile* is a result of cutting a DTM surface along a horizontal alignment using a "saw." As shown in Figure 8-1, the profile is made of the line segments or the intersecting edges of the DTM surface and the vertical plane of the horizontal alignment (hereafter referred to as the horizontal alignment plane).

The profile is stretched along the horizontal alignment to become an elevation/station graph (see Figure 8-2). The X-axis in the graph represents the stations along a horizontal alignment, while the Y-axis represents the corresponding elevations of the existing surface. The elevation/station graph shows the terrain characteristics along the horizontal alignment.

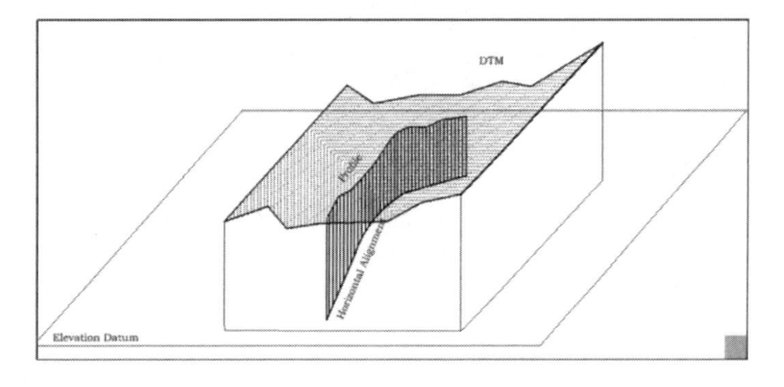

Figure 8-1 Profile Resulting from DTM Surface and Horizontal Alignment Plane

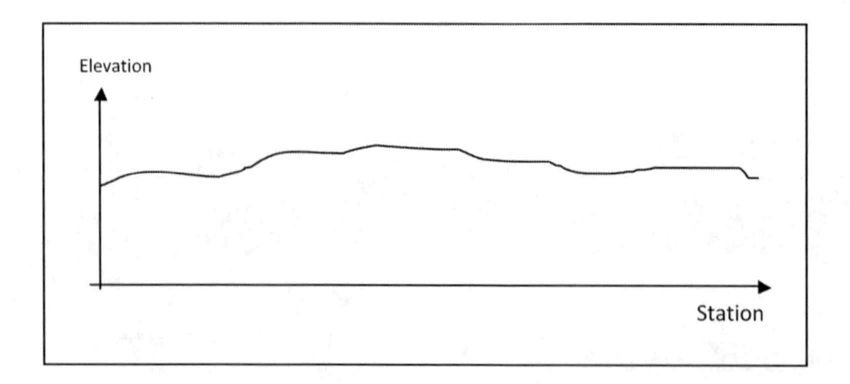

Figure 8-2 Elevation/Station Graph

This chapter describes how to use InRoads to create profiles for the tutorial project. The DTM surface **OG.dtm** and the geometry project **Tutorial.alg** (which contains horizontal alignments for Route 8, Laguna Canyon Road, and ramps) will be used as inputs (see Figure 8-3) for the profiles.

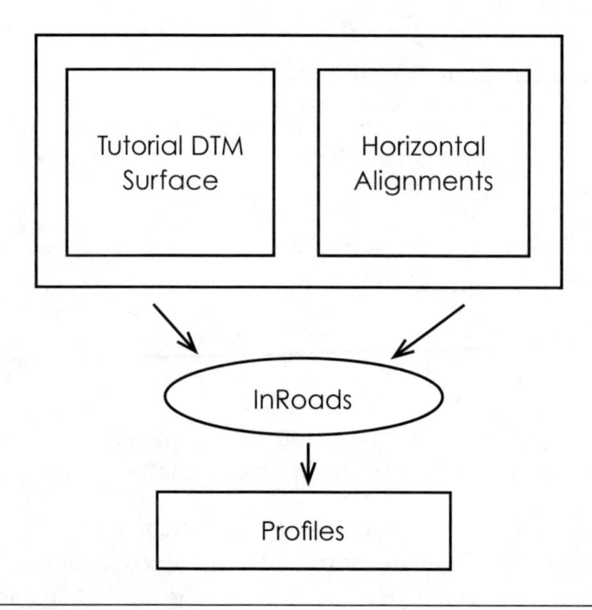

Figure 8-3 Profile-Creating Process

8.1 Load DTM Surface and Horizontal Alignment into InRoads

Before we can create profiles for the tutorial project, you need to have MicroStation and InRoads installed on your computer. You also need to copy **tutorial_2D_HA.dgn, OG.dtm** and **tutorial.alg** from **C:/Temp/Book/Chp8/Master Files** folder to the **C:/Temp/Tutorial/Chp8/Master Files** folder or you can download the files from the book website www.csupomona.edu/~xjia/GeometricDesign/Chp8/Master Files/. You may use your own **tutorial_2D_HA.dgn, OG.dtm**, and **Tutorial.alg** files you created when you worked with the previous chapters.

Follow these steps to open the **tutorial_2D_HA.dgn** design file from the **C:/Temp/Tutorial/Chp8/Master Files** folder:

1. Start MicroStation and InRoads.
2. When MicroStation prompts you to open a MicroStation file, select **tutorial_2D_HA.dgn.**
   a. Keep the **MicroStation → View 1** window open and close other view windows if multiple view windows are open.

      You will see the InRoads design environment similar to Figure 8-4.

   b. If you do not see the contours in the **View 1** window, from the **View 1** Window Toolbar, select the ***MicroStation → Fit View*** tool.

      Note that **tutorial_2D_HA.dgn** is a 2-D design file.

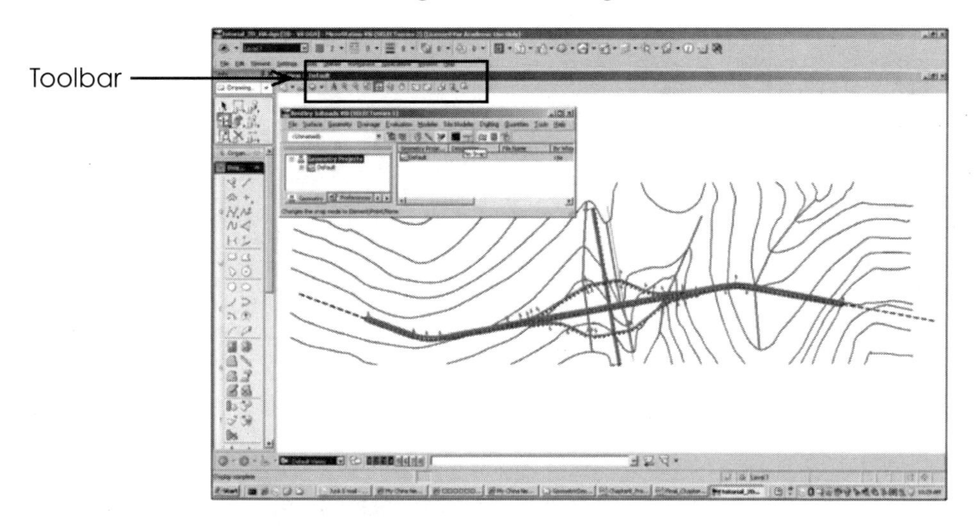

Figure 8-4 Tutorial_2D_HA file

3. Go to ***InRoads → File → Open.***

The **Open** window appears.

   a. Browse the **OG.dtm** file and select the file.
   b. Select ***Open*** and ***Cancel.***

      You will see the **OG.dtm** file loaded into the InRoads **Workspace Bar** window under the **Surface** tab (see Figure 8-5).

   c. Right click the **OG** in the **Workspace Bar** window and select **Set Active.**

      The **OG** DTM surface is selected to be active. A red rectangle encloses the surface icon associated with **OG**.

4. Go to ***InRoads → File → Open.***

The **Open** window appears.

   a. Browse and select the **Tutorial.alg** file.
   b. Select ***Open*** and ***Cancel.***

      You will see the **Tutorial.alg** file loaded into the InRoads **Workspace Bar** window under the **Geometry** tab (see Figure 8-6).

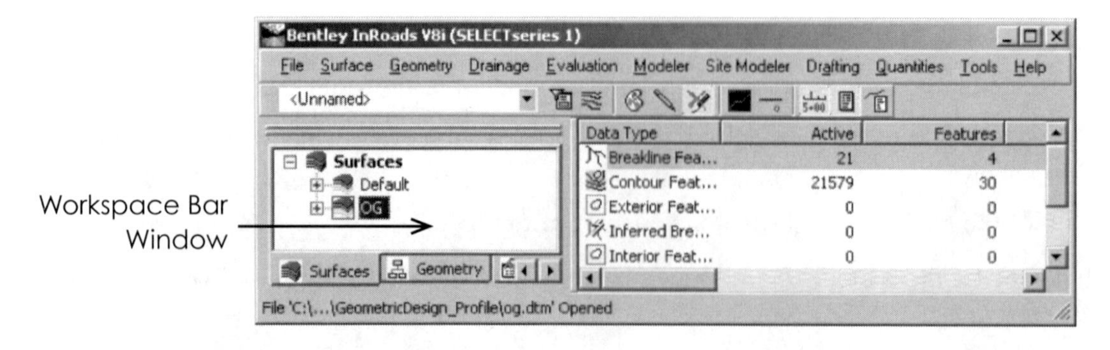

Figure 8-5 Set OG to be Active DTM Surface

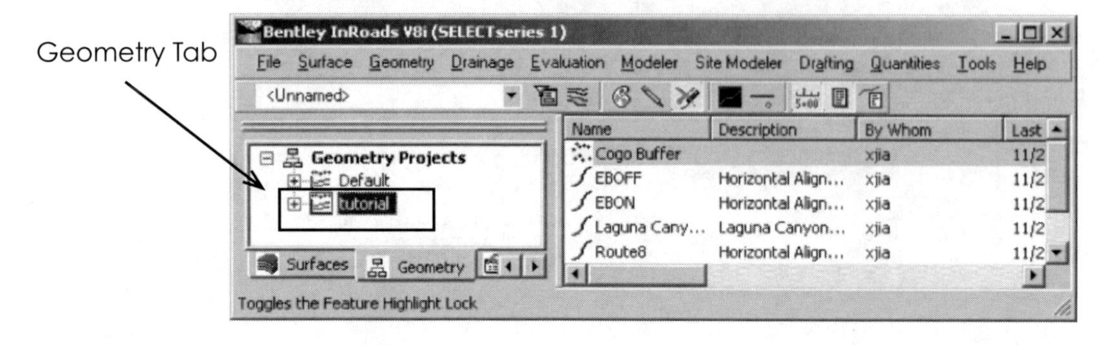

Figure 8-6 The Loaded Tutorial.alg

c. Right click **Tutorial** (the geometry project in the **Workspace Bar** window) and select **Set Active.**

The **Tutorial** geometry project is selected to be active. A red rectangle encloses the surface icon associated with **Tutorial.**

8.2 Create the Profile for Route 8

Now we need to create the profile for Route 8 following these steps:

1. Select ***InRoads → Workspace Bar → Geometry Tab.***
2. Expand the **Tutorial** geometry project by clicking on the + sign next to it.

You will see six horizontal alignments listed under **Tutorial.**

3. Right click ***Route 8*** and set it to be active.

   a. Select ***InRoads → Evaluation → Profile → Create Profile.***

   The **Create Profile** window appears (see Figure 8-7).

   b. Do the following in the **Create Profile** window. Under the **General** entry:

Set Name:	***Route 8***	
Direction:	***Left to Right*** checked	
Surfaces:	**OG** selected	
Exaggeration		
	Vertical:	***5.0000***
	Horizontal:	***1.0000***

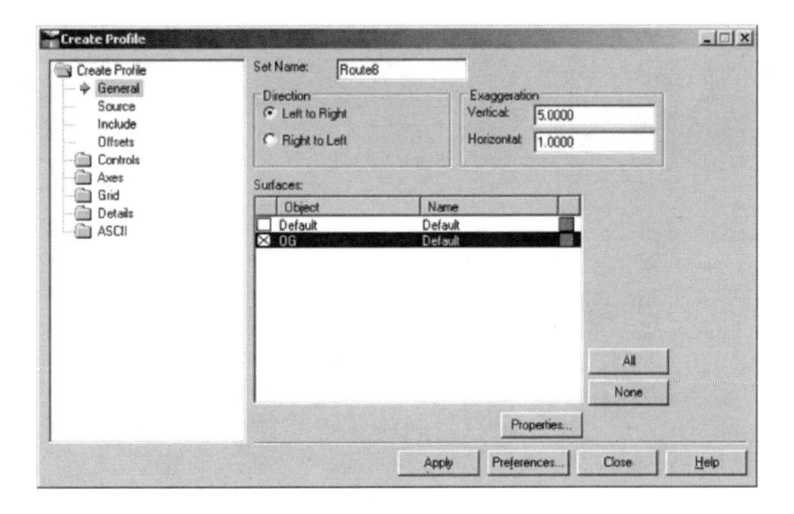

Figure 8-7 Create Profile Window

Note that the **vertical exaggeration** of 5.0000 indicates that 1 ft of actual terrain elevation is represented by 5 ft of screen units in the profile. The **horizontal exaggeration** of 1.0000 means the station scale for horizontal alignments and for profiles is the same.

The **Left to Right** option indicates that the profile is displayed from left to right.

The selection of **OG** surface indicates that the profile for Route 8 will be created from the cut of the **OG** DTM surface along the Route 8 horizontal alignment. InRoads allows us to select more than one surface for creating profiles. If two DTM surfaces are selected, the profile will have two elevation/station graphs. In this project, we only use the **OG** DTM surface to create one elevation/station chart for Route 8.

Example 8-1:

Assume a profile with the vertical exaggeration of 5.0000 is created for Route 44 (see Figure 8-8). The actual elevation of a point on the Route 44 horizontal alignment is 2,300.00 ft. What is the elevation value of this point (expressed by screen units) on the profile?

Solution

Given the vertical exaggeration of 5.0000, the elevation on the profile (represented by screen units) is 2,300 $\times$ 5 = 11,500 ft.

When you use the MicroStation Measure Distance command to measure the segment A-A', the result of the measure is 11,500 ft.

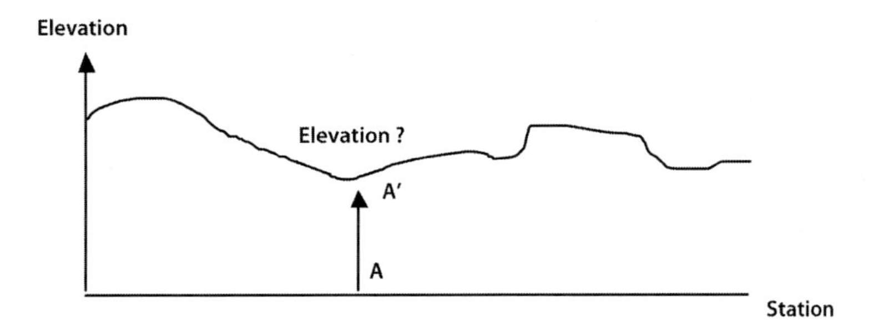

Figure 8-8 Profile for Example 8-1

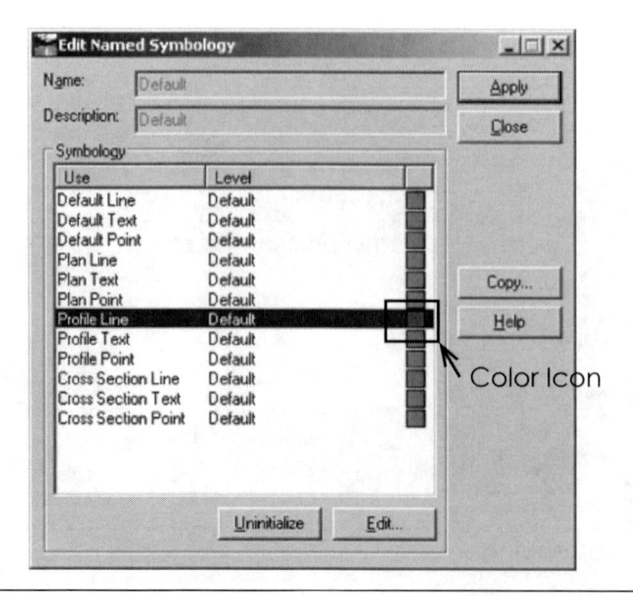

Figure 8-9 Edit Named Symbology Window

c. Still under the **General** entry, double click the Color Icon (see Surface Panel in Figure 8-7) for the **OG** DTM surface.

   The **Edit Named Symbology** window appears (see Figure 8-9).

d. In the **Edit Named Symbology** window, double click the Color Icon for the Profile Line (see Figure 8-9).

   The **Line Symbology** window appears.

e. In the **Line Symbology** window, select the color of your choice. Select OK.

   The profile line color is changed to your choice.

f. In the **Edit Named Symbology** window, select ***Apply*** and ***Close***.

   You have finished the work for the **General** entry.

g. Under the **Source** entry, make sure:

Create:	***Window and Data***
Alignment:	***Route 8***
All other options:	***Un-checked***

h. Under the **Details** entry, make sure:

Title Text:	***Route 8 Profile***
Relative to Bottom Axis:	***Unchecked***
All other options:	***Default***

i. Select **Apply** and place a point to locate the lower left corner of the profile.

   Figure 8-10 shows the profile of Route 8.

j. Select ***Close***.

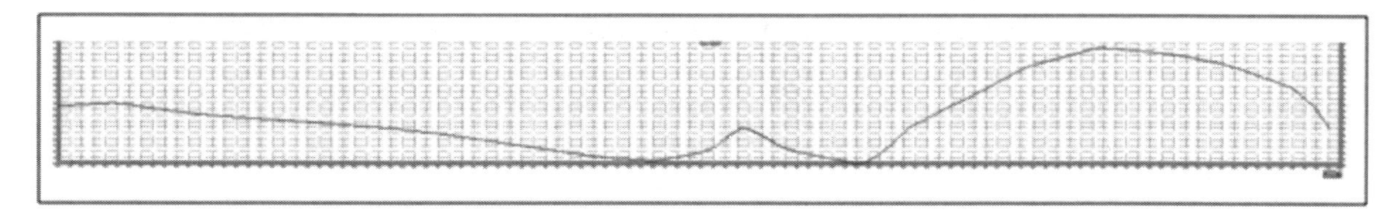

Figure 8-10 Route 8 Profile

Example 8-2:

In Figure 8-11, we measure the distance between P and P' points and get 450 ft. Why?

Solution

From the profile, we know that the actual distance from P to P' is shown to be 3,080 − 2,990 = 90 ft. The vertical exaggeration for this profile is 5.0000. It means that MicroStation uses 5 × 90 = 450 ft. (screen units) to represent the elevation of Point P on Route 8.

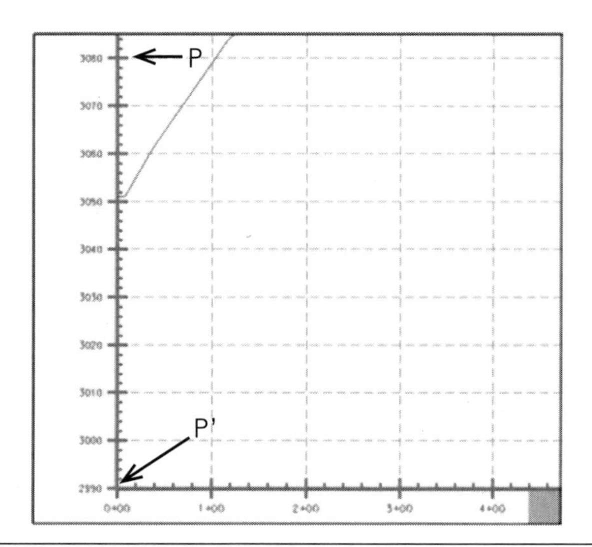

Figure 8-11 Vertical Exaggeration in Example 8-2

8.3 Create the Profile for the Laguna Canyon Road

Now we need to create the profile for Laguna Canyon Road following these steps:

1. Select ***InRoads → Workspace Bar → Geometry Tab.***
2. Expand the **Tutorial** geometry project.

You will see six horizontal alignments listed under **Tutorial**.

3. Right click ***Laguna Canyon Road*** and set it to be active.

4. Select ***InRoads → Evaluation → Profile → Create Profile.*** The **Create Profile** window appears.

   a. Do the following in the **Create Profile** window. Under the **General** entry,

Set Name:	***Laguna Canyon Road***
Direction:	***Left to Right*** checked
Surfaces:	**OG** selected
Exaggeration	
	Vertical: ***5.0000***
	Horizontal: ***1.0000***

   b. Double click the Color Icon for the **OG** DTM surface.

      The **Edit Named Symbology** window appears.

   c. In the **Edit Named Symbology** window, double click the Color Icon for the Profile Line.

      The **Line Symbology** window appears.

   d. In the **Line Symbology** window, select the color of your choice. Select OK.

      The profile line color is changed to your choice.

   e. In the **Edit Named Symbology** window, select ***Apply*** and ***Close.***

   f. Under the **Source** entry, make sure:

Create:	***Window and Data***
Alignment:	***Laguna Canyon Road***
All other options:	***Un-checked***

   g. Under the **Details** entry, make sure:

Title Text:	***Laguna Canyon Road Profile***
Relative to	
Bottom Axis:	***Unchecked***
All other options:	***Default***

   h. Select ***Apply*** and place a point to locate the lower left corner of the profile.

      Figure 8-12 shows the profile of Laguna Canyon Road.

   i. Select ***Close.***

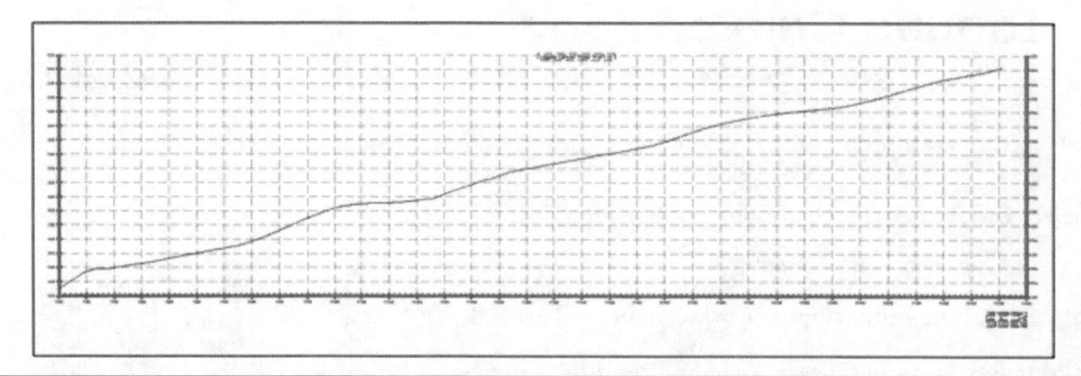

Figure 8-12 Laguna Canyon Road Profile

8.4 Create the Profile for the Eastbound Off Ramp

Now we need to create the profile for the Eastbound Off Ramp following these steps:

1. Select ***InRoads → Workspace Bar → Geometry Tab.***
2. Expand the **Tutorial** geometry project.

You will see six horizontal alignments listed under **Tutorial**.

3. Right click ***EBOFF*** and set it to be active.
4. Select ***InRoads → Evaluation → Profile → Create Profile.***

The **Create Profile** window appears.

a. Do the following in the **Create Profile** window. Under the **General** entry:

Set Name:	***EBOFF***	
Direction:	***Left to Right*** checked	
Surfaces:	**OG** selected	
Exaggeration		
	Vertical:	***5.0000***
	Horizontal:	***1.0000***

b. Double click the Color Icon for the **OG** DTM surface. The **Edit Named Symbology** window appears.

c. In the **Edit Named Symbology** window, double click the Color Icon for the Profile Line.

   The **Line Symbology** window appears.

d. In the **Line Symbology** window, select the color of your choice. Select OK.

   The profile line color is changed to your choice.

e. In the **Edit Named Symbology** window, select ***Apply*** and ***Close***.

f. Under the **Source** entry, make sure:

Create:	***Window and Data***
Alignment:	***EBOFF***
All other options:	***Un-checked***

g. Under the **Controls** entry, make sure under the Elevation frame:

Use:	***Checked***
High:	***3080.00***
Low:	***2970.00***

h. Under the **Details** entry, make sure:

Title Text:	***Eastbound Off Profile***
Relative to	
Bottom Axis:	***Unchecked***
All other options:	***Default***

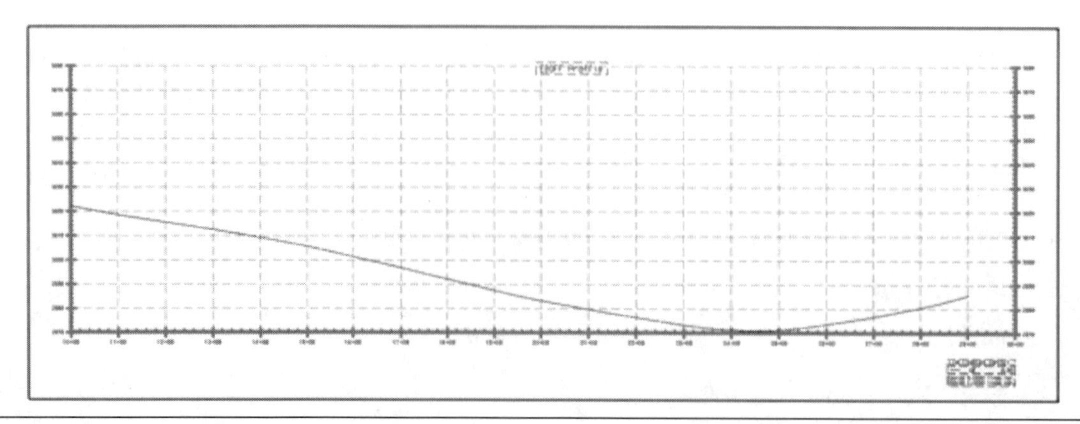

Figure 8-13 Eastbound Off Ramp Profile

i. Select **Apply** and place a point to locate the lower left corner of the profile.

   Figure 8-13 shows the profile of the Eastbound Off Ramp.

j. Select ***Close.***

8.5 Create the Profile for the Eastbound On Ramp

Now we need to create the profile for the Eastbound On Ramp following these steps:

1. Select ***InRoads → Workspace Bar → Geometry Tab.***
2. Expand the **Tutorial** geometry project.

You will see six horizontal alignments listed under **Tutorial**.

3. Right click ***EBON*** and set it to be active.
4. Select ***InRoads → Evaluation → Profile → Create Profile.***

The **Create Profile** window appears.

a. Do the following in the **Create Profile** window. Under the **General** entry:

   Set Name: ***EBON***
   Direction: **Left to Right** checked
   Surfaces: **OG** selected
   Exaggeration
   Vertical: ***5.0000***
   Horizontal: ***1.0000***

b. Double click the Color Icon for the **OG** DTM surface.

   The **Edit Named Symbology** window appears.

c. In the **Edit Named Symbology** window, double click the Color Icon for the Profile Line.

   The **Line Symbology** window appears.

d. In the **Line Symbology** window, select the color of your choice. Choose OK.

The profile line color is changed to your choice.

e. In the **Edit Named Symbology** window, select ***Apply*** and ***Close***.

f. Under the **Source** entry, make sure:

Create:	***Window and Data***
Alignment:	***EBON***
All other options:	***Un-checked***

g. Under the **Details** entry, make sure:

Title Text:	***Eastbound On Profile***
Relative to	
Bottom Axis:	***Unchecked***
All other options:	***Default***

h. Select ***Apply*** and place a point to locate the lower left corner of the profile.

Figure 8-14 shows the profile of the Eastbound On Ramp.

i. Select ***Close***.

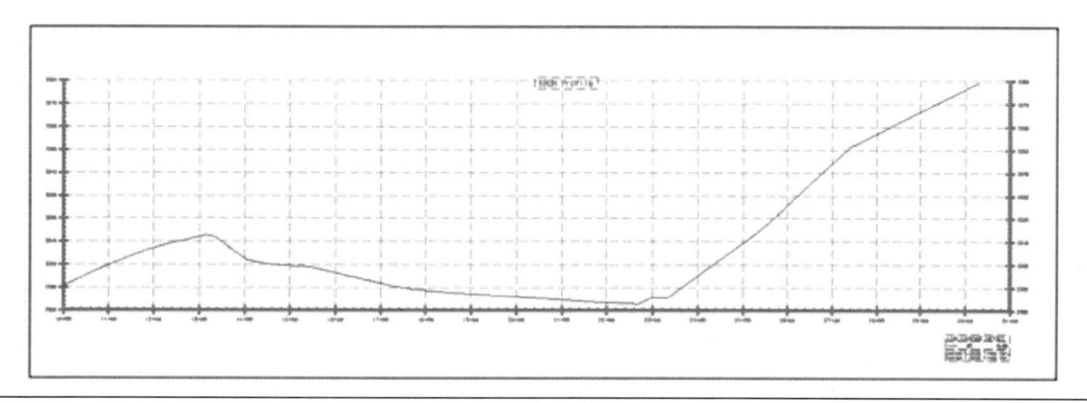

Figure 8-14 Eastbound On Ramp Profile

8.6 Create the Profile for the Westbound Off Ramp

Now we need to create the profile for the Westbound Off Ramp following these steps:

1. Select ***InRoads → Workspace Bar → Geometry Tab.***
2. Expand the **Tutorial** geometry project.

You will see six horizontal alignments listed under **Tutorial**.

3. Right click ***WBOFF*** and set it to be active.
4. Select ***InRoads → Evaluation → Profile → Create Profile.***

The **Create Profile** window appears.

a. Do the following in the **Create Profile** window. Under the **General** entry:

Set Name:	***WBOFF***
Direction:	***Left to Right*** checked
Surfaces:	**OG** selected
Exaggeration	
	Vertical: ***5.0000***
	Horizontal: ***1.0000***

b. Double click the Color Icon for the **OG** DTM surface.

The **Edit Named Symbology** window appears.

c. In the **Edit Named Symbology** window, double click the Color Icon for the Profile Line.

The **Line Symbology** window appears.

d. In the **Line Symbology** window, select the color of your choice. Select OK.

You will see the color for the profile line is changed to your choice.

e. In the **Edit Named Symbology** window, select ***Apply*** and ***Close***.

f. Under the **Source** entry, make sure:

Create:	***Window and Data***
Alignment:	***WBOFF***
All other options:	***Un-checked***

g. Under the **Details** entry, make sure:

Title Text:	***Westbound Off Profile***
Relative to	
Bottom Axis:	***Unchecked***
All other options:	***Default***

h. Select ***Apply*** and place a point to locate the lower left corner of the profile.

Figure 8-15 shows the profile of the Westbound Off Ramp.

i. Select ***Close***.

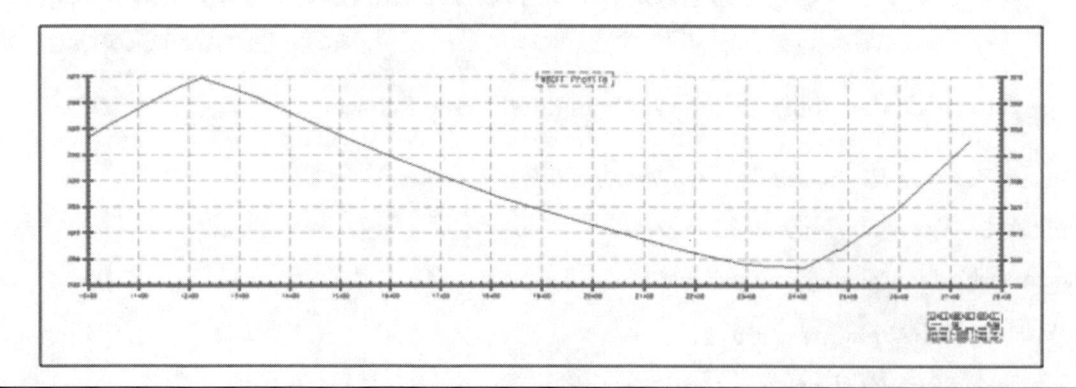

Figure 8-15 Westbound Off Ramp Profile

8.7 Create the Profile for Westbound On Ramp

Now we need to create the profile for Westbound On Ramp using the following steps:

1. Select ***InRoads → Workspace Bar → Geometry Tab.***
2. Expand the **Tutorial** geometry project.

You will see six horizontal alignments listed under **Tutorial**.

3. Right click ***WBON*** and set it to be active.
4. Select ***InRoads → Evaluation → Profile → Create Profile.***

The **Create Profile** window appears.

a. Do the following in the **Create Profile** window. Under the General entry:

Set Name:	***WBON***	
Direction:	***Left to Right*** checked	
Surfaces:	**OG** selected	
Exaggeration		
	Vertical:	***5.0000***
	Horizontal:	***1.0000***

b. Double click the Color Icon for the **OG** DTM surface.

The **Edit Named Symbology** window appears.

c. In the **Edit Named Symbology** window, double click the Color Icon for the Profile Line.

The **Line Symbology** window appears.

d. In the **Line Symbology** window, select the color of your choice. Select OK.

The profile line color is changed to your choice.

e. In the **Edit Named Symbology** window, select ***Apply*** and ***Close***.

f. Under the **Source** entry, make sure:

Create:	***Window and Data***
Alignment:	***WBON***
All other options:	***Un-checked***

g. Under the Controls entry, make sure under the Elevation frame:

Use:	***Checked***
High:	***3080.00***
Low:	***3010.00***

h. Under the **Details** entry, make sure:

Title Text:	***Westbound On Profile***
Relative to	
Bottom Axis:	***Unchecked***
All other options:	***Default***

i. Select ***Apply*** and place a point to locate the low-left corner of the profile.

Figure 8-16 shows the profile of the Westbound On Ramp.

j. Select ***Close***.

k. Save the working **tutorial_2D_HA.dgn** file as **C:/Temp/Tutorial/Chp8/Master Files/tutorial_2D_Profile.dgn**.

The difference between the two files is that the **tutorial_2D_Profile.dgn** has profiles, while the **tutorial_2D_HA.dgn** does not. In later chapters, the **tutorial_2D_Profile.dgn** will be used to create vertical alignments.

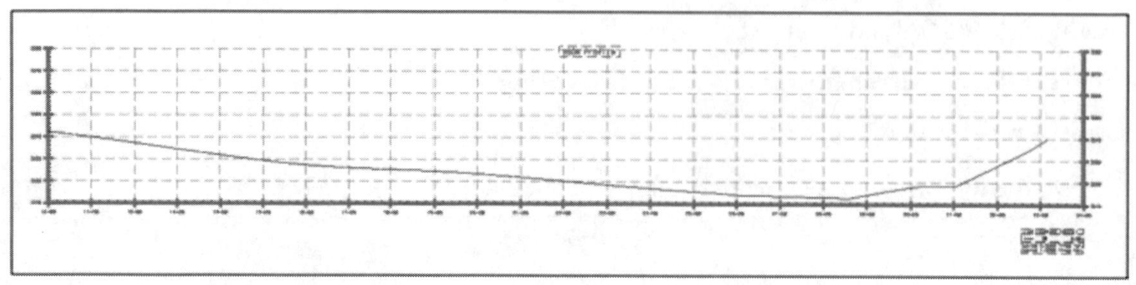

Figure 8-16 Westbound On Ramp Profile

Name____ Date________________

8.8 Questions

Q8.1 What is a profile?

Q8.2 Is a profile a vertical alignment?
To create a profile, we need to have a few things active in the InRoads environment. What are the two primary things that should be set active?

Q8.3 The X-axis in a profile represents the stations along a horizontal alignment, while the Y-axis represents the elevations of an existing surface. Normally, the scale (or the horizontal exaggera tion) for the X-axis is 1:1 and the scale (or the vertical exaggeration) for the Y-axis is 1:5. Explain how to set these two scales in the InRoads environment.

Q8.4 How do you know which surface is the active surface? How do you know which geometry project is active? How do you know which horizontal alignment is active in the active geometry project?

Q8.5 Which level are the profiles (created in this chapter) located on? Level 1 or Default?

Q8.6 How do you change the title for a profile? If you want to enlarge the title, how do you do that?

Q8.7 Are there any Caltrans standards on profiles?
Assume a profile with the vertical exaggeration of 10.0000 is created for Route 44. The starting elevation of the profile is 2,000 ft. The actual elevation of a point on Route 44 horizontal alignment is 2,300.00 ft. What is the distance measured from this point on the profile to the X axis?

Chapter 9

Alignments

LEARNING OBJECTIVES

After completing this chapter you should know:

1. Design controls related to the vertical alignments of the tutorial project;
2. Caltrans standards for the vertical alignments of the tutorial project;
3. How to create six vertical alignments;
4. How to check the vertical alignments for conformance to Caltrans design standards.

SECTIONS

This chapter concentrates on vertical alignment. Design of a vertical alignment is a trade-off process that determines a series of grades and vertical curves in order to provide safe, smooth, continuous operation at the design speed of the facility.

9.1 Understanding the Tutorial Project

Before we start using MicroStation and InRoads to create the vertical alignments for the tutorial project, let's review the project data and have a good understanding of the design controls associated with the vertical alignment design.

Design Speed In this tutorial project, the design speed is as follows:

Route 8: 60 mph
Ramps 25: 50 mph
Laguna Canyon Road: 45 mph

Geometric Control Points There are three control points that govern the vertical alignment design of Route 8 and Laguna Canyon Road: elevation, grade, and coordinates. These control points are Point A and Point B on Route 8 and Point C on Laguna Canyon Road. The elevation and the northing and easting values of these three points are provided in Chapter 4.

The vertical alignment between Points A and B shall be consistent with vertical alignments leading to Point A and coming out of Point B. The grades at Point A (2.0%, West to East) and Point B (1.5%, West to East) are the design "agreement" among design teams for Route 8. Each design team shall comply with this agreement.

Points A and B can be used *only* as the beginning of vertical curves (BVCs), ending of vertical curves (EVCs), or points on grade lines outside of any vertical curves.

The vertical alignment for Laguna Canyon Road shall pass through Point C and shall be in the grade of 4% up to the North.

The **tutorial.dgn** file in the **C:/Temp/Book/Chp9/Master Files** folder (or you can download the file from the book website www.csupomona.edu/~xjia/GeometricDesign/Chp9/Master Files/) contains these three points. You need to verify the control points in MicroStation.

9.2 Design Standards Applied to Vertical Alignments

Caltrans criteria and standards that are applicable to the vertical alignment design of the tutorial project are adopted from the Caltrans Highway Design Manual (*HDM*) and are described as follows:

1. Standards on grade lines

Grade lines are reference lines by which the elevation of the pavement and other features of a highway are established. The grade lines are typically controlled by topography, type of highway, horizontal alignment, performance of heavy vehicles, right-of-way costs, safety, sight distance, construction costs, cultural development, drainage, and pleasing appearance.

All portions of grade lines must meet safety requirements (such as sight distances) for the design speed of roads.

Grade lines should generally coincide with the axis of rotation for superelevation. Their relation to cross section should be as follows:

a. **Undivided Highways.** The grade line should coincide with the highway centerline (see Figure 9-1a).

b. **Ramps.** The grade line is usually positioned at the left edge of traveled way (ETW; see Figure 9-1b).

c. **Divided Highways.** The grade line should be positioned at the centerline of the median for paved medians with the width of 65 ft or less, thus avoiding a "saw tooth" section (see Figure 9-1c). The grade line may be positioned at the ultimate median edge of traveled way when:

   - The median edges of traveled way of two roadways are at equal elevation (see Figure 9-1d).
   - Two roadways are at different elevations (see Figure 9-1e).

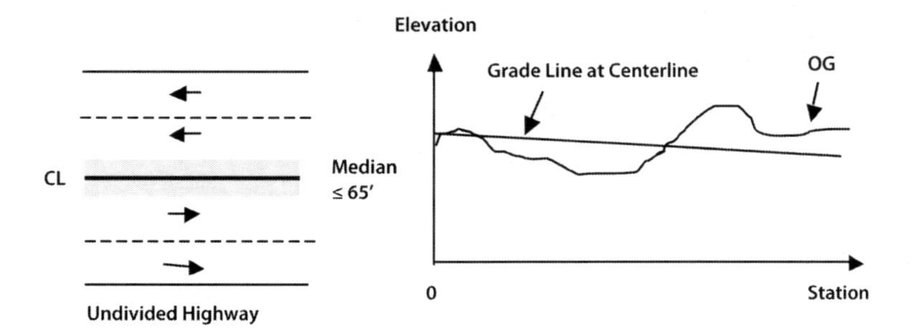

Figure 9-1a Grade Line for Undivided Highway

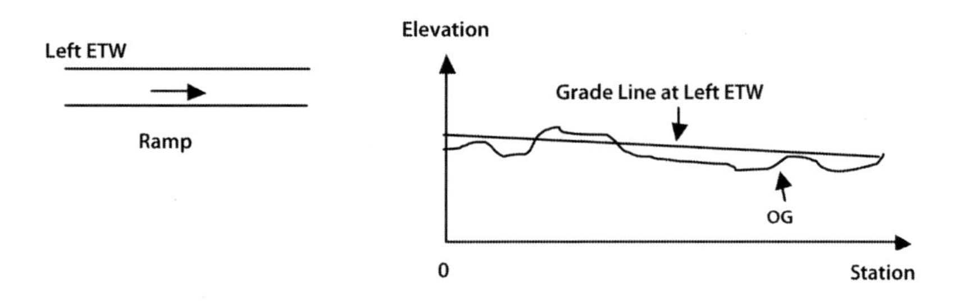

Figure 9-1b Grade Line for Ramp

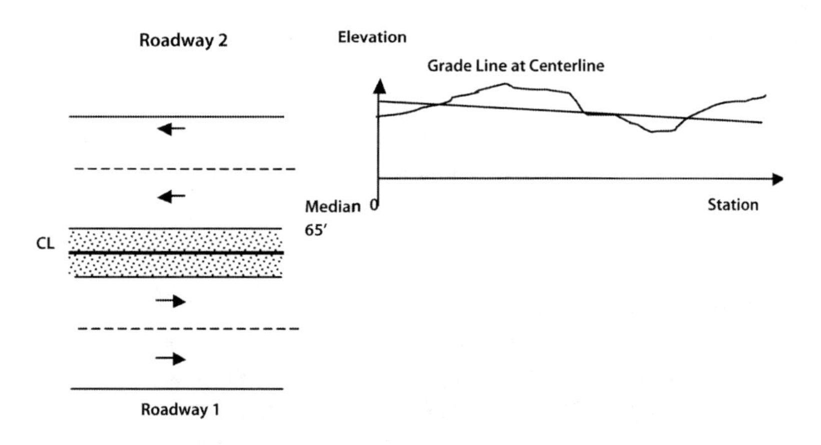

Figure 9-1c Grade Line for Divided Highway

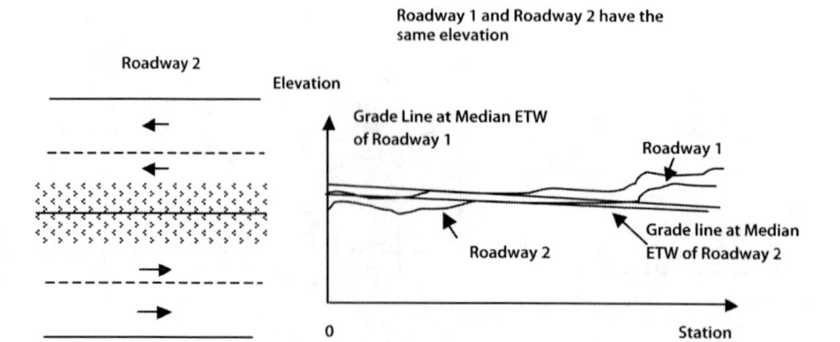

Figure 9-1d Grade Line for Divided Highway

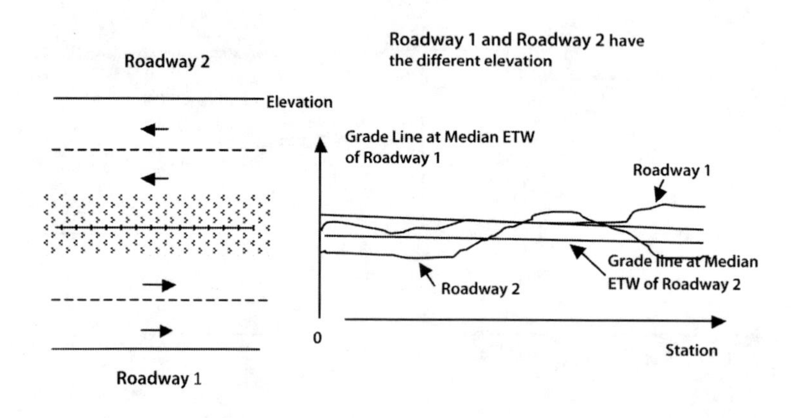

Figure 9-1e Grade Line for Divided Highway

2. Standards on grades

Grades affect vehicle speeds and overall capacity. They also cause operational problems at ramp intersections.

a. Maximum Grades

Table 204.3 of Caltrans *HDM* shows the maximum grades that shall not be exceeded in the design of vertical alignments. For example, the maximum grades for freeways and expressways are 3%, 4%, and 6% for level, rolling, and mountainous terrain, respectively. The maximum grades for rural highways are 4%, 5%, and 7% for level, rolling, and mountainous terrain, respectively. The maximum grades for urban highways are 6%, 7%, and 9% for level, rolling, and mountainous terrain, respectively.

Ramp grades should not exceed 8%. On descending on-ramps and ascending off-ramps, 1% steeper is allowed (see Index 504.2(5) of Caltrans *HDM*).

Maximum grade is not a complete design control. The length of an uphill grade is important as well, because it affects capacity, level of service, and delay when slow-moving trucks, buses, and recreational vehicles are present.

A common criterion for all types of highways is to consider the addition of a climbing lane where the running speed of trucks falls 10 miles per hour or more below the running speed of remaining traffic. Figure 9-2 (or Figure 204.5 of Caltrans *HDM*) shows the speed reduction curves for a 200 lb/hp truck, which is representative of large trucks operating near maximum gross weight. The 10 miles per hour reduction criterion may be used as one way to determine needs for climbing lanes.

Example 9-1:

Assume the grade is 4% for a segment on a freeway in the rolling area. The length of the segment is 1,000 ft. Is this grade OK to meet the Caltrans standards? Is the length of the segment allowed for keeping the designed level of service without adding a climbing lane?

Solution

Based on the roadway type and the terrain characteristics, the maximum grade is set at 4% according to Caltrans standards. The design grade is 4%, equal to the maximum grade, so the grade is OK.

Given the grade of 4%, the critical length is 1,800 ft., greater than the designed segment length of 1,000 ft. The climbing lane is not needed.

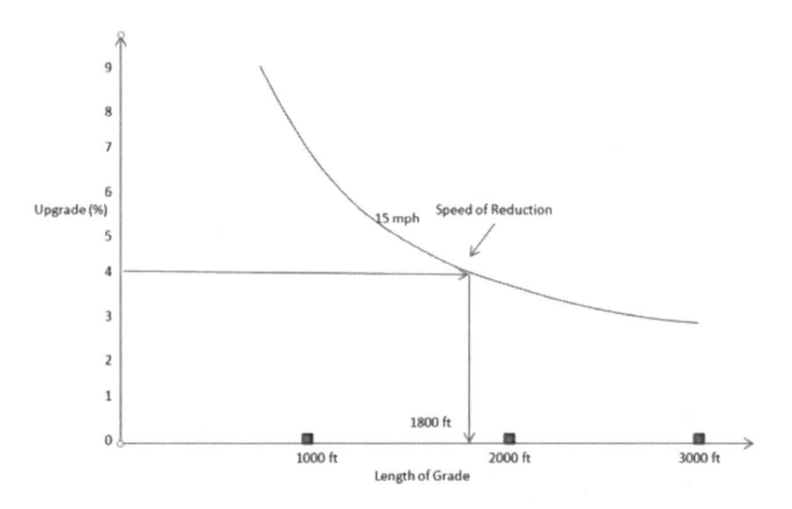

Figure 9-2 Critical Lengths of Grade for Design

b. Minimum Grades

Minimum grades should be 0.5% in snow country and 0.3% at other locations. Except for conventional highways in urban or suburban areas, a level grade line is permissible in level terrain where side fill slopes are 4:1 or flatter, and dikes are not needed to carry water in the roadbed. Flat grades are not permissible in superelevation transitions due to flat spots that cause ponds on the roadbed.

For the tutorial project, the minimum grade is set at 0.5% for Route 8 and ramps.

3. Standards for vertical curves

Vertical curves (either crest or sag curves) are parabolic curves that provide a gradual transition of a grade line from one tangent grade to another. The reasons to use parabolic curves instead of circular curves are that the circular curves will make people in vehicles feel sudden "drop" or "roller coaster" effects due to gravitational forces (see Figure 9-3). Parabolic curves are the trajectory of a particle or body in motion under the influence of a uniform gravitational field without air resistance. When a roadway is built with a parabolic curve, vehicles are supported by the curve. People in the vehicles do not have gravitational "drop" and feel comfortable in traveling.

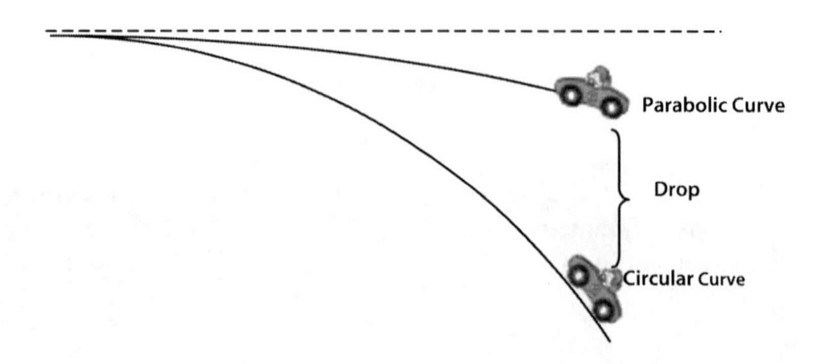

Figure 9-3 Parabolic Curve

Properly designed vertical curves should ensure adequate sight distance, safety, comfortable driving, good drainage, and pleasing appearance. Given two tangent grades, G_1 and G_2 as shown in Figure 9-4, designers should design a vertical curve whose length is sufficient for the grade change.

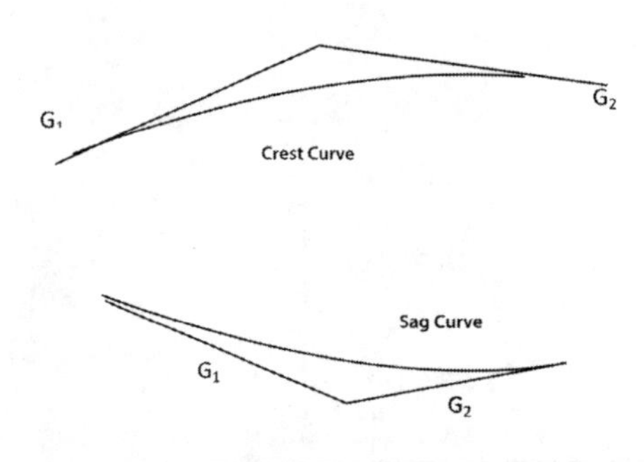

Figure 9-4 Length of Curve for Grade Change

The word *sufficient* means that the length of a vertical curve (designed) should be greater than or equal to the minimum length of the vertical curve for the grade transition from G_1 to G_2 in Figure 9-4.

a. Algebraic grade difference ≥ 2% and design speed ≥ 40 mph

The minimum length of vertical curve should be equal to 10V, where V = design speed.

The minimum length of vertical curve is also controlled by the stopping sight distance criteria. Figure 9-5 shows the relationship between the stopping sight distance and the length of crest curves for V = 45 mph, 60 mph, 65 mph, and 70 mph, respectively. Figure 9-6 shows the relationship between the stopping sight distance and the length of sag curves for V = 45 mph, 60 mph, 65 mph, and 70 mph, respectively.

Example 9-2:

Assume the design speed of a freeway is 65 mph. G_1 = 3.0%, G_2 = –2.0%. What is the algebraic grade difference or the total grade change? What is the minimum length of vertical curve?

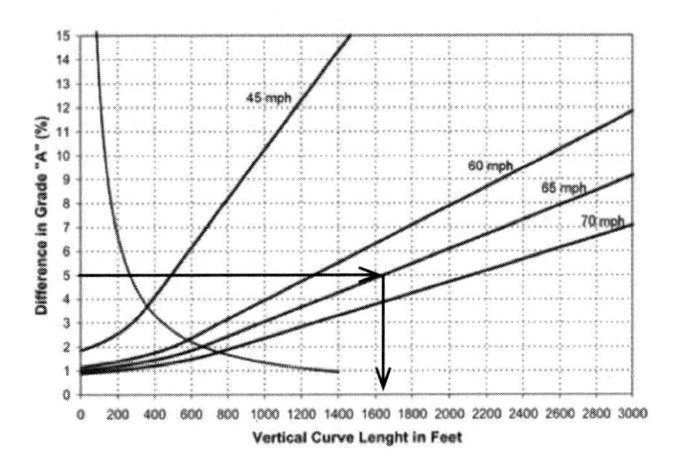

Figure 9-5 Stopping Sight Distance and Length of Crest Curves

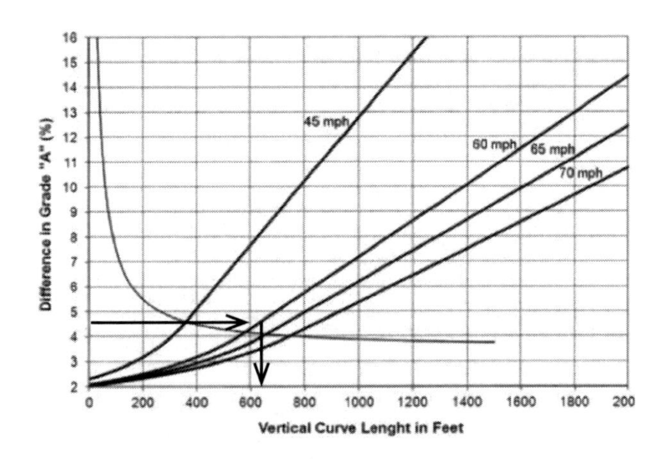

Figure 9-6 Stopping Sight Distance and Length of Sag Curves

Solution

This is a crest curve. The algebraic grade difference is $|-2.0\% - 3.0\%| = 5\%$. Following the equation $L_{min} = 10V$, the minimum length of the vertical curve is 10×65 mph = 650 ft.

Using Figure 9-5 (for crest curves), the minimum length of the vertical curve is 1,620 ft.

Since 1,620 ft > 650 ft, the minimum length of the vertical curve is 1,620 ft.

Example 9-3:

Assume the design speed of a freeway is 60 mph. $G_1 = -2.5\%$, $G_2 = 2.0\%$. What is the algebraic grade difference or the total grade change? What is the minimum length of the vertical curve?

Solution

This is a sag curve. The algebraic grade difference is $|2.0\% - (-2.5\%)| = 4.5\%$.

Following the equation $L_{min} = 10V$, the minimum length of the vertical curve is 10×60 mph = 600 ft.

Using Figure 9-6 (for sag curves), the minimum length of the vertical curve is 650 ft.

Since 650 ft > 600 ft, the minimum length of the vertical curve is 650 ft.

b. Algebraic grade difference < 2% or design speed < 40 mph

The vertical curve length should be a minimum of 200 ft.

Example 9-4:

Assume the design speed of a freeway is 60 mph. $G_1 = 0.5\%$, $G_2 = -0.5\%$. What is the algebraic grade difference or the total grade change? What is the minimum length of vertical curve?

Solution

This is a crest curve. The algebraic grade difference is $|-0.5\% = 0.5\%| = 1.0\% < 2\%$.

Figure 9-5 (for crest curves) is not applicable in this case because the intersection between the 1% line and the 60 mph curve does not exist.

The minimum length of the vertical curve therefore is 200 ft.

c. Algebraic grade difference ≤ 0.5%

Vertical curves are not required.

Grade breaks should not be too close. The length of breaks should be smaller than 50 ft. A total of all grade breaks in 200 ft should not exceed 0.5%.

Since flat vertical curves may develop poor drainage at the level section, adjusting the gutter grade or shortening the vertical curve may overcome drainage problems.

Broken-back vertical curves consist of two vertical curves in the same direction separated by a short grade tangent. A profile with such curvature should be avoided, particularly in sags where the view of both curves is unpleasing.

4. Grade line of structures

When an overcrossing structure (a local road running on the top of a state highway facility) or an undercrossing structure (a local road running under a state highway facility) is considered for a highway project, the grade line of the structure should be carefully determined.

There are five factors that affect the grade line of structures:

a. Vertical Clearances

Vertical clearance should be provided for overcrossing and undercrossing bridge structures.

For bridges crossing freeways and expressways, all construction except overlay projects shall provide a minimum vertical clearance of **16.5 ft** over the roadbed of the state facility (e.g., main lanes, shoulders, ramps, collector-distributor roads, speed change lanes, etc.). For overlay projects, the vertical clearance shall be provided at a minimum of **16 ft** over the roadbed of the state facility.

For bridges crossing conventional highways, parkways, and local facilities, all projects shall have a minimum vertical clearance of **15 ft** over the traveled way and a minimum vertical clearance of **14.5 ft** over the shoulders of all portions of the roadbed.

Example 9-5:

An undercrossing bridge is designed for Route 8. Laguna Canyon Road goes under the bridge (see Figure 9-7). What is the vertical clearance from the pavement surface of Laguna Canyon Road to the bottom of the bridge deck?

Solution

The undercrossing bridge needs to provide vertical clearance for vehicles on Laguna Canyon Road. Since Laguna Canyon Road is a conventional highway, the minimum vertical clearance is 15 ft. The minimum vertical clearance over the shoulders of Laguna Canyon Road is 14.5 ft.

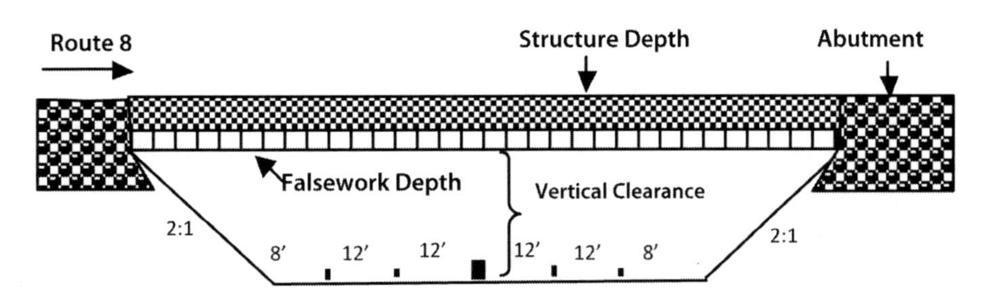

Figure 9-7 Structure Layout for Example 9-5

b. Structure depth

The depth of a highway structure is controlled by the structure's span and the depth to span ratio (d/s) associated with the structure. Caltrans defines the d/s based on the type of construction, aesthetics, costs, falsework limitations, and vertical clearance limitations.

Caltrans defines the d/s to be 0.06, 0.045, 0.055, and 0.004 for single span with span length $\leq$ 100 ft, single span with span length $>$ 100 and $\leq$ 180 ft, continuous structures with multiple spans with span length $\leq$ 100 ft, and continuous structures with multiple spans with span length $>$ 100 ft, respectively.

Example 9-6:

Assume an undercrossing structure is considered to allow Laguna Canyon Road to go under Route 8. Its layout is shown in Figure 9-7. The vertical clearance is 15 ft. from the bottom of the bridge deck to the pavement surface of Laguna Canyon Road. Falsework is not considered. What is the depth of the structure?

Solution

This structure is a singlespan structure. The vertical clearance above Laguna Canyon Road is 15 ft. The structure span is $s = (8 + 12 + 12 + 12 + 12 + 8) + 15 \times 2 + 15 \times 2 = 124$ ft.

The span is longer than 100 ft, so the d/s is 0.045.

The depth of the structure $= 0.045 \times 124 = 5.58$ ft.

c. Falsework

Falsework refers to temporary structures used in construction to support permanent structures. These temporary structures hold components in place until the permanent structures can support themselves. Falsework molds concrete to form a desired shape and provides scaffolding to give workers access to the permanent structure being constructed.

Falsework over traffic is often economically justified during bridge construction to provide a support-free open area (falsework opening or traffic opening) beneath the permanent structure.

To establish the grade of a permanent structure to be constructed with a falsework opening, allowance must be made for the depth of the falsework. Steps used to determine the allowance are:

Step 1 Determine the type of facility to be spanned by the structure

Four types of facilities are grouped in Table 204.8 of Caltrans *HDM* (www.dot.ca.gov/hq/oppd/hdm/pdf/english/chp0200.pdf). These types are freeways, nonfreeways, special, and other roadways for fire or utility access, or quasipublic roads with very light traffic.

Step 2 Determine the width of the falsework opening

The width of the falsework opening or traffic opening, which significantly affects construction costs, is controlled by factors such as staging and traffic handling requirements, the width of the approach roadbed (that will exist at the time the bridge is constructed), traffic volumes, local agency desires, controls in the form of existing facilities, and the practical problems of falsework construction.

For the normal minimum width of traffic openings and required falsework spans for various lane and shoulder combinations, refer to Table 204.8 of Caltrans *HDM*.

After the falsework opening is determined, a field review of the bridge site should be made to ensure that existing facilities (drainage, other bridges, or roadways) will not conflict with the falsework.

Step 3 Determine the normal span for the falsework

The normal span for the falsework is determined based on the traffic opening selected in Step 2.

Step 4 Determine the falsework depth

The falsework depth is determined based on the normal span selected in Step 3 and the depth of the permanent structure.

Example 9-7:

Assume an undercrossing structure is considered to allow Laguna Canyon Road to go under Route 8. Its layout is shown in Figure 9-7. The vertical clearance is 15 ft from the bottom of the bridge deck to the pavement surface of Laguna Canyon Road. Falsework is considered. What is the falsework depth?

Solution

Step 1 Since the falsework is to provide supports for the permanent bridge that spans Laguna Canyon Road, the facility type for this example is nonfreeway.

Step 2 Laguna Canyon Road has 4 lanes + 2-8' shoulders, and the traffic opening is 64 ft.

Step 3 The normal span for the falsework is 72 ft, given the normal span is 64 ft. (according to Page 200-24, Caltrans *HDM*, www.dot.ca.gov/hq/oppd/hdm/pdf/english/chp0200.pdf).

Step 4 Given the span of the falsework is 72 ft and the depth of the permanent structure is 5.58 ft, the depth of the falsework is 3'5" = 3.42 ft.

The elevation difference between the pavement surface of Laguna Canyon Road and the pavement surface of Route 8 is 24 ft. (or 15 ft for vertical clearance + 3.42 for falsework depth + 5.58 ft for structure depth). In other words, the grade line of Route 8 on the bridge is 24 ft over the grade line of Laguna Canyon Road.

Note that the minimum vertical falsework clearance over freeways and nonfreeways shall be 15 ft. Worker safety must be considered when determining vertical falsework clearance. Requests for approval of temporary vertical clearances less than 15 ft should discuss the impact on worker safety.

d. Steel or precast concrete structures

When vertical clearance cannot be easily provided during geometric design, falsework depth requirements can be removed by considering steel and precast concrete girders in lieu of cast-in-place concrete. Such falsework elimination may permit lower grade lines and reduce approach fill heights.

e. Grade line on bridge decks

A minimum fall of 0.05-foot per station should be provided when vertical curves are placed on bridge decks. This fall should not extend over a length greater than 100 ft.

When a tangent grade is considered on a bridge deck, the flattest allowable tangent grade should be 0.3%.

5. Coordination of horizontal and vertical alignments

Horizontal and vertical alignments should be properly balanced. When possible, vertical curves should be superimposed on horizontal curves. In doing so, sight restrictions can be reduced, changes in profile can be made less apparent, and a pleasing appearance can be achieved.

Give special care to the areas where horizontal and vertical curves are superimposed. The combination of superelevation and profile grades may result in distortion in the outer pavement edges and create drainage concerns or confuse drivers at night. In these areas, edge of pavement profiles should be plotted, and smooth curves need to be introduced to eliminate irregularities or distortions.

Give special care as well to long open curves. A uniform grade line should be used because a rolling profile results in a poor appearance.

9.3 Vertical Alignment Design Using InRoads

Design of a vertical alignment is a process of creating a series of graphical elements (tangent grades and vertical curves) over the profile of a horizontal alignment in MicroStation (see Figure 9-8). These graphical elements, once conforming to design standards, are imported into InRoads to form a vertical alignment.

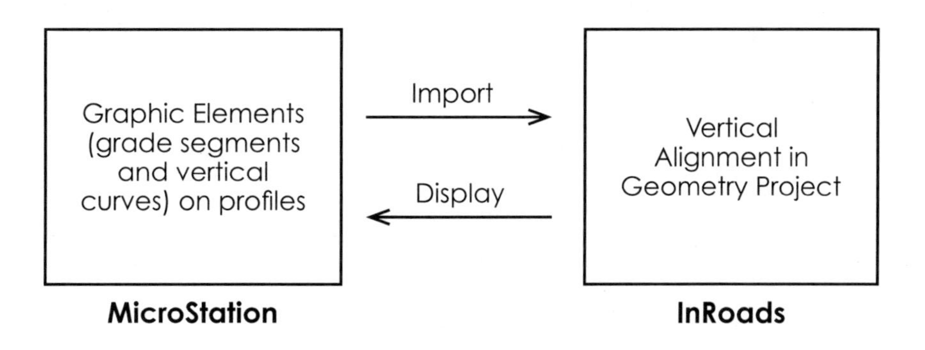

Figure 9-8 Vertical Alignment Design Process

9.3.1 Vertical Alignment for Laguna Canyon Road

Before we create the vertical alignments for the tutorial project, you need to have MicroStation and InRoads installed on your computer. You also need to copy the **tutorial_2D_Profile.dgn, OG.dtm**, and **tutorial.alg** files from the **C:/Temp/Book/Chp9/Master Files** folder to **C:/Temp/Tutorial/Chp9/Master Files** folder or you can download the files from the book website (www.csupomona.edu/~xjia/GeometricDesign/Chp9/Master Files/). You may use your own **tutorial_2D_Profile.dgn, OG.dtm**, and **Tutorial.alg** files you created when you worked with the previous chapters.

We start working on the design of the vertical alignment for Laguna Canyon Road. Steps involved in the design are:

1. Open the **tutorial_2D_Profile.dgn** file from the **C:/Temp/Tutorial/Chp9/Master Files** folder.
   a. Go to ***MicroStation → Settings →Levels → Manager.***

      The **Level Manager** window appears.
2. Go to ***Levels → New*** in the **Level Manager** window.
   a. Name the new level **Laguna Canyon Road_VA.**
   b. Select the **Color, Line Style,** and **Line Weight** of your choice.
   c. Set the level to be active.
3. Go to ***MicroStation → Laguna Canyon Road Profile.***
   a. Draw a line segment with the two points below:

      Station/Elevation: 10+00 / 2969.16 ft.

      Station/Elevation: 44+12.00 / 3105.64 ft.
   b. Go to the ***MicroStation → Move/Copy Parallel*** command.
   c. Do the following in the **Move/Copy Parallel** window:

      Distance: ***Check 45.80***

      Keep Original: ***Check***
   d. Draw a construction line to select the 2,960 gridline and make a parallel copy (see Figure 9-9).

      Note that the difference between the design elevation (2,969.16 ft) at Point C and the 2,960 ft gridline is 9.16 ft. Also note that the elevation exaggeration is 5, therefore the offset used in the **Move/Copy Parallel Copy** command is 9.16 × 5 = 45.80 ft from the 2,960 ft. gridline.

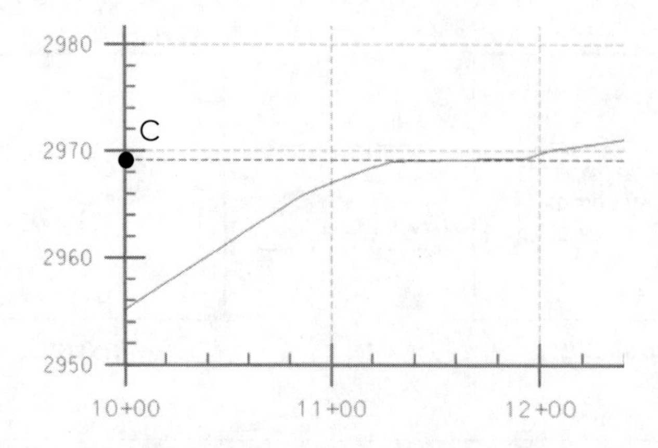

Figure 9-9 Determine Point C on Laguna Canyon Road Profile

e. Do the following in the **Move/Copy Parallel** window:

Distance:	***Check 28.20***
Keep Original:	***Check***

f. Select the 3,010 gridline and make a parallel copy (see Figure 9-10).

Note that the difference between the design elevation (3,105.64 ft) at Point C' and the 3,100 ft gridline is 5.64 ft. Also note that the elevation exaggeration is 5, therefore the offset used in the **Move/Copy Parallel Copy** command is 5.64 × 5 = 28.20 ft. from the 3,100 gridline.

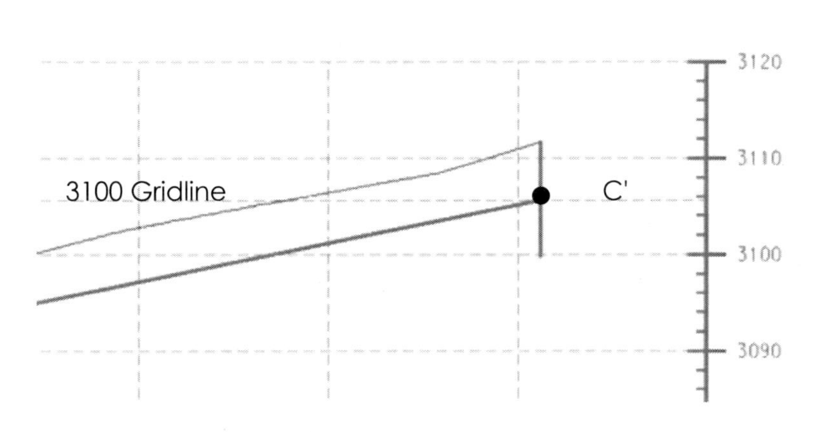

Figure 9-10 Determine Point C' on Laguna Canyon Road Profile

g. Select the ***MicroStation → Place Line*** command at the end of the profile line in the **Laguna Canyon Road** Profile. The angle is 270°. Point C' is the intersection of the parallel line and the vertical line (see Figure 9-10).

h. Select the ***MicroStation → Place Line*** command to draw a line segment between the starting and ending points (Point C and Point C').

i. Delete all the temporary line segments.

Figure 9-11 shows the final line segment on the **Laguna Canyon Road** profile.

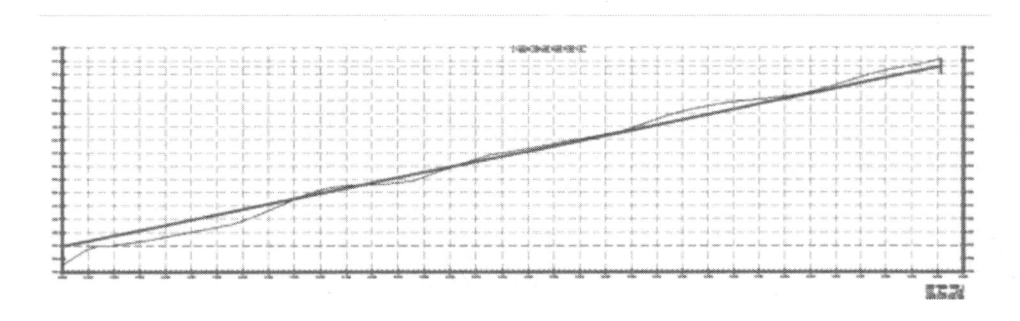

Figure 9-11 Vertical Alignment of Laguna Canyon Road

4. Caltrans Standards Conformance Check

Table 9-1 shows the result of the conformance check

Table 9-1 Conformance Check		
Element Type	**Length**	**Conformance Check**
Segment	3,412 ft. Grade 4%	This is a local road. The grade is 4% from South to North. OK.

5. Import the graphical element into InRoads to create the vertical alignment of Laguna Canyon Road
   a. Make sure the **Laguna Canyon Road_VA** level is active in MicroStation.
   b. Make sure the **tutorial** geometry project is active. If it is not active, set it to be active.
   c. Make sure the horizontal alignment **Laguna Canyon Road** is set to be active in the **tutorial geometry** project.
   d. Go to ***InRoads → File → Import → Geometry ....***

   The **Import Geometry** window appears (see Figure 9-12).

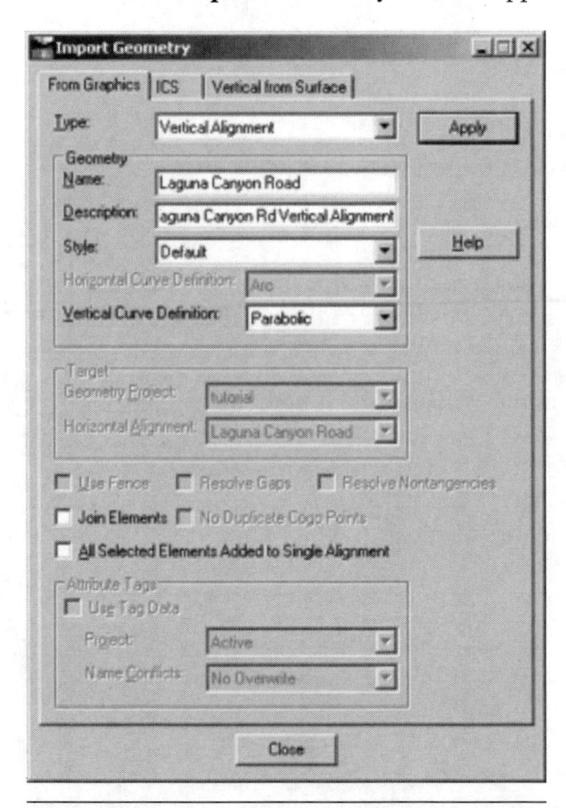

Figure 9-12 Import Geometry Window

   e. Select the **From Graphics** tab and type the following:

Type:	**Vertical Alignment**
Name:	**Laguna Canyon Road**
Description:	**Laguna Canyon Rd Vertical Alignment**
Style:	**Default**
Vertical Curve Definition:	**Parabolic**
Other fields:	**Unchecked**

   f. Select ***Apply***.

   InRoads prompts you to select a graphical element.

   g. Select the line segment in the **Laguna Canyon Road** Profile.
   h. Select ***Accept*** and ***Reset***.
   i. Select ***Close***.
   j. Check in the **Workspace Bar** window to make sure the vertical alignment Laguna Canyon Road was created.

You will see a **Workspace Bar** window similar to Figure 9-13.

6. Save the **tutorial** geometry project
   a. Make sure the **tutorial** geometry project is set to be active.
   b. Go to ***InRoads → File → Save → Geometry Project***.
   c. Save the **tutorial** geometry project as ***tutorial.alg*** in the **C:/Temp/Tutorial/Chp9/Master Files** folder.

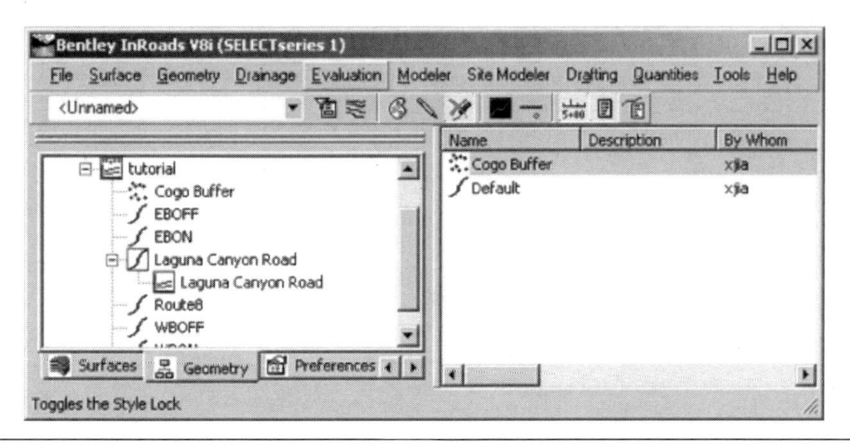

Figure 9-13 Laguna Canyon Road Vertical Alignment Created

7. The vertical alignment **Laguna Canyon Road** is created and the **tutorial** geometry project is saved. Now we can view the vertical alignment **Laguna Canyon Road** using InRoads tools.

   a. Make sure the **tutorial** geometry project is set to be active.

   b. Make sure the horizontal alignment **Laguna Canyon Road** and the vertical alignment **Laguna Canyon Road** are set to be active.

   c. Go to ***InRoads → Geometry → View Geometry → Active Vertical.***

   d. Go to ***InRoads → Geometry → View Geometry → Vertical Annotation ....***

   The **Vertical Annotation** window appears.

   e. Select the **Tangents** tab and type the following (see Figure 9-14):

Grade:	**Check on**
Format:	**50%**
Other fields:	**Unchecked**

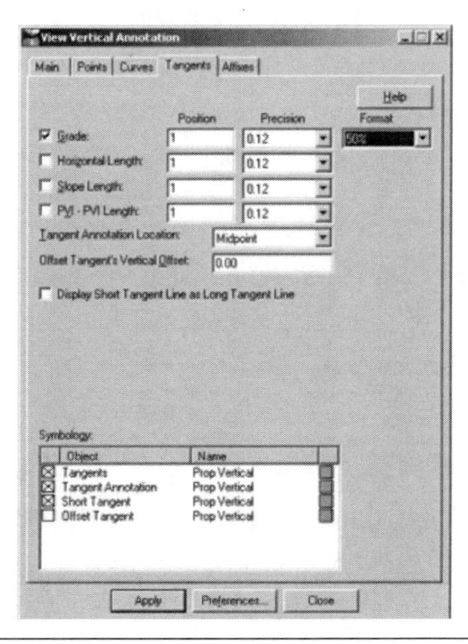

Figure 9-14 View Vertical Annotation Window

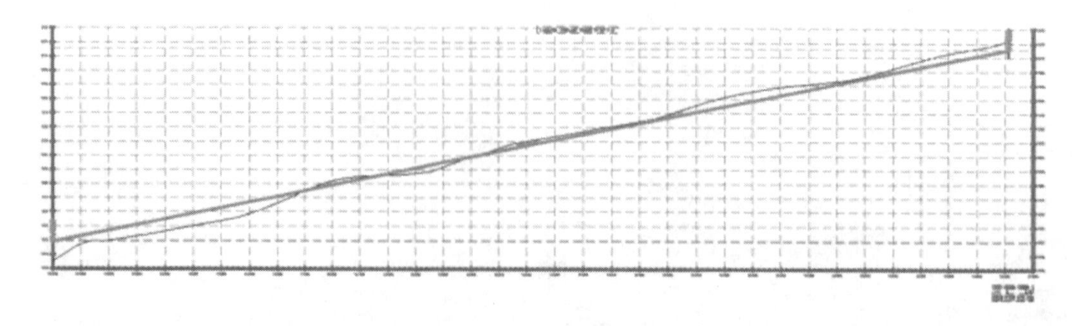

Figure 9-15 Vertical Alignment for Laguna Canyon Road

f. In the **View Annotation** window, select ***Apply*** and ***Close***.

Vertical annotations are displayed along with the vertical alignment of Laguna Canyon Road (see Figure 9-15).

9.3.2 Vertical alignment for Route 8

Now we design the Route 8 vertical alignment. The design involves four tasks: 1) entrance grade design, 2) exit grade design, 3) controlling feature identification, and 4) grade line design to link the entrance and exit grades.

Follow these steps in the design process:

1. Go to ***MicroStation → Settings → Levels → Manager.***

   The **Level Manager** window appears.

2. Go to ***Levels → New*** in the **Level Manager** window.

   a. Name the new level **Route8_VA.**

   b. Select the **Color, Line Style,** and **Line Weight** of your choice.

   c. Set the ***Route8_VA*** level as the active level.

   d. Close the **Level Manager** window.

3. Create graphical elements for Route 8 by doing the following:

   Entrance Grade Design: Create a tangent grade of 2.0% to connect to Point A (see Figure 9-16).

   a. Go to the ***MicroStation → Move/Copy Parallel*** command. Do the following in the **Move/Copy Parallel** window:

Distance:	***Check 5.9***
Keep Original:	***Check***

   b. Select the 3,050 gridline and make a parallel copy.

   c. The intersection of the parallel line and the Y-axis (Point A) is the design elevation.

   Note that the difference between the design elevation (3,051.18 ft.) at Point A and the 3,050 gridline is 1.18 ft. Also note that the elevation exaggeration is 5, therefore the offset used in the **Move/Copy Parallel Copy** command is $1.18 \times 5 = 5.9$ ft. from the 3,050 gridline.

   d. Draw a horizontal line segment of **200 ft.** (angle = 0°; see Figure 9-16).

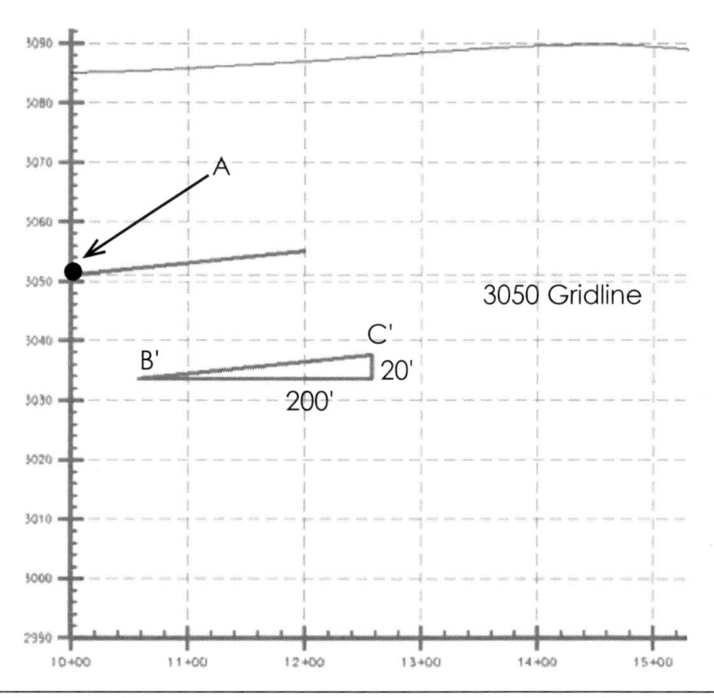

Figure 9-16 Design Elevation

e. At the end of the horizontal line segment, draw a vertical line segment of **20 ft.** (angle = 90°; see Figure 9-16).

   As the elevation exaggeration is 5, the raise for a run of 200 ft is 2.0% × 200 × 5 = 20 ft if the grade is 2.0%.

f. Draw the grade segment connecting Point B' and Point C' (see Figure 9-16).

g. Copy the grade segment and move Point B' to Point A (see Figure 9-16).

h. Delete all temporary line segments.

   Exit Grade Design: Create a tangent grade of 1.5% to connect to Point B.

a. Go to the ***MicroStation → Move/Copy Parallel*** command.

b. Type the following in the **Move/Copy Parallel** window (Figure 9-17):

   Distance: ***Check 19.95***

   Keep Original: ***Check***

c. Select the 3,080 gridline and make a parallel copy.

   The intersection of the copied line and the Y-axis is the design elevation (see Figure 9-17).

   Note that the difference between the design elevation (3,083.99 ft.) and the 3,080 gridline is 3.99 ft. Also note that the elevation exaggeration is 5, therefore the offset used in the **Move/Copy Parallel Copy** command is 3.99 × 5 = 19.95 ft from the 3,080 gridline.

d. Draw a vertical segment passing through the end of the Route 8 profile line and intersecting the 19.95-ft offset line.

   The intersection is Point B.

e. Draw a horizontal line segment of **200 ft.** (angle = 0°; see Figure 9-17).

f. At the end of the horizontal line segment, draw a vertical line segment of **15 ft.** (angle = 90°; see Figure 9-17).

As the elevation exaggeration is 5, the raise for a run of 200 ft is 1.5% × 200 × 5 = 15 ft if the grade is 1.5%.

g. Draw the grade segment connecting Point B' and Point C' (see Figure 9-17).

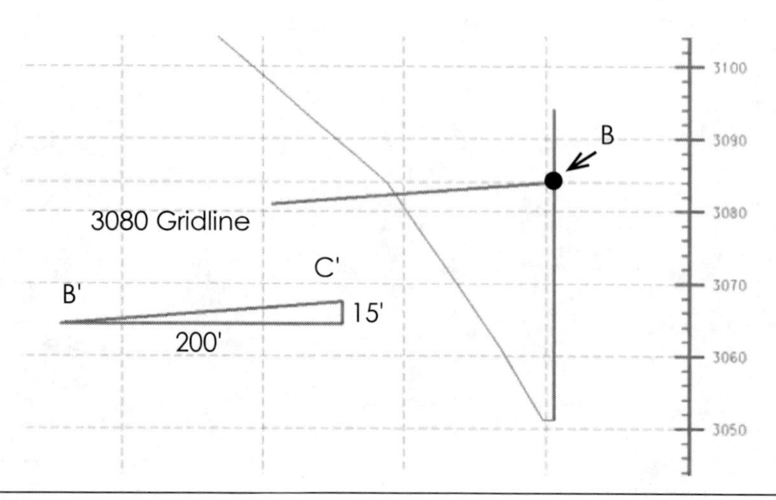

Figure 9-17 Design Elevation at Point B

h. Copy the grade segment and move Point C' to Point B (see Figure 9-17).

i. Delete all temporary line segments.

Controlling feature identification (bridges and culverts)

The grade line for Route 8 runs over **bridges** under which Laguna Canyon Road passes and **culverts** by which water passes under Route 8 from North to South. The elevation of the grade line is controlled by the location and elevation of the bridges and the culverts.

Step 1 Identify the locations, sizes, and elevations of the bridges

a. Go to ***MicroStation → Window Area*** and zoom in to an area that covers the intersection of Route 8 and Laguna Canyon Road (see Figure 9-18).

b. Go to ***MicroStation → Settings → Levels → Manager.***

The **Level Manager** window appears.

c. Go to ***Levels → New*** in the **Level Manager** window.

d. Name the new level **Bridge** and select the **Color, Line Style,** and **Line Weight** of your choice.

e. Set the ***Bridge*** level as the active level.

f. Close the **Level Manager** window.

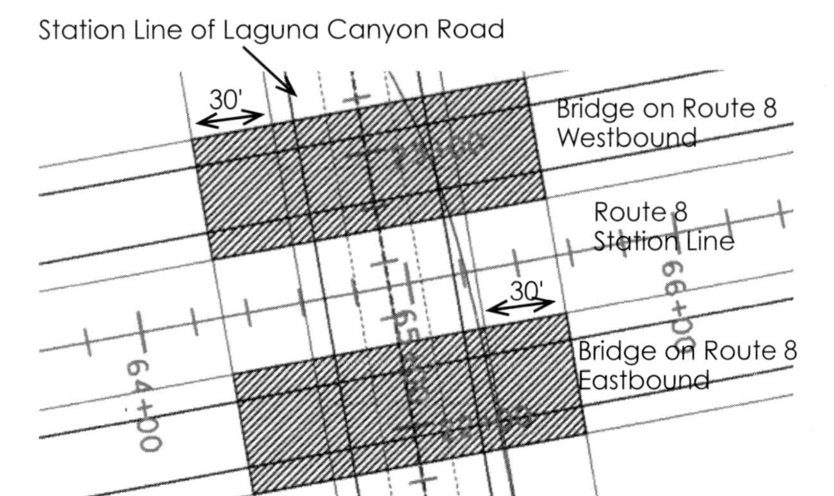

Figure 9-18 Locations of Bridges

g. Select the ***MicroStation → Place Shape*** command to draw two rectangle shapes to represent the bridges.

   Each bridge is enclosed by two Route 8 shoulder lines and two line segments that are **30 ft** from the shoulder lines of Laguna Canyon Road. The 30 ft offset is the run required to make the vertical clearance of 15 ft, given the side slope under the bridge, is 2:1 (run:raise).

h. Go to the ***MicroStation → Measure Distance*** command to measure the starting and end stations of the bridges on Route 8.

   Starting station: 64+30.67
   End station: 65+54.67

i. Go to the ***MicroStation → Measure Distance*** command to measure the starting and end stations of the bridges on Laguna Canyon Road.

   Bridge on Route 8 Westbound:

   Starting station: 22+72.14
   End station: 23+16.14

   Bridge on Route 8 Eastbound:

   Starting station: 21+86.14
   End station: 22+30.14

   The elevation of Route 8 grade line is calculated based on the vertical clearance of the bridges, the falsework depth, the bridge structure depth, and the elevation of the critical point on Laguna Canyon Road.

   The critical point on Laguna Canyon Road is at 23+16.14 since the road has an up-grade of 4% from South to North.

j. Use the vertical alignment designed for Laguna Canyon Road to determine the elevation of the critical point (see Figure 9-19).

   The elevation at 23+16.14 is measured to be 3,021.81 ft. The falsework depth for Route 8 bridge is 3.42 ft (see Example 9-7). The vertical clearance of the bridges is 15 ft and the structure depth is 5.58 ft. The elevation of the Route 8 grade line is 3,021.81 ft + 15 ft + 3.42 ft + 5.58 ft = 3,045.81 ft (see Figure 9-20).

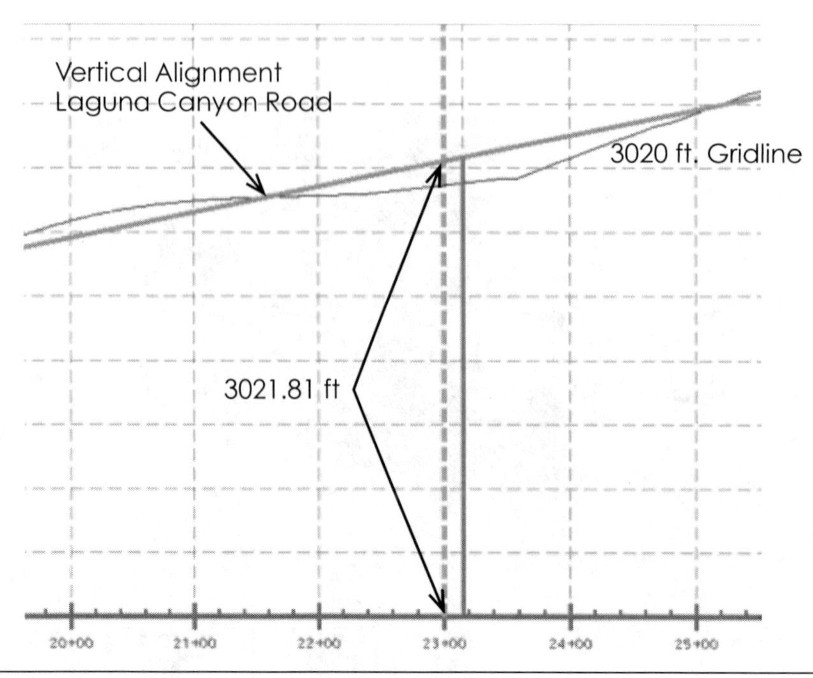

Figure 9-19 Vertical Clearance of the Critical Point

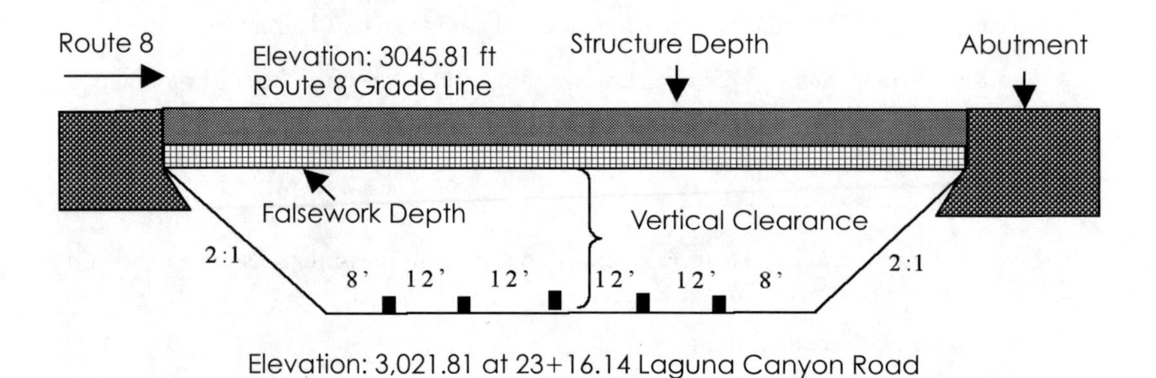

Figure 9-20 Structure Layout

Step 2 Draw the bridge on Route 8 Profile.

a. Go to the ***MicroStation* → *Window Area*** command and zoom in to the Route 8 Profile created in Chapter 8.

b. Go to the ***MicroStation* → *Move/Copy Parallel*** command to get vertical offset lines (passing through Station 64+60.67 and Station 65+24.67) and two horizontal offset lines at elevations 3,045.81 ft and 3,021.81 ft, respectively.

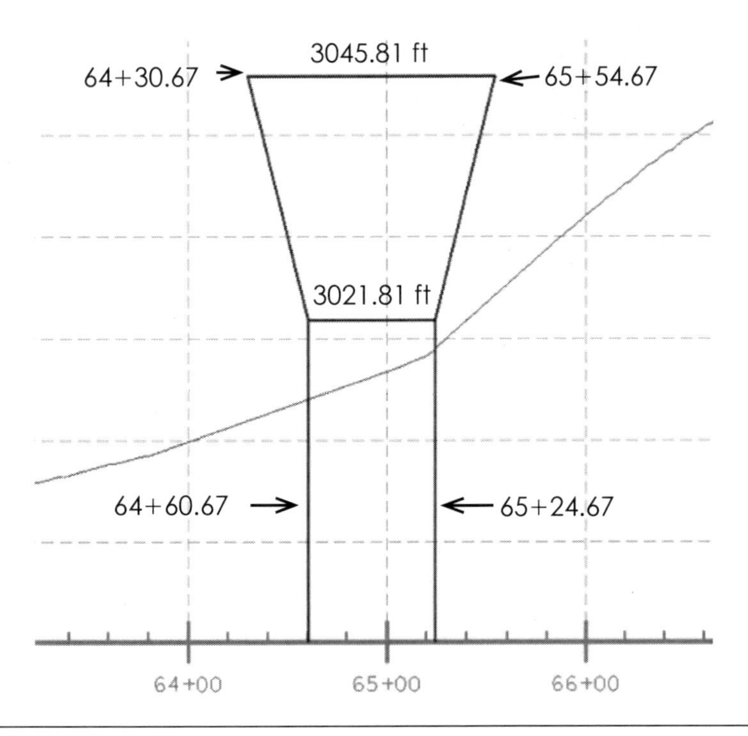

Figure 9-21 Bridge Location, Size, and Elevation

c. Select the ***MicroStation*** → ***Place Shape*** command to draw the cross section of bridges on the Route 8 Profile (see Figure 9-21).

Step 3 Identify location, size, and elevation of culverts

a. Go to ***MicroStation*** → ***Settings*** → ***Levels*** → ***Manager.***

The **Level Manager** window appears.

b. Go to ***Levels*** → ***New*** in the **Level Manager** window.

c. Name the new level **Culvert** and select the **Color, Line Style,** and **Line Weight** of your choice.

d. Set the ***Culvert*** level as the active level.

e. Close the **Level Manager** window.

f. Do the following to identify the location of culverts. Determine which breaklines are valleys or ridge lines (see Figure 9-22).

Example 9-8:

Given the breakline that runs through the contours shown in Figure 9-23, is this breakline a valley line?

Solution

Draw a dashed line that crosses contours and the breakline. We will see the cross section of the contours along the dashed line (see Figure 9-24). This indicates the breakline is a valley line.

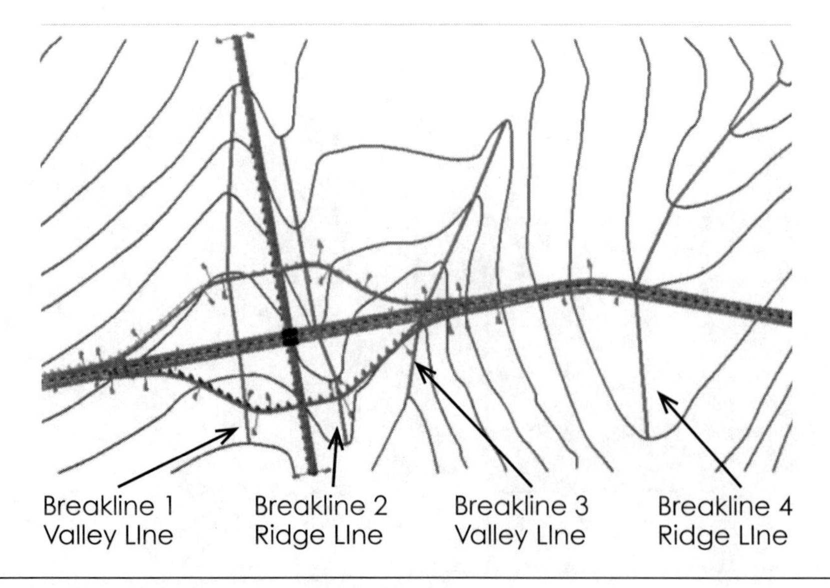

Figure 9-22 Location of Valleys and Ridge Lines

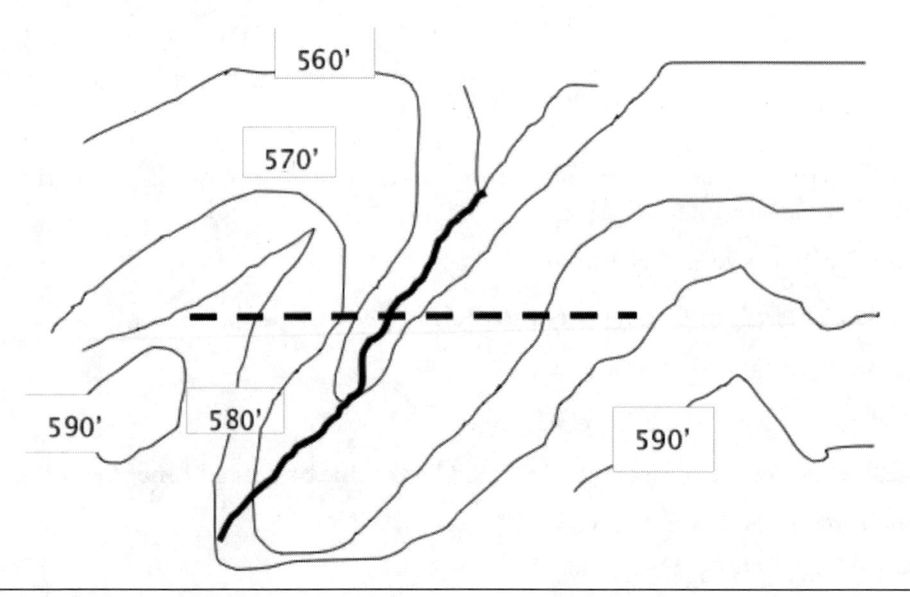

Figure 9-23 Contours for Example 9-8

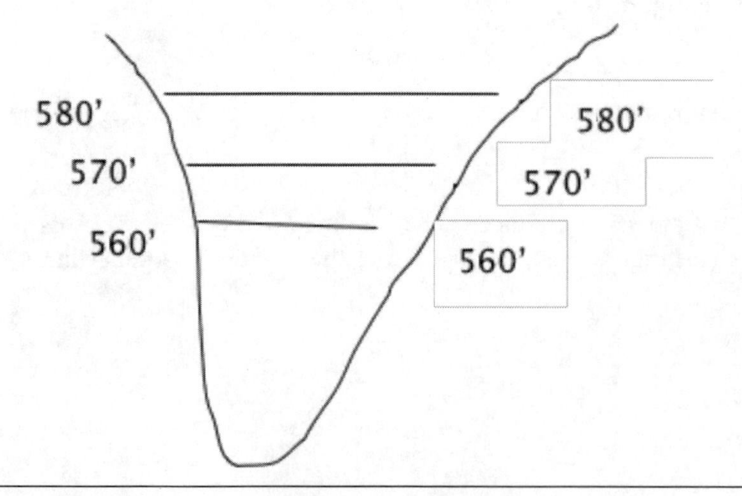

Figure 9-24 Cross Section Showing the Valley

The two valley lines (Breakline 1 and Breakline 3) run across Route 8, which requires two culverts to carry water under Route 8.

g. Check the intersections of Route 8 and the valley lines.

   We find that the two culverts are located at Stations 59+89.87 and 77+54.56, respectively.

h. Check the existing elevations of Route 8 Profile at stations 59+89.87 and 77+54.56.

   We find the existing elevations are 2,994.64 ft. and 2,991.14 ft. (see Figures 9-25a and 9-25b). The culverts are made of pipes. The size of the pipes as described in the previous chapters is 4 ft. in diameter.

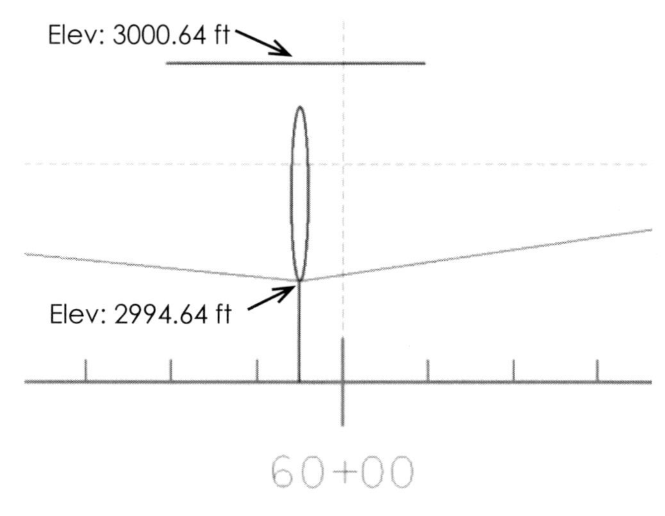

Figure 9-25a Culvert Design at Station 59+89.87

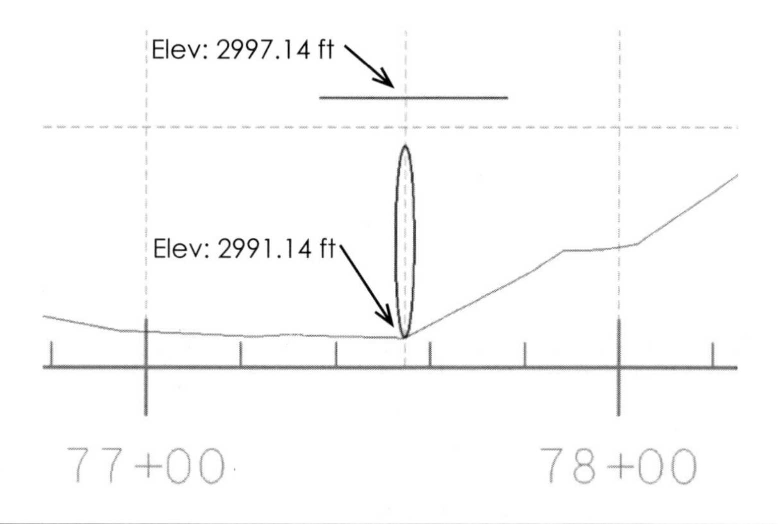

Figure 9-25b Culvert Design at Station 77+54.56

In summary, there are three control features (one bridge and two culverts) to consider in the vertical alignment design of Route 8. The locations, elevations, and sizes of these features are shown in Figure 9-26. Note that the culverts could be discarded since their elevations are significantly lower than that for Route 8 bridges.

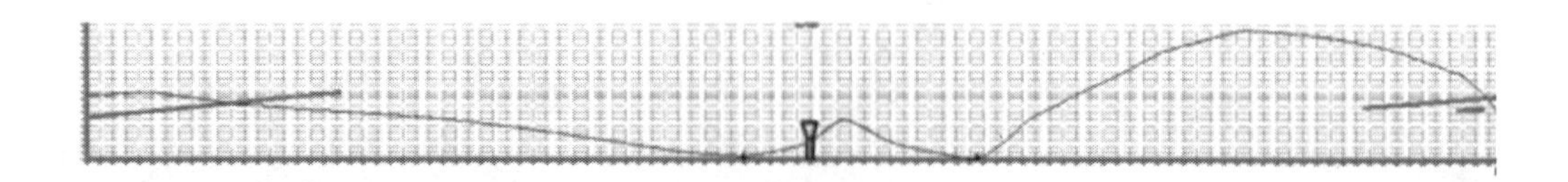

Figure 9-26 Control Points for Route 8 Vertical Alignment

Fill the Gap: Create a grade segment over Route 8 bridges and fill the gap between Points A and B.

There are unlimited grade segments available for connecting the entrance and exit grades. Create one grade segment that meets the Caltrans standards.

Follow these steps for the grade segments:

a. Set the ***MicroStation → Route8_VA*** level as the active level.

b. Select the ***MicroStation → Place Line*** command to create a grade segment of −1% (see Figure 9-27).

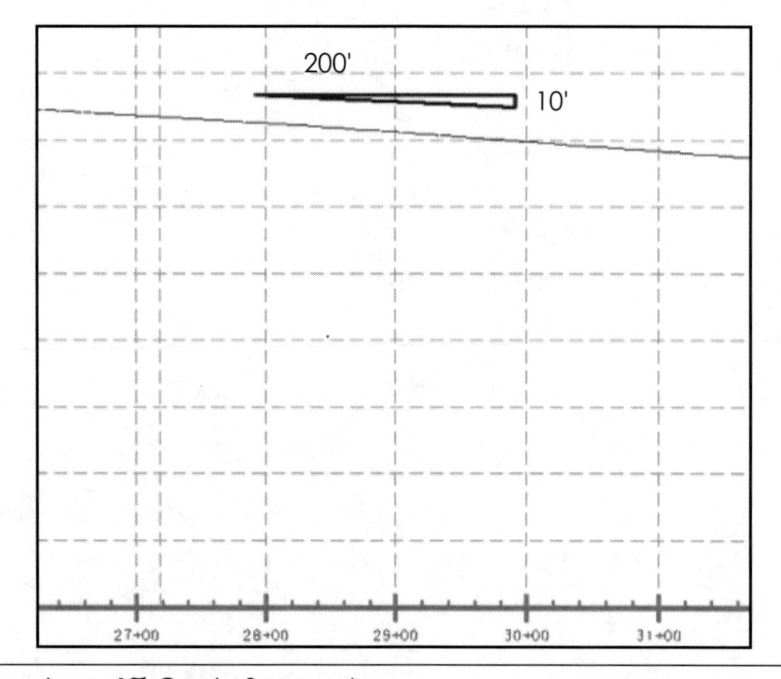

Figure 9-27 Create a −1% Grade Segment

c. Go to the ***MicroStation → Move*** command to move the grade segment to intersect the entrance grade and the exit grade. Set it to be above the top side of the bridges (see Figure 9-28).

The stations and elevations of the two intersecting points are measured as:

Station/Elevation:	27+18.00/3,085.54 ft.
Station/Elevation:	81+72.71/3,030.99 ft.

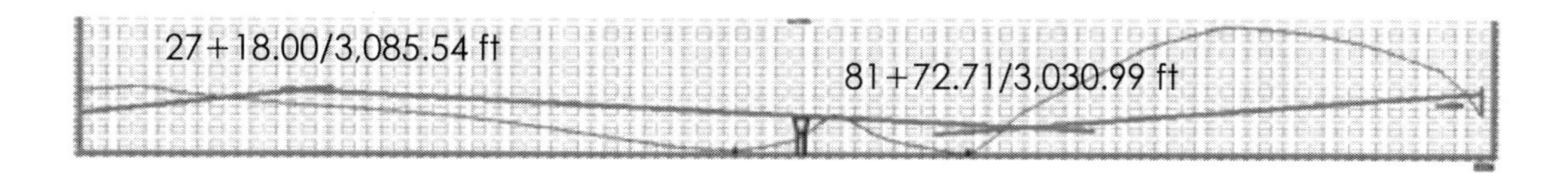

Figure 9-28 Grade Segment for Route 8 Vertical Alignment

d. Go to ***MicroStation → Tools → Tool Boxes ....***

e. Check on the **Fillets** option in the **Tool Boxes** window.

The **Fillets** toolbox appears.

f. Select the **Construct Parabolic Fillet** tool in the **Fillets** toolbox.

The **Construct Parabolic Fillet** window appears.

g. Type the following in the **Construct Parabolic Fillet** window:

Distance:	***3000.0000***
Type:	***Horizontal***
Truncate:	***None***

The distance of 3,000.0000 specifies the length of the vertical curve. The Horizontal type indicates the parabolic curve is aligned with the horizontal view axis. This setting is used for highway design to join intersecting grade lines.

h. Select the entrance grade and the grade segment over the bridges.

The parabolic curve of 3,000 ft is filleted into the grade segments.

i. Again select the **Construct Parabolic Fillet** tool in the **Fillets** toolbox.

j. Type the following in the **Construct Parabolic Fillet** window:

Distance:	***3000.0000***
Type:	***Horizontal***
Truncate:	***None***

k. Select the exit grade and the grade segment over the bridges.

The parabolic curve of 3,000 ft is filleted into the grade segments (see Figure 9-29).

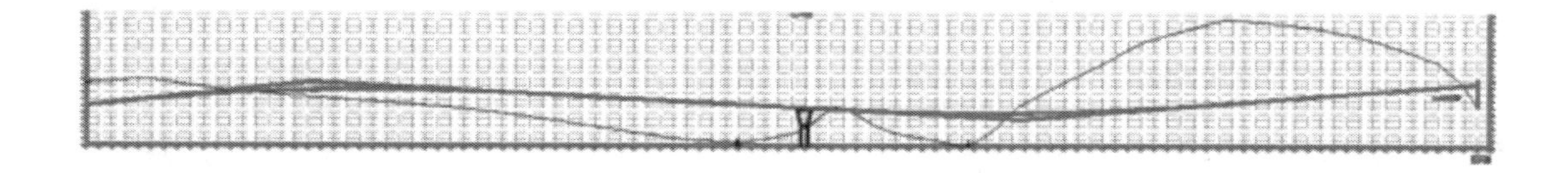

Figure 9-29 Parabolic curves on Route 8 Profile

4. Caltrans Standards Conformance Check

Table 9-2 shows the graphical elements that form the vertical alignment of Route 8.

5. Import the graphical elements into InRoads to create Route 8 vertical alignment
   a. Make sure the **Route8_VA** level is active in MicroStation.
   b. Make sure the **tutorial** geometry project is active. If it is not active, set it to be active.

Table 9-2 Conformance Check

Element Type	Length (L)	Conformance Check
Grade 1 or Entrance Grade	218.00' 2.00%	The entrance grade matches the grade agreed with other design teams. It keeps the grade consistency for Route 8. The grade of 2.00% is within the range of 0.3% (min) and 4% (max).
Parabolic Curve 1	3,000.00'	The algebraic grade difference is |–1.0% – 2%) | = 3.0%. Following the equation L_{min} = 10V, the minimum length of the vertical curve is 10 × 60 mph = 600 ft. The minimum length of the vertical curve is 750 ft (see Figure 9-5). 750 ft > 600 ft. Select 750 ft. 3,000 ft > 750 ft. OK.
Grade 2	2454.71' –1.0%	The grade of –1.0% is within the range of –0.3% (min) and –4% (max).
Parabolic Curve 2	3,000.00'	The algebraic grade difference is | 1.5% – (-1.0%) | = 2.5%. Following the equation L_{min} = 10V, the minimum length of the vertical curve is 10 × 60 mph = 600 ft. The minimum length of the vertical curve is less than 200 ft. (see Figure 9-6). 600 ft. > 200 ft. Select 600 ft. 3,000.00 ft. > 600 ft. OK.
Grade 3 or Exit Grade	2,033.14' 1.5%	The exit grade matches the grade agreed with among design teams. It keeps the grade consistency for Route 8. The grade of 1.5% is within the range of 0.3% (min) and 4% (max).

   c. Make sure the **Route 8** horizontal alignment is active.
   d. In MicroStation, create a complex chain using **Create Complex Chain** command.
   e. Select the **three tangent grades** for Route 8 and create a chain for Route 8 (see Figure 9-30).

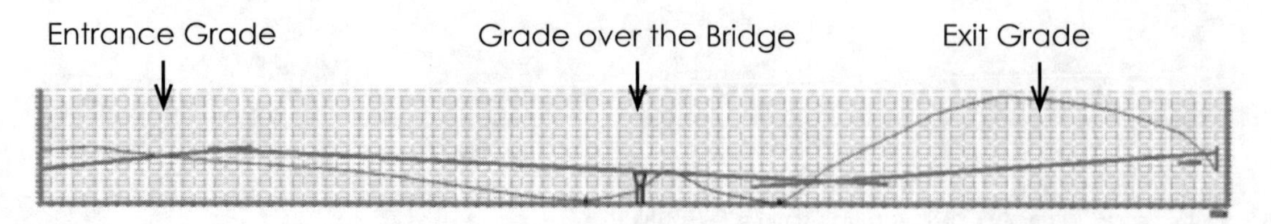

Figure 9-30 Use of Three Segments to Create a Chain

You may wonder why the two parabolic curves are not selected to make a complex chain. The reason is that the vertical curves are made up of small line segments. When the vertical curves are imported into InRoads, the small line segments, instead of two vertical curves, are used to form the vertical alignment.

Recognizing the limitations of InRoads, we first use the three tangent grades to make a vertical alignment. We then complete the vertical alignment by adding two vertical curves in InRoads.

a. Go to ***InRoads → Import → Geometry …***.

The **Import Geometry** window appears (see Figure 9-31).

b. Select the **From Graphics** tab and type the following:

Type:	**Vertical Alignment**
Name:	**Route 8**
Description:	**Route 8 Vertical Alignment**
Style:	**Default**
Vertical Curve Definition:	**Parabolic**
Other fields:	**Default**

c. Select ***Apply***.

InRoads prompts you to select the graphical elements.

d. Select the chain you just created.

e. Select ***Accept*** and ***Reset***.

f. Select ***Close***.

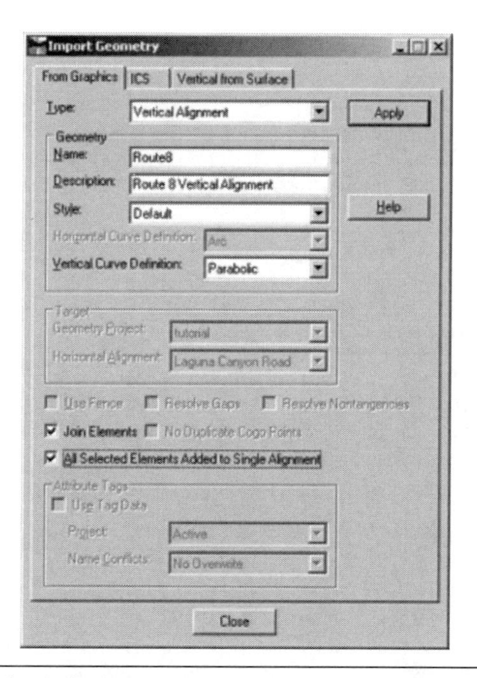

Figure 9-31 Import Geometry Window

g. Check the **Workspace Bar** window to see if the vertical alignment Route 8 was created.

You will see a Workspace Bar window similar to Figure 9-32.

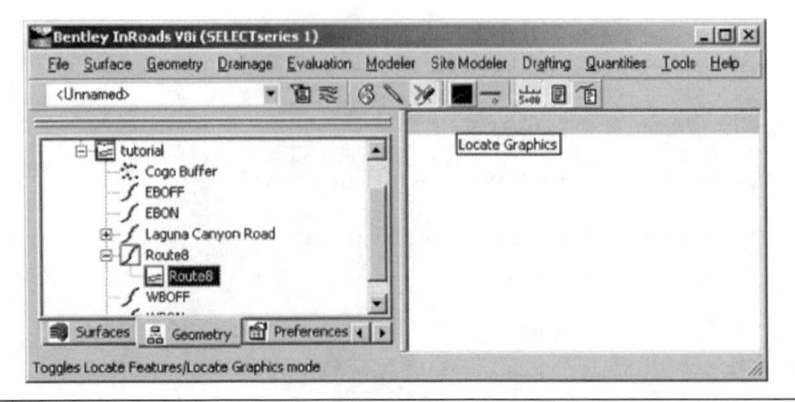

Figure 9-32 Route 8 Created

h. Go to ***InRoads → Geometry → View Geometry → Active Vertical.***

i. Go to ***InRoads → Geometry → Vertical Curve Set → Define Curve ....***

The **Define Curve** window appears (see Figure 9-33).

j. In the **Length** field, type ***3000.00.***

k. Select ***Apply***.

l. Select ***Next***.

m. In the **Length** field, type ***3000.00.***

n. Select ***Apply*** and ***Close***.

6. Save the **tutorial** geometry project

a. Make sure the **tutorial** geometry project is set to be active.

b. Go to ***InRoads → File → Save → Geometry Project.***

c. Save the **tutorial** geometry project as ***tutorial.alg*** in the **C:/Temp/Tutorial/Chp9/Master Files** folder.

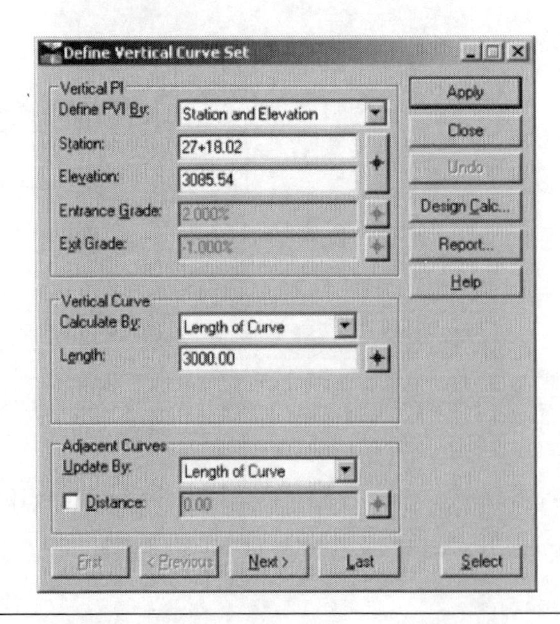

Figure 9-33 Define Curve Window

7. The **Route 8** vertical alignment is created and the **tutorial** geometry project is saved. Now we can view the **Route 8** vertical alignment using InRoads tools.

   a. Make sure the **tutorial** geometry project is set to be active.

   b. Make sure **Route 8** vertical alignment is set to be active.

   c. Go to ***InRoads → Geometry → View Geometry → Vertical Annotation ....***

      The **Vertical Annotation** window appears.

   d. Select the **Points** tab and make sure the following appears in the Symbology Listbox:

PVC Text:	**Check on**
PVI Text:	**Check on**
PVT Text:	**Check on**
High Point Text:	**Check on**
Low Point Text:	**Check on**
PVC Point:	**Check on**
PVI Point:	**Check on**
PVT Point:	**Check on**
High Point:	**Check on**
Low Point:	**Check on**
All other fields:	**Check off**

   e. Select the **Curves** tab and make sure all the fields are the same as shown in Figure 9-34.

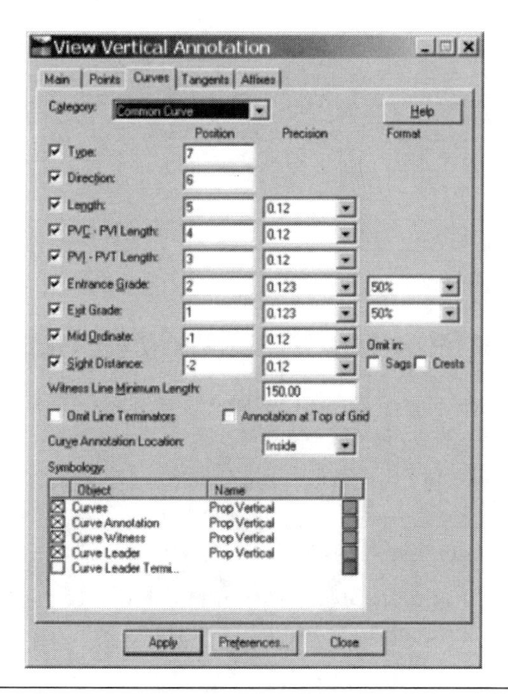

Figure 9-34 Curve Settings

   f. Select the **Tangents** tab and do the following (see Figure 9-35):

Grade:	**Check on**
Other fields:	**Default**

g. In the **View Vertical Annotation** window, select ***Apply*** and ***Close***.

Vertical annotations are displayed along with the vertical alignment of Route 8 (see Figure 9-36).

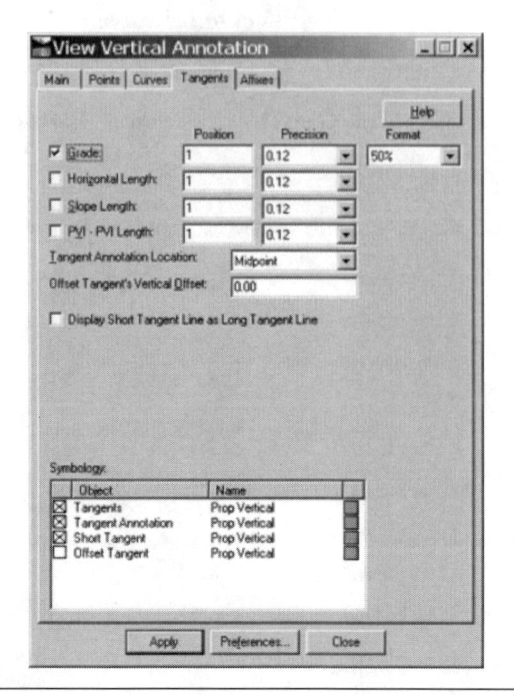

Figure 9-35 Tangent Settings

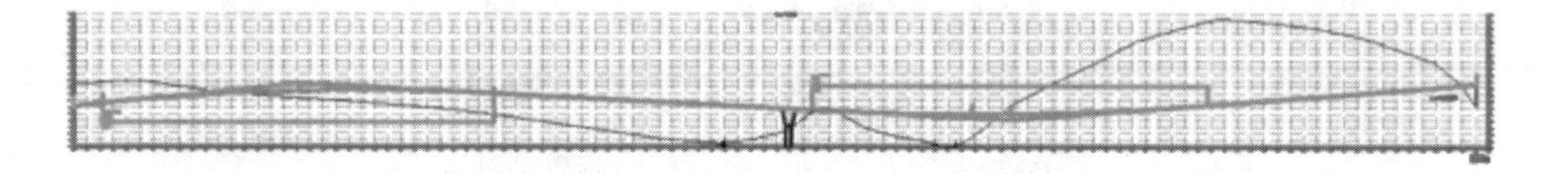

Figure 9-36 Route 8 Vertical Alignment

9.3.3 Vertical alignment for the Eastbound Off Ramp

The vertical alignment design for the Eastbound Off Ramp is a process that lays out the grade line in the ramp's profile. It consists of three components: freeway-side grade, local-road-side grade, and a series of grades and curves that connect to Route 8 and Laguna Canyon Road.

1. Location and design of freeway-side grade

The freeway-side grade segment should ensure elevation and grade consistency in the transition area (or the gore area) where drivers change direction from Route 8 to the ramp.

The ramp elevation at the freeway side should be the same as at the intersection (Point A') between the Route 8 right side ETW line and the ramp station line (see Figure 9-37). No sudden drop of elevation is allowed when drivers exit from Route 8.

The intersection A' is 55' (or 31' + 12' + 12') away from the Route 8 station line at Station 47+92.66.

Use the ***MicroStation → Measure Distance*** command to measure the elevation of Route 8 at Station 47+92.66 in the Route 8 Profile.

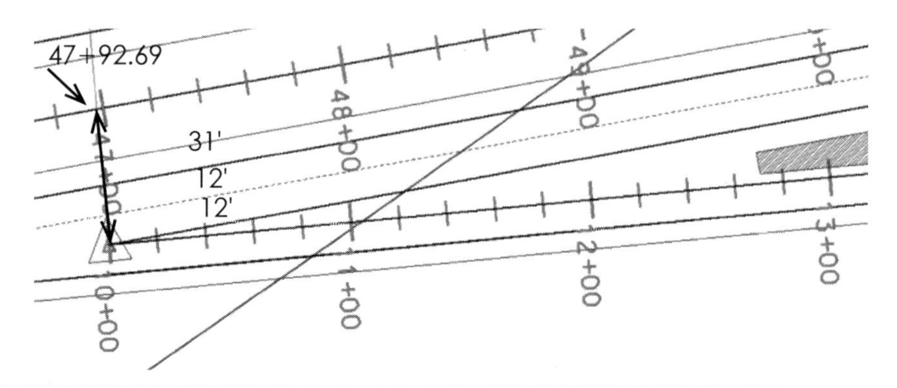

Figure 9-37 Exit Station (at Route 8) for Eastbound Off Ramp

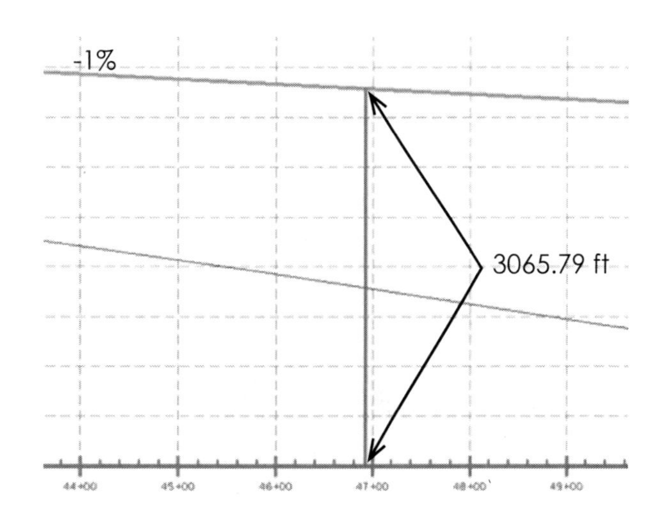

Figure 9-38 Elevation of Route 8 at Station 46+92.66

The elevation is measured to be 3,065.79 ft on the Route 8 vertical alignment (see Figure 9-38). The elevation at the freeway side (or Point A') can be calculated (see Figure 9-39):

= 3,065.79' + 2% × 31 − 2% × 24

= 3,065.93'

Note that Caltrans standards require the cross slopes in the median area and on the pavement surface be –2% (from median side ETW to right side ETW), which keeps water away from the lanes.

Note that Figure 9-39 shows the grade of Route 8 at the place where drivers exit to the ramp is –1% from West to East.

The design of the freeway-side grade segment for the Eastbound Off Ramp is shown in Figure 9-40. Note that the grade on the Profile is –1% from West to East.

a. Go to ***MicroStation → Settings → Levels → Manager.***

   The **Level Manager** window appears.

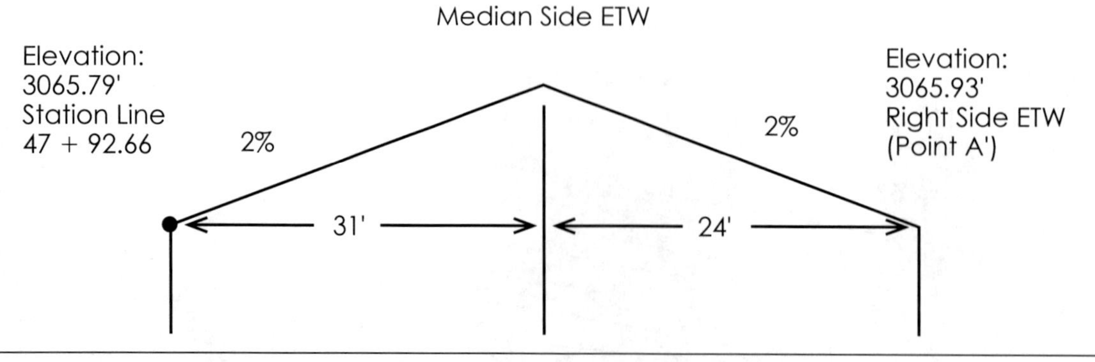

Figure 9-39 Elevation of the Eastbound Off Ramp at Freeway Side

b. Go to ***Levels → New*** in the **Level Manager** window.

c. Name the new level **EBOFF_VA.** Select the **Color, Line Style,** and **Line Weight** of your choice.

d. Set the ***EBOFF_VA*** level as the active level.

e. Close the **Level Manager** window.

f. Go to the ***MicroStation → Window Area*** command and zoom in to the Eastbound Off Ramp Profile created in Chapter 8.

g. Go to the ***MicroStation → Move/Copy Parallel*** command to get a horizontal offset line at elevation 3,065.93 ft.

Note the offset from the 3,065.00 gridline is 29.65 ft due to the vertical exaggeration of 5, 29.65 ft/5 = 5.93 ft.

h. Select the ***MicroStation → Place Line*** command to draw the –1% grade segment on the Eastbound Off Ramp Profile (see Figure 9-40).

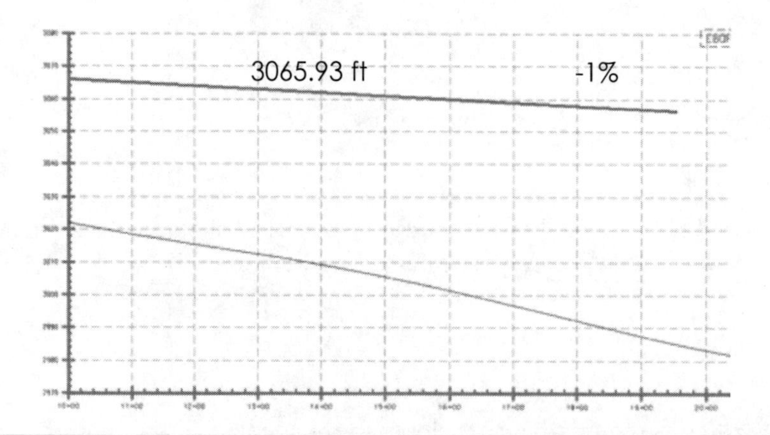

Figure 9-40 Freeway-Side Grade Design of the Eastbound Off Ramp

2. Location and design of local-road-side grade

The local-road-side grade segment provides a connection to Laguna Canyon Road. Its elevation and grade should be designed to ensure that drivers can change direction comfortably and safely from the ramp to the local road.

The elevation of the local-road-side grade segment and the elevation at the intersection (Point A') between Laguna Canyon Road's ETW line and the ramp station line (see Figure 9-41) should be the same. The intersection (Point A') is 24 ft away from the centerline or station line of Laguna Canyon Road at 16+96.08.

Use the ***MicroStation → Measure Distance*** command to measure the elevation of Laguna Canyon Road at Station 16+96.08. The elevation is measured to be 2,997.00 ft (see Figure 9-42).

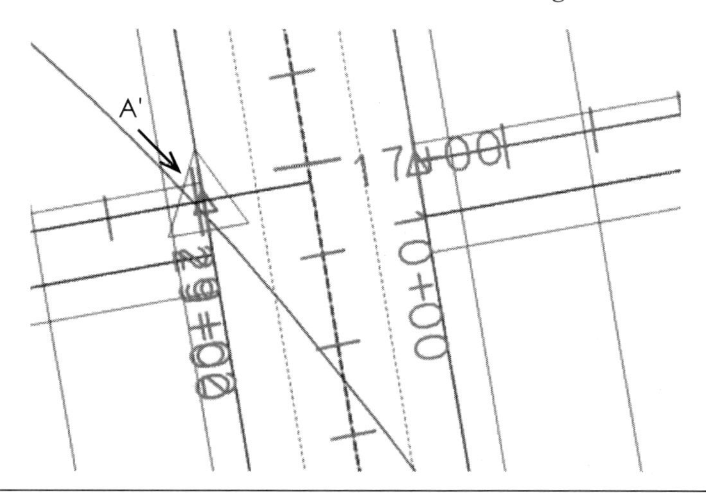

Figure 9-41 Location of Eastbound Off Ramp on Laguna Canyon Road

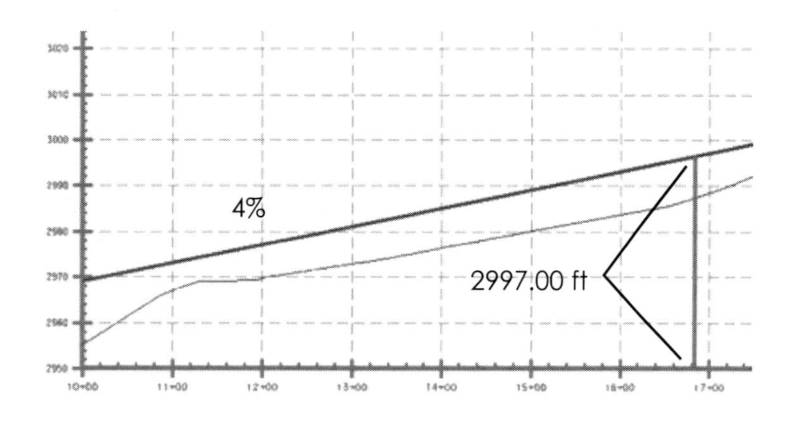

Figure 9-42 Elevation at Laguna Canyon Road at Station 16+96.08

We assume Laguna Canyon Road is used as the primary roadway to control the cross section of the intersection. The local-road-side elevation of the ramp is then calculated (see Figure 9-43):

$= 2{,}997.00' - 24 \times 2\%$

$= 2{,}996.52$ ft.

In Figure 9-43, water in the intersection flows away from the 2% down slope. The local-road-side grade of the Eastbound Off Ramp can be in the range of –2% to 2%. Any grade greater than 2% will make drivers turn onto Laguna Canyon Road with uncomfortable "dipping" impacts. In this tutorial project, we select 2% for the grade (from the view of West to East) or –2% from the view of the Laguna Canyon Road centerline to the end of the Eastbound Off Ramp.

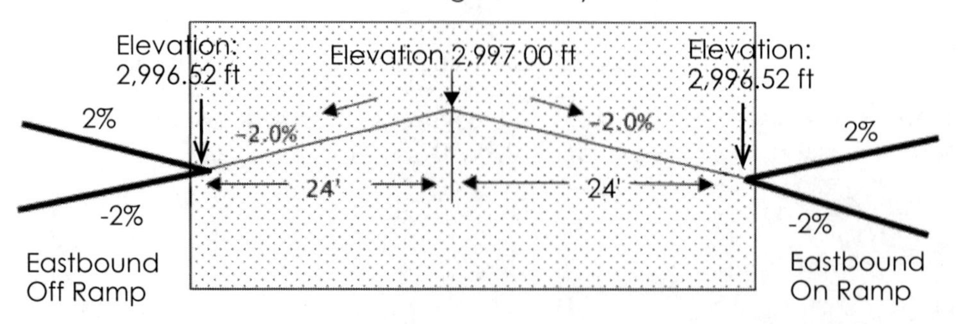

Figure 9-43 Local-Road-Side Elevation of the Eastbound Off Ramp

The design of local-road-side grade segment is shown in Figure 9-44.

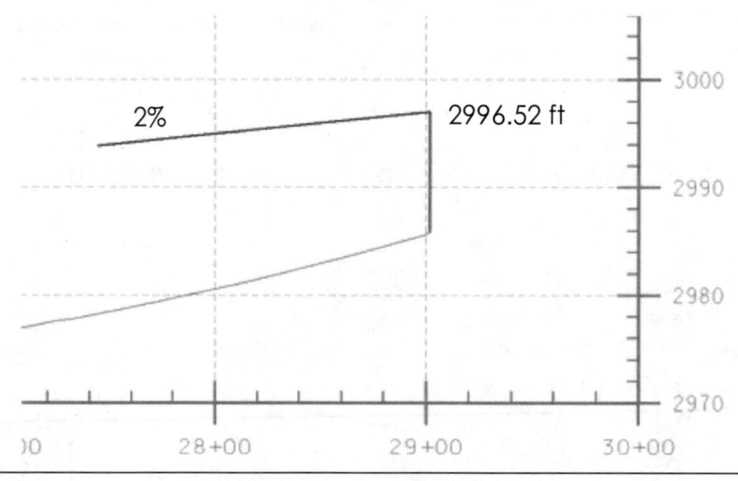

Figure 9-44 Local-Road-Side Grade Design of Eastbound Off Ramp

3. Location and design of grade line in the middle

Before we design the grade line to link to Route 8 and Laguna Canyon Road, we need to identify features that control the elevation of the grade line.

The breakline 1 as shown in Figure 9-22 runs through the Eastbound Off Ramp. A culvert is needed to carry water cross the ramp.

Use the ***MicroStation → Measure Distance*** command to identify the culvert location. The culvert is located at Station 23+91.52, where the breakline1 intersects the station line of the Eastbound Off Ramp (see Figure 9-45).

The elevation of the culvert is 2,971.36 ft at Station 23+91.52 (see Figure 9-46). The diameter of the culvert is given to be 4 ft. The elevation of the culvert top side therefore is 2,975.36 ft. We assume that a 2-ft buffer is needed between the culvert and the ramp's pavement surface. The minimum elevation that the ramp's grade line can be lowered is 2,977.36 ft.

From Figure 9-46, you can see the culvert is below the 2% grade line. Therefore the culvert is not a controlling feature affecting the ramp design.

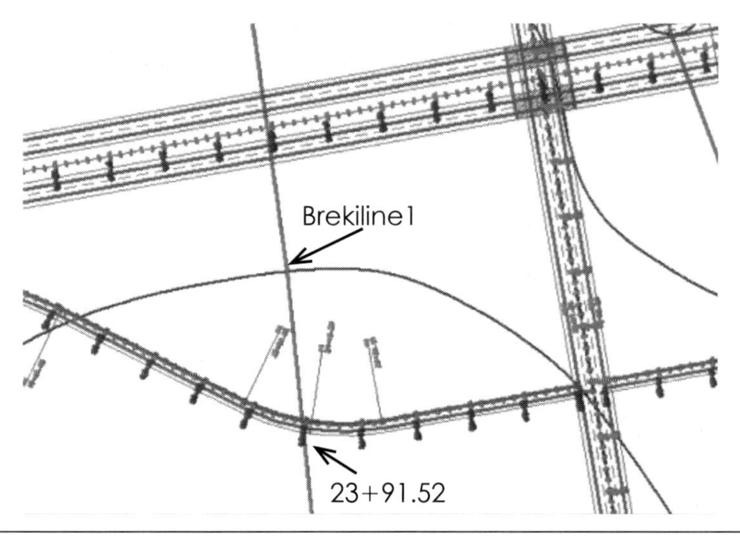

Figure 9-45 Location of Culvert on Eastbound Off Ramp

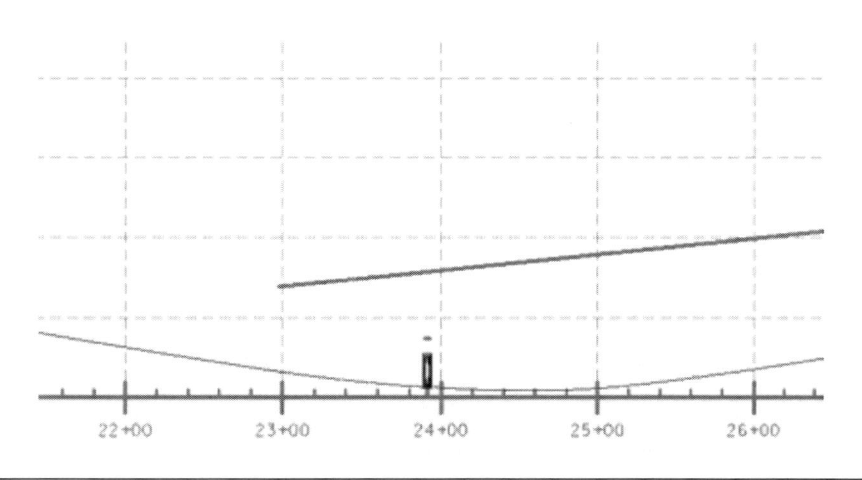

Figure 9-46 Culvert on Eastbound Off Ramp

Now we need to create the grade line that connects to the two end grade segments. Considering the maximum grade of 8% that can be used for ramps and after some trial and error, we use 7% (see Figure 9-47).

a. Go to the ***MicroStation → Place Line*** command to create a segment with a 7% grade.

b. Use the ***MicroStation → Move*** command to move the grade segment to intersect the two end grade segments (see Figure 9-47).

The stations and elevations of the two intersecting points are measured as:

Station/Elevation: 15+26.13/3,060.67 ft.
Station/Elevation: 25+44.54/2,989.38 ft.

c. Go to ***MicroStation → Tools → Tool Boxes ....***

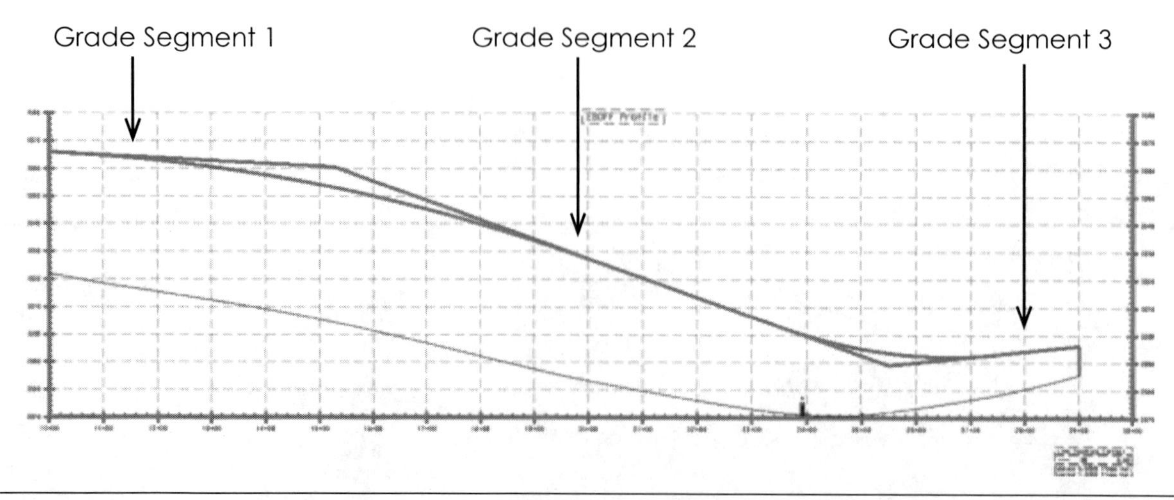

Figure 9-47 Vertical Alignment Design of Eastbound Off Ramp

d. Check on the **Fillets** option in the **Tool Boxes** window.

The **Fillets** toolbox appears.

e. Select the **Construct Parabolic Fillet** tool in the **Fillets** toolbox.

The **Construct Parabolic Fillet** window appears.

f. Type the following in the **Construct Parabolic Fillet** window:

Distance: ***1000.0000***
Type: ***Horizontal***
Truncate: ***None***

g. Select the grade segment 1 and the grade segment 2 (see Figure 9-47).

The parabolic curve of 1,000 ft. is filleted into the grade segments.

h. Again select the **Construct Parabolic Fillet** tool in the **Fillets** toolbox.

i. Enter the following in the **Construct Parabolic Fillet** window:

Distance: ***400.0000***
Type: ***Horizontal***
Truncate: ***None***

j. Select the grade segment 2 and the grade segment 3 (see Figure 9-47).

The parabolic curve of 400 ft. is filleted into the grade segments.

4. Caltrans Standards Conformance Check

Table 9-3 shows the graphical elements that form the vertical alignment of the Eastbound Off Ramp.

5. Import the graphical elements into InRoads to create the vertical alignment of the Eastbound Off Ramp

a. Make sure the **EBOFF_VA** level is active in MicroStation. If **EBOFF_VA** is not available, create the level using MicroStation commands.

b. Make sure the **tutorial** geometry project is active. If it is not active, set it to be active.

c. Make sure the horizontal alignment **EBOFF** is active.

Table 9-3 Conformance Check

Element Type	Length (L)	Conformance Check
Grade Segment 1	26.13' −1.0%	The grade matches the grade on Route 8. OK.
Parabolic Curve 1	1,000.00'	The algebraic grade difference is $\lvert -7.0\% - (-1\%) \rvert = 6.0\%$. Following the equation $L_{min} = 10V$, the minimum length of the vertical curve is 10 × 50 mph = 500 ft. The minimum length of the vertical curve is 850 ft. (see Figure 9-5). 850 ft > 500 ft. Select 850 ft. 1,000 ft > 850 ft. OK.
Grade Segment 2	318.41' −7%	7% < 8%. OK.
Parabolic Curve 2	400.00'	The algebraic grade difference is $\lvert 2\% - (-7.0\%) \rvert = 9.0\%$. The minimum length of the vertical curve is 200 ft given V = 30 mph. The minimum length of the vertical curve is 350 ft (see Figure 9-6). 350 ft > 200 ft. Select 350 ft. 400 ft > 350 ft. OK.
Grade Segment 3	157.61' 2.0%	The grade of 2.0% extends the cross slope of the ramp intersection, eliminating the "dipping" effects.

d. In MicroStation, create a complex chain using the **Create Complex Chain** command.

e. Select the **three tangent grades** (in the order of Segment 1, Segment 2, and Segment 3) for Eastbound Off Ramp and create a chain (see Figure 9-47).

f. Go to ***InRoads → Import → Geometry …***.

The **Import Geometry** window appears (see Figure 9-48).

g. Select the **From Graphics** tab and type the following:

Type:	**Vertical Alignment**
Name:	**EBOFF**
Description:	**EBOFF Vertical Alignment**
Style:	**Default**
Vertical Curve Definition:	**Parabolic**
Other fields:	**see Figure 9-48**

h. Select ***Apply***.

InRoads prompts you to select the graphical elements.

i. Select the chain you just created.

j. Select ***Accept*** and ***Reset***.

k. Select ***Close***.

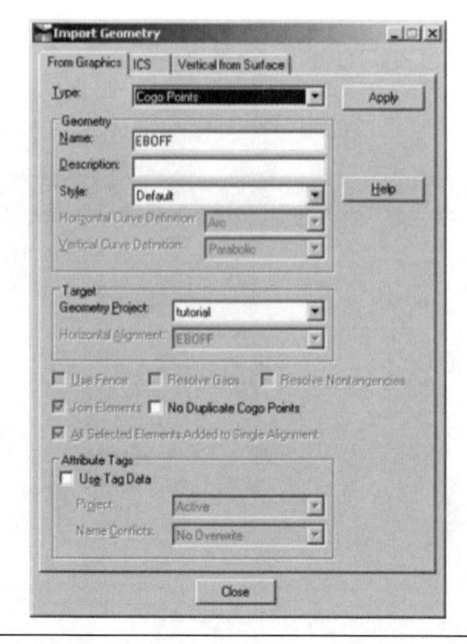

Figure 9-48 Import Graphics for EBOFF Vertical Alignment

l. Check in the **Workspace Bar** window to make sure the vertical alignment **EBOFF** is created. You will see a Workspace Bar window similar to Figure 9-49.

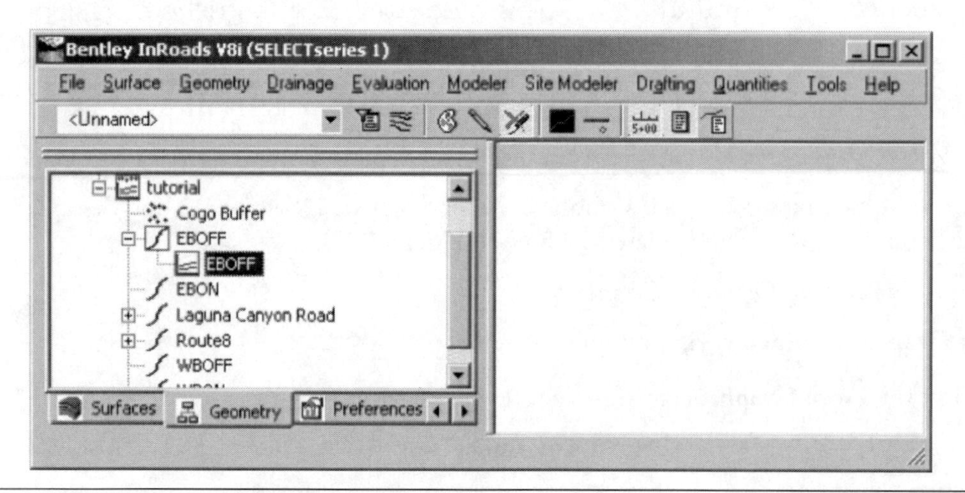

Figure 9-49 Vertical Alignment of Eastbound Off Ramp

m. Go to ***InRoads → Geometry → View Geometry → Active Vertical.***

n. Go to ***InRoads → Geometry → Vertical Curve Set → Define Curve ....***

The **Define Curve** window appears (see Figure 9-50).

o. In the **Length** field, type ***1000.00***. Select ***Apply***.

p. Select ***Next***.

q. In the **Length** field, type ***400.00.***

r. Select ***Apply*** and ***Close***.

6. Save the **tutorial** geometry project

   a. Make sure the **tutorial** geometry project is set to be active.

   b. Go to ***InRoads → File → Save → Geometry Project.***

   c. Save the **tutorial** geometry project as ***tutorial.alg*** in the **C:/Temp/Tutorial/Chp9/Master Files** folder.

7. The **EBOFF** vertical alignment is created and the tutorial geometry project is saved. Now we can view **EBOFF** vertical alignment using InRoads tools.

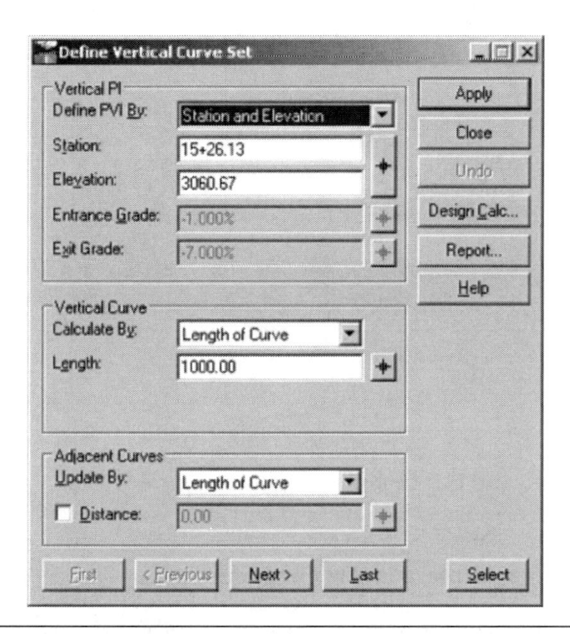

Figure 9-50 Define Curve Window

   a. Make sure the **tutorial** geometry project is set to be active.

   b. Make sure the **EBOFF** vertical alignment is set to be active.

   c. Go to ***InRoads → Geometry → View Geometry → Vertical Annotation ....***

      The **Vertical Annotation** window appears.

   d. Select the **Points** tab and make sure the following appears in the Symbology listbox:

PVC Text:	**Check on**
PVI Text:	**Check on**
PVT Text:	**Check on**
High Point Text:	**Check on**
Low Point Text:	**Check on**
PVC Point:	**Check on**
PVI Point:	**Check on**
PVT Point:	**Check on**
High Point:	**Check on**
Low Point:	**Check on**
All other fields:	**Check off**

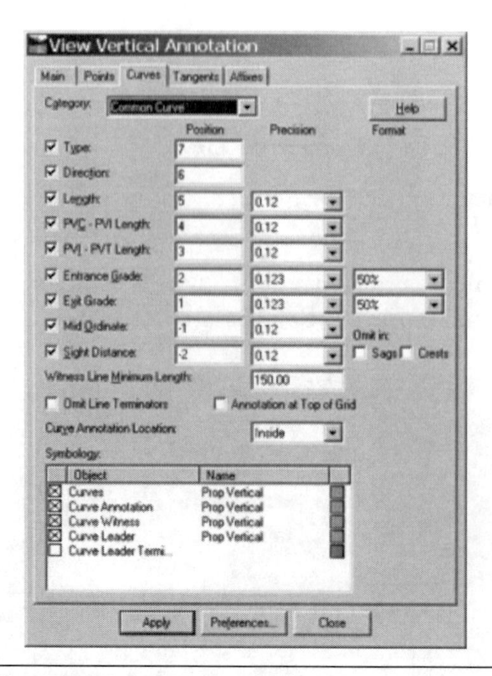

Figure 9-51 Curve Tab in the View Vertical Annotation

e. Select the **Curves** tab and make sure all the fields are the same as shown in Figure 9-51.

f. Select the **Tangents** tab and do the following (see Figure 9-52):

Grade:	**Check on**
Other fields:	**Same as shown in Figure 9-52**

g. In the **View Vertical Annotation** window, select ***Apply*** and ***Close***.

Vertical annotations are displayed along the vertical alignment of the Eastbound Off Ramp (see Figure 9-53).

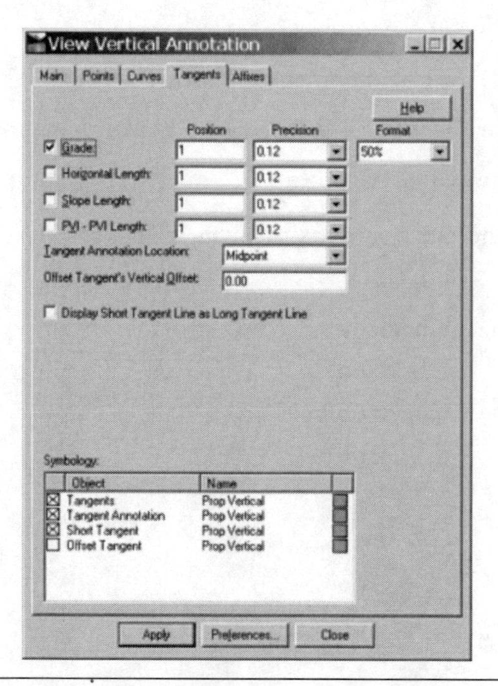

Figure 9-52 Tangents Tab of the View Vertical Annotation

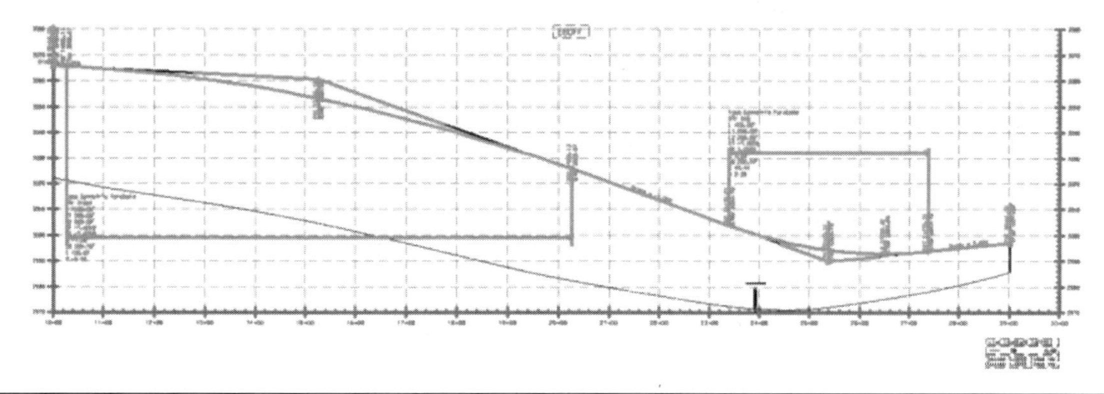

Figure 9-53 Vertical Alignment for Eastbound Off Ramp

9.3.4 Vertical Alignment for Eastbound On Ramp

The vertical alignment design for the Eastbound On Ramp consists of three components: local-road-side grade, freeway-side grade, and a series of grades and curves that connect to Route 8 and Laguna Canyon Road.

1. Location and design of local-road-side grade

The local-road-side grade segment of the Eastbound On Ramp connects to Laguna Canyon Road. Its elevation and grade should be designed so that drivers can change direction comfortably from Laguna Canyon Road to the ramp.

The ramp elevation should be the same as that at the intersection (Point A') between Laguna Canyon Road's ETW line and the ramp station line (see Figure 9-54). The intersection (Point A') is 24 ft away from the centerline or station line of Laguna Canyon Road at 16+96.08.

Use the ***MicroStation → Measure Distance*** command to measure the elevation of Laguna Canyon Road at Station 16+96.08. The elevation is measured to be 2,997.00 ft.

The local-road-side elevation of the ramp is then calculated:

$= 2997.00' - 24 \times 2\%$

$= 2996.52$ ft.

The local-road-side grade of the Eastbound On Ramp is –2% (from West to East) for this tutorial project.

The design of local-road-side grade segment is shown in Figure 9-55. Note that the grade is 1% from West to East direction. The grade is in the range of 2% to –2%.

a. Go to ***MicroStation → Settings → Levels → Manager.***

   The **Level Manager** window appears.

b. Go to ***Levels → New*** in the **Level Manager** window.

c. Name the new level to be **EBON_VA**. Select the **Color, Line Style**, and **Line Weight** of your choice.

d. Set the ***EBON_VA*** level as the active level.

e. Close the **Level Manager** window.

f. Go to the ***MicroStation → Window Area*** command and zoom in to the Eastbound On Ramp Profile created in Chapter 8.

g. Go to the ***MicroStation → Move/Copy Parallel*** command to get a horizontal offset line at elevation 2,996.52 ft.

Note the offset from 2,990.00 gridline is 32.60 ft due to the vertical exaggeration of 5, 32.60 ft./ 5 = 6.52 ft.

h. Select the ***MicroStation → Place Line*** command to draw the +1% grade segment on the Eastbound On Ramp Profile (see Figure 9-55).

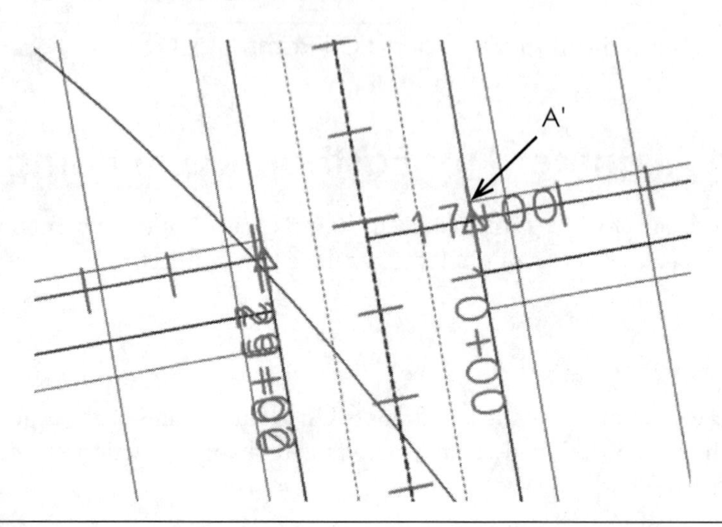

Figure 9-54 Location of Eastbound On Ramp on Laguna Canyon Road

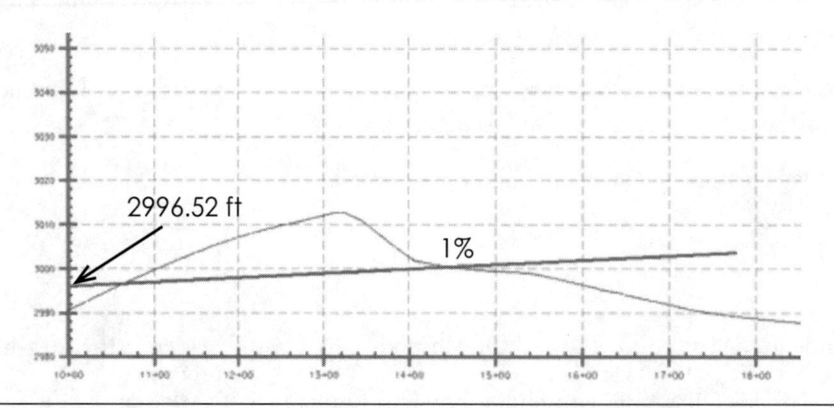

Figure 9-55 Local-Road-Side Grade Design of Eastbound On Ramp

2. Location and design of freeway-side grade

The ramp's freeway-side grade segment should have the same elevation and grade as those on Route 8 in the transition area (or the gore area) where drivers change direction from the ramp to Route 8.

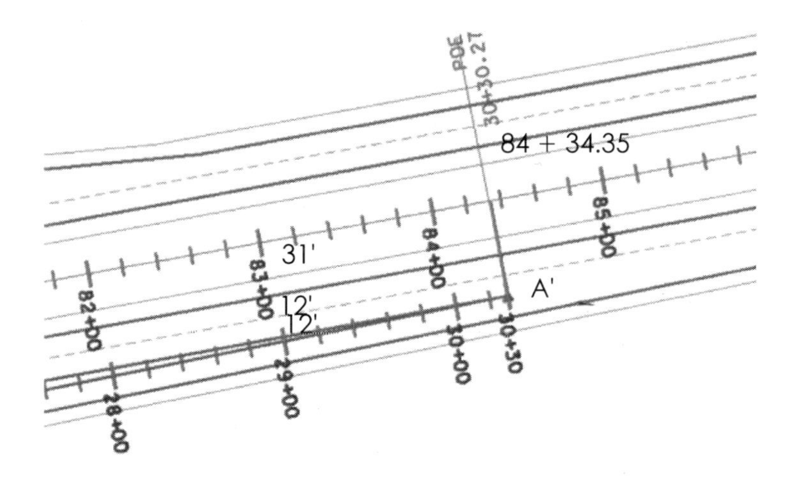

Figure 9-56 Elevation Calculation

Go to the ***MicroStation → Place Line*** command to create a line segment perpendicular to the Route 8 Station Line and connect to the intersection (Point A') between the Route 8 right side ETW line and the ramp station line (see Figure 9-56).

The intersection A' is 55' (or 31' + 12' + 12') away from the Route 8 station line at Station 84+34.35.

Use the ***MicroStation → Measure Distance*** command to measure the elevation of Route 8 at Station 84+34.35 in the Route 8 Profile.

The elevation is measured to be 3,041.31 ft. (see Figure 9-57). The elevation at the freeway side (or Point A') can be calculated (see Figure 9-58):

$= 3041.31' + 2\% \times 31 - 2\% \times 24$

$= 3041.45'$

Station 84+34.35 is located in the sag curve. Its grade is to be approximately 1% from West to East. To ensure the grade consistency, the grade for the ramp entrance segment is to be 1% from West to East.

The design of the freeway-side grade segment for the Eastbound On Ramp is shown in Figure 9-59. Note that the grade for the segment is 1% from West to East.

a. Go to the ***MicroStation → Window Area*** command and zoom in to the Eastbound On Ramp Profile.

b. Go to the ***MicroStation → Move/Copy Parallel*** command to get a horizontal offset line at elevation 3,041.45 ft.

   Note the offset from 3,040.00 ft. gridline is 7.25 ft. due to the vertical exaggeration of 5, 7.25 ft./5 = 1.45 ft.

c. Select the ***MicroStation → Place Line*** command to draw the –1% grade segment on the Eastbound On Ramp Profile (see Figure 9-59).

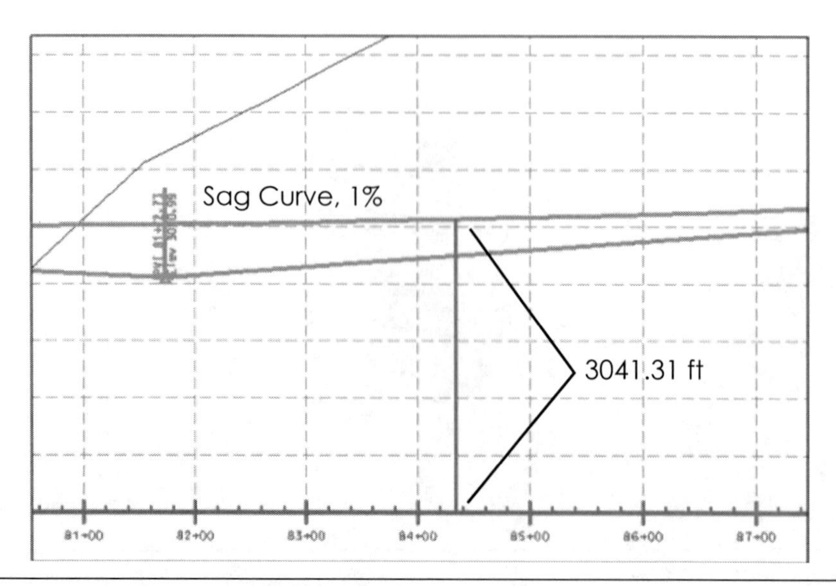

Figure 9-57 Elevation of Route 8 at Station 84+34.38

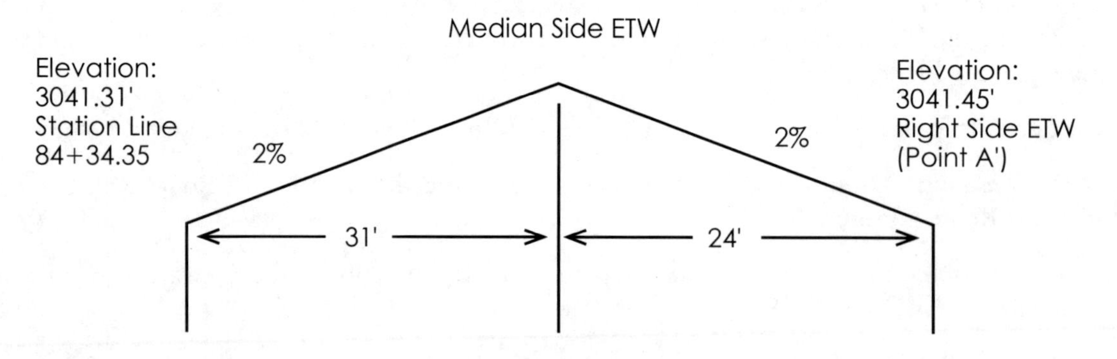

Figure 9-58 Elevation of the Eastbound On Ramp at Freeway Side

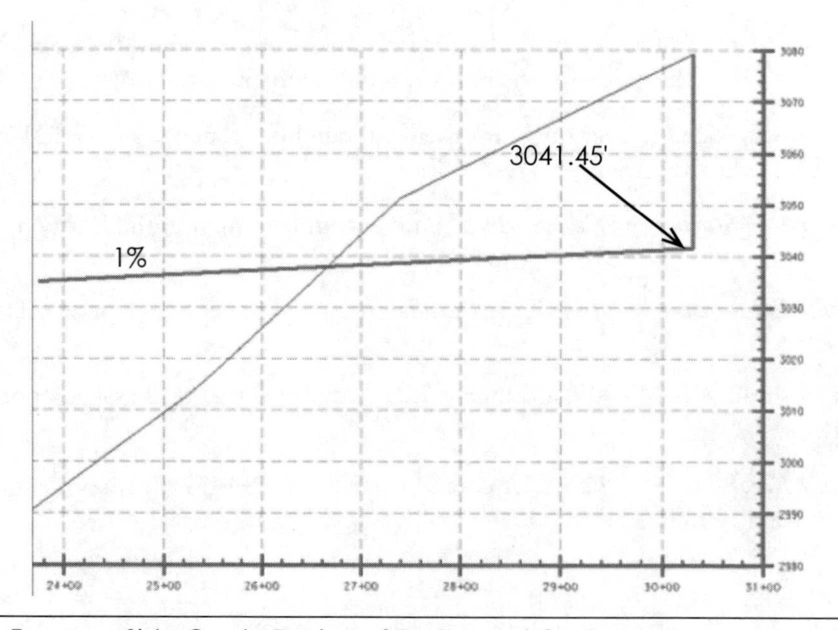

Figure 9-59 Freeway-Side Grade Design of Eastbound On Ramp

3. Location and design of grade line between two ramp ends

The grade line design needs to consider the location and elevation of the culvert that passes through the ramp.

Go to the ***MicroStation → Measure Distance*** command to measure the location of the culvert or the intersection point as shown in Figure 9-22. The culvert is located at Station 22+69.44, where the breakline 3 intersects the station line of the Eastbound On Ramp (see Figure 9-22).

The elevation of the culvert is 2,982.95 ft at Station 22+69.44. The diameter of the culvert is given to be 4 ft. The elevation of the culvert top side therefore is 2,986.95 ft. We assume that a 2-ft buffer is needed between the culvert and the ramp's pavement surface. The minimum elevation that the grade line of the ramp can be lowered is 2,988.95 ft.

From Figure 9-60a, you can see the culvert is significantly below the grades at the two ends (or local-road-side and freeway-side grade segments). Therefore, the culvert is not a controlling feature affecting the ramp design.

Now we need to create the grade line that connects to the two end grade segments.

a. Go to the ***MicroStation → Place Line*** command to create a segment with a 4% grade.
b. Go to the ***MicroStation → Move*** command to move the grade segment to intersect the two end grade segments (see Figure 9-60b).

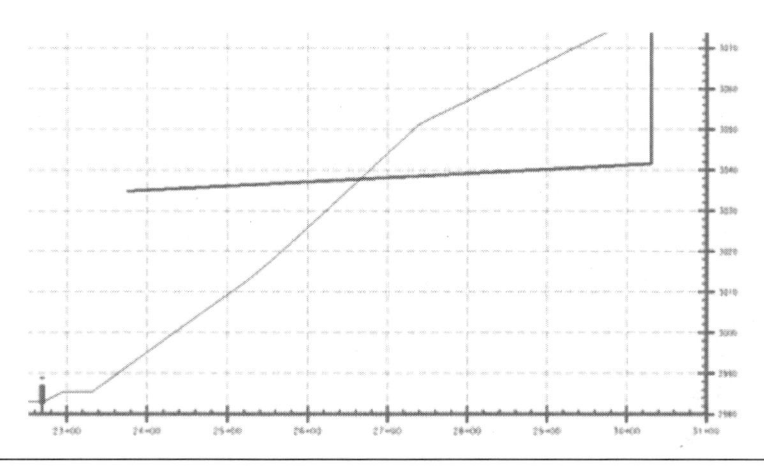

Figure 9-60a Location and Elevation of Culvert at Eastbound On Ramp

The stations and elevations of the two intersecting points are measured as:

Station/Elevation: 14+82.30/3,001.34 ft.
Station/Elevation: 22+79.97/3,033.25 ft.

c. Go to ***MicroStation → Tools → Tool Boxes ...***.
d. Check on the **Fillets** option in the **Tool Boxes** window.

   The **Fillets** toolbox appears.
e. Select the **Construct Parabolic Fillet** tool in the **Fillets** toolbox.

   The **Construct Parabolic Fillet** window appears.

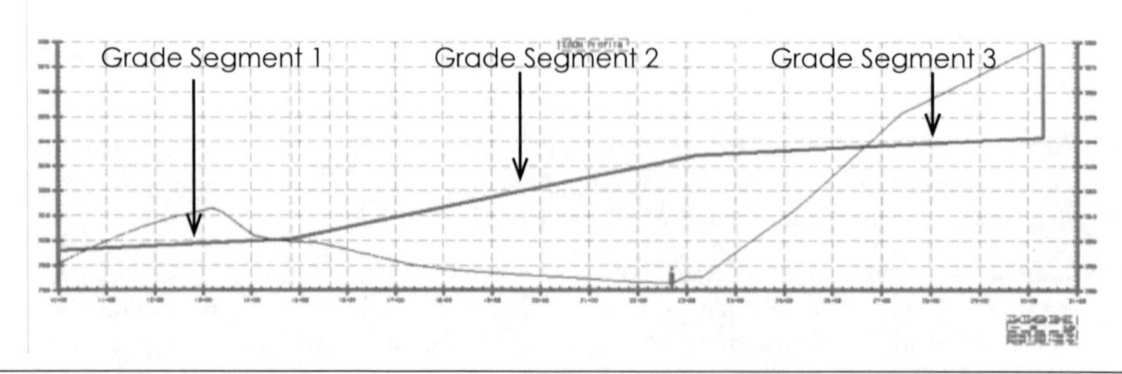

Figure 9-60b Vertical Alignment Design of Eastbound On Ramp

f. Type the following in the **Construct Parabolic Fillet** window:

Distance:	***600.0000***
Type:	***Horizontal***
Truncate:	***None***

g. Select the grade segment 1 and the grade segment 2.

The parabolic curve of 600 ft. is filleted into the grade segments.

h. Select the **Construct Parabolic Fillet** tool in the **Fillets** toolbox.

i. Type the following in the **Construct Parabolic Fillet** window:

Distance:	***500.0000***
Type:	***Horizontal***
Truncate:	***None***

j. Select the grade segment 2 and the grade segment 3.

The parabolic curve of 500 ft. is filleted into the grade segments.

4. Caltrans Standards Conformance Check

Table 9-4 shows the graphical elements that form the vertical alignment of the Eastbound On Ramp.

Table 9-4 Conformance Check

Element Type	Length (L)	Conformance Check
Grade 1 Segment	182.30' 1.0%	The grade of 1.0% is within the range of –2% to 2%. OK.
Parabolic Curve 1	600.00'	The algebraic grade difference is |3.5% – (1.0%)| = 2.5%. Following the equation L_{min} = 10V, the minimum length of the vertical curve is 10 × 30 mph = 300 ft. The minimum length of the vertical curve is about 200 ft (see Figure 9-6 or Figure 201.5 of Caltrans *HDM*). 300 ft > 200 ft. Select 300 ft. 600.00 ft > 300 ft. OK.

Grade Segment 2	247.67 ft 4%	4% < 8%. OK.
Parabolic Curve 2	500.00' ft.	The algebraic grade difference is | 1.0% – (3.5%) | = 2.5%. Following the equation L_{min} = 10V, the minimum length of the vertical curve is 10 × 50 mph = 500 ft. The minimum length of the vertical curve is about 300 ft (see Figure 9-6). Select 500 ft. 500 ft. ≥ 500 ft. OK.
Grade Segment 2	570.03' 1.0%	The grade matches the grade on Route 8. OK

5. Import the graphical elements into InRoads to create the vertical alignment of the Eastbound On Ramp
   a. Make sure the **EBON_VA** level is active in MicroStation. If **EBON_VA** is not available, create the level using MicroStation commands.
   b. Make sure the **tutorial** geometry project is active. If it is not active, set it to be active.
   c. Make sure the horizontal alignment **EBON** is active.
   d. In MicroStation, create a complex chain using the **Create Complex Chain** command.
   e. Select the **three tangent grades** (in the order of Segment 1, Segment 2, and Segment 3) for Eastbound On Ramp and create a chain.
   f. Go to ***InRoads → Import → Geometry ...***.

      The **Import Geometry** window appears.
   g. Select the **From Graphics** tab and type the following:

      Type: **Vertical Alignment**
      Name: **EBON**
      Description: **EBON Vertical Alignment**
      Style: **Default**
      Vertical
      Curve Definition: **Parabolic**
      Other fields: **Default values**
   h. Select ***Apply***.

      InRoads prompts you to select the graphical elements.
   i. Select the chain you just created.
   j. Select ***Accept*** and ***Reset***.
   k. Select ***Close***.
   l. Check in the **Workspace Bar** window to make sure the vertical alignment **EBON** is created.

      You will see a Workspace Bar window similar to Figure 9-61.

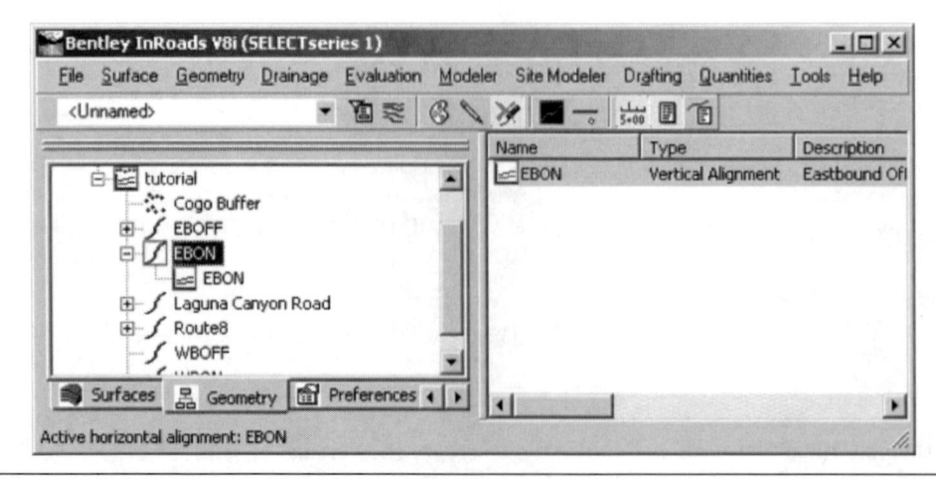

Figure 9-61 Vertical Alignment of Eastbound On Ramp

m. Go to ***InRoads → Geometry → View Geometry → Active Vertical.***

n. Go to ***InRoads → Geometry → Vertical Curve Set → Define Curve ….***

The **Define Curve** window appears (see Figure 9-62).

o. In the **Length** field, type ***600.00.***

p. Select ***Apply***. Select ***Next***.

q. In the **Length** field, type ***500.00.***

r. Select ***Apply*** and ***Close***.

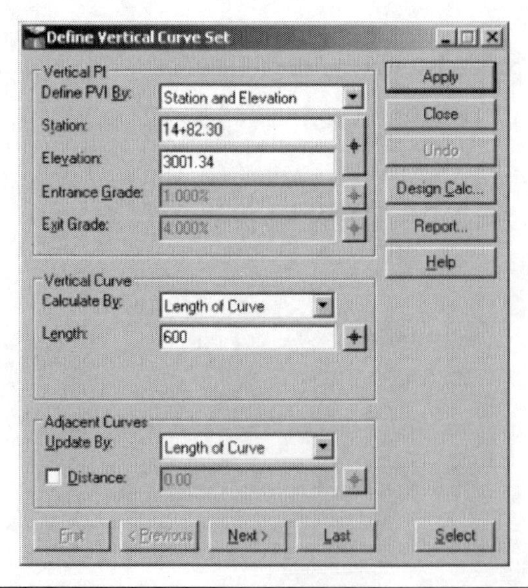

Figure 9-62 Define Curve Window

6. Save the **tutorial** geometry project

   a. Make sure the **tutorial** geometry project is set to be active.

   b. Go to ***InRoads → File → Save → Geometry Project.***

   c. Save the **tutorial** geometry project as ***tutorial.alg*** in the **C:/Temp/Tutorial/Chp9/Master Files** folder.

7. The **EBON** vertical alignment is created and the **tutorial** geometry project is saved. Now we can view **EBON** vertical alignment using InRoads tools.
   a. Make sure the **tutorial** geometry project is set to be active.
   b. Make sure the **EBON** vertical alignment is set to be active.
   c. Go to ***InRoads → Geometry → View Geometry → Vertical Annotation ...***.

      The **Vertical Annotation** window appears.
   d. Select the **Points** tab and make sure the following appears in the Symbology listbox:

PVC Text:	**Check on**
PVI Text:	**Check on**
PVT Text:	**Check on**
High Point Text:	**Check on**
Low Point Text:	**Check on**
PVC Point:	**Check on**
PVI Point:	**Check on**
PVT Point:	**Check on**
High Point:	**Check on**
Low Point:	**Check on**
All other fields:	**Check off**

   e. Select the **Curves** tab and make sure all the fields are the same as shown in Figure 9-63.
   f. Select the **Tangents** tab and do the following (see Figure 9-64):

Grade:	**Check on**
Other fields:	**Same as shown in Figure 9-64**

   g. In the View **Vertical Annotation** window, select ***Apply*** and ***Close***.

      The vertical annotations are displayed along with the vertical alignment of Eastbound On Ramp (see Figure 9-65).

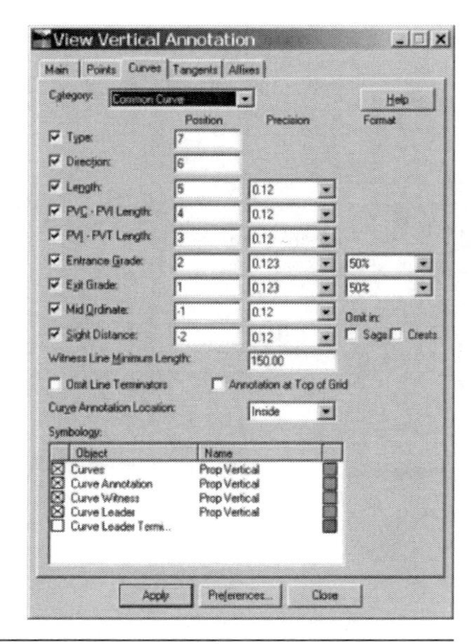

Figure 9-63 Curve Tab in the View Vertical Annotation

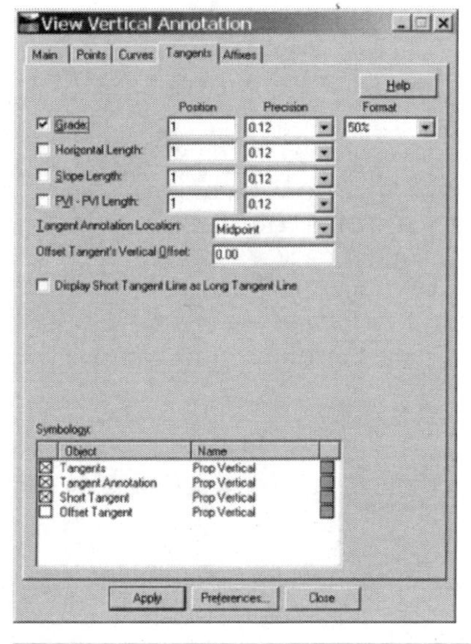

Figure 9-64 Tangents Tab of the View Vertical Annotation

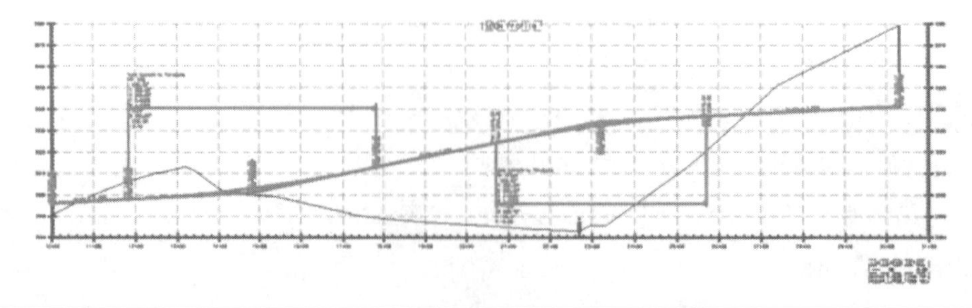

Figure 9-65 Vertical Alignment for Eastbound On Ramp

9.3.5 Vertical Alignment for Westbound Off Ramp

The vertical alignment design for the Westbound Off Ramp consists of three components: local-road-side grade, freeway-side grade, and a series of grades and curves that connect to Route 8 and Laguna Canyon Road.

1. Location and design of local-road-side grade

The local-road-side grade segment connects to Laguna Canyon Road. Its elevation should be the same as that at the intersection (Point A') between Laguna Canyon Road's ETW line and the ramp station line (see Figure 9-66). The intersection (Point A') is 24 ft away from the centerline or station line of Laguna Canyon Road at 28+06.19.

Use the ***MicroStation* → *Measure Distance*** command to measure the elevation of Point A' at Station 28+06.19 of the Laguna Canyon Road Profile. The elevation is measured to be 3,041.41 ft.

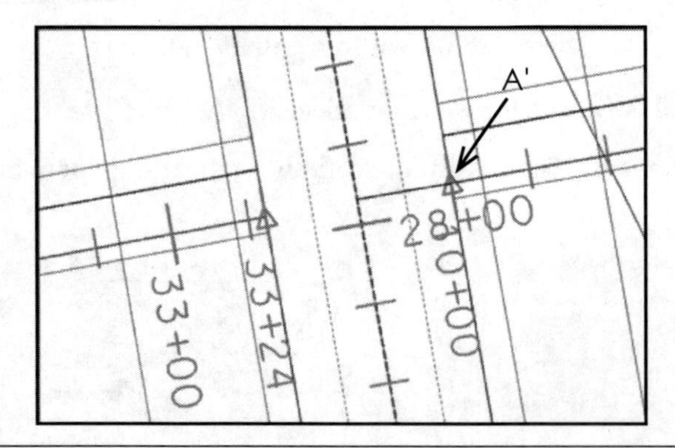

Figure 9-66 Location of Westbound Off Ramp on Laguna Canyon Road

The local-road-side elevation of the ramp is then calculated (see Figure 9-67):

$= 3{,}041.41' - 24 \times 2\%$

$= 3{,}040.93$ ft.

The local-road-side grade of the Westbound Off Ramp is –2% (from West to East) for this tutorial project.

The design of local-road-side grade segment is shown in Figure 9-68. Note that the grade is –2% from West to East direction, the same cross slope as the ramp intersection.

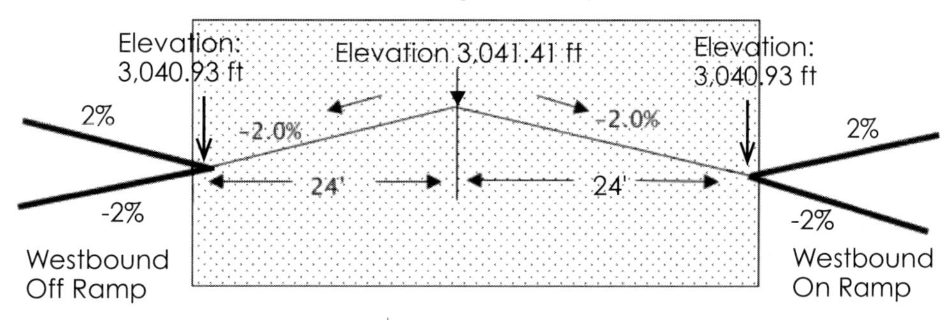

Figure 9-67 Local-Road-Side Elevation of the Westbound Off Ramp

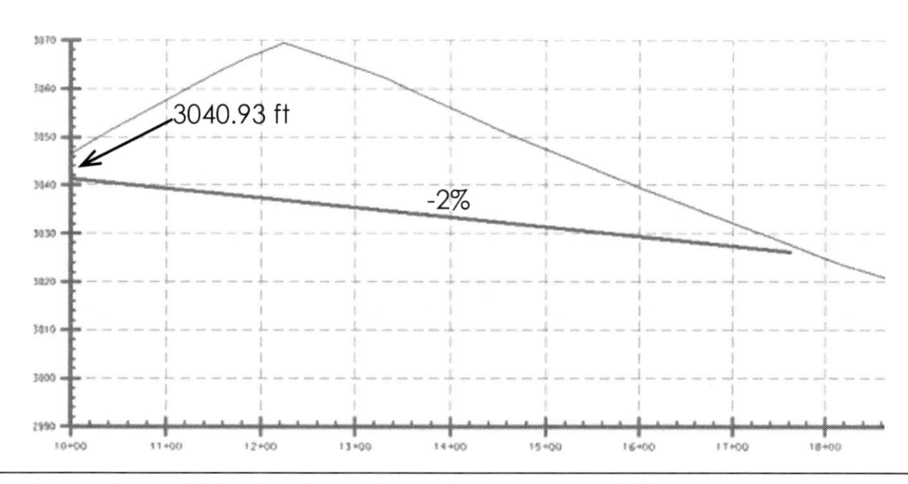

Figure 9-68 Local-Road-Side Grade Design of Westbound Off Ramp

a. Go to ***MicroStation*** → ***Settings*** → ***Levels*** → ***Manager.***

   The **Level Manager** window appears.

b. Go to ***Levels*** → ***New*** in the **Level Manager** window.

c. Name the new level **WBOFF_VA**. Select the **Color, Line Style,** and **Line Weight** of your choice.

d. Set the ***WBOFF_VA*** level as the active level.

e. Close the **Level Manager** window.

f. Go to the ***MicroStation*** → ***Window Area*** command and zoom in to the Westbound Off Ramp Profile.

g. Go to the ***MicroStation*** → ***Move/Copy Parallel*** command to get a horizontal offset line at elevation 3,040.93 ft.

Note the offset from 3,040 ft gridline is 4.65 ft due to the vertical exaggeration of 5, 4.65 ft/5 = 0.93 ft.

h. Select the ***MicroStation → Place Line*** command to draw the –2% grade segment on the Westbound Off Ramp Profile (see Figure 9-68).

2. Location and design of freeway-side grade

The freeway-side grade segment should have the same elevation and grade as those on Route 8 to ensure a safe change of direction from Route 8 to the ramp in the transition area (or the gore area).

Go to the ***MicroStation → Place Line*** command to create a line segment perpendicular to the Route 8 Station Line and connect to the intersection (Point A') between the Route 8 right side ETW line and the ramp station line (see Figure 9-69). The intersection A' is 55' (or 31' + 12' + 12') away from the Route 8 station line at Station 81+33.31.

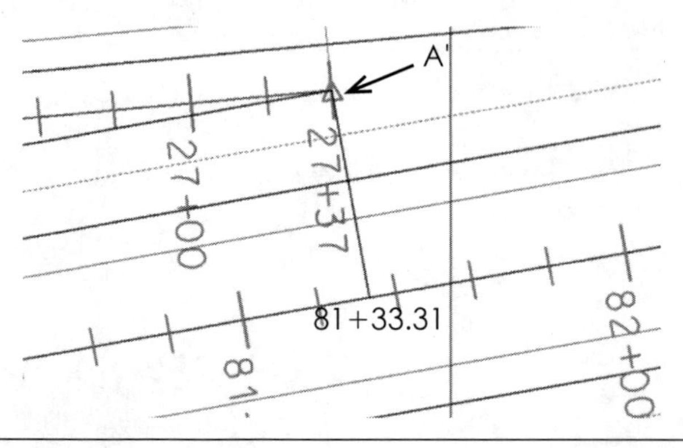

Figure 9-69 Elevation Calculation

Use the ***MicroStation → Measure Distance*** command to measure the elevation of Route 8 at Station 81+33.31 in the Route 8 Profile. The elevation is measured to be 3,040.28 ft. (see Figure 9-70). The elevation at the freeway side (or Point A') can be calculated (see Figure 9-71):

= 3,040.28' + 2% × 31 − 2% × 24

= 3,040.42'

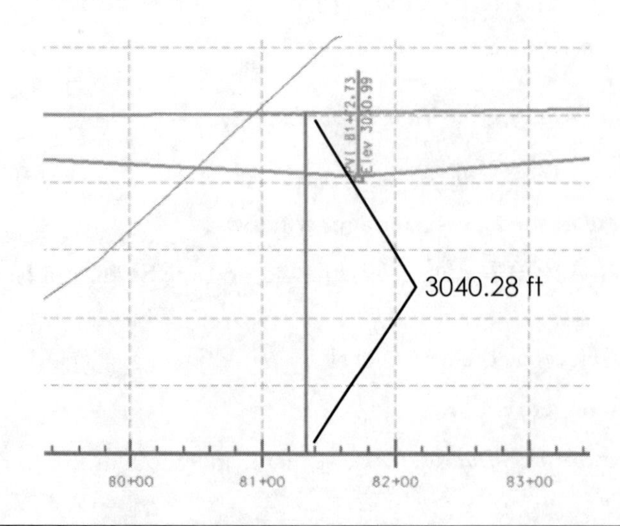

Figure 9-70 Elevation of Route 8 at Station 81+33.31

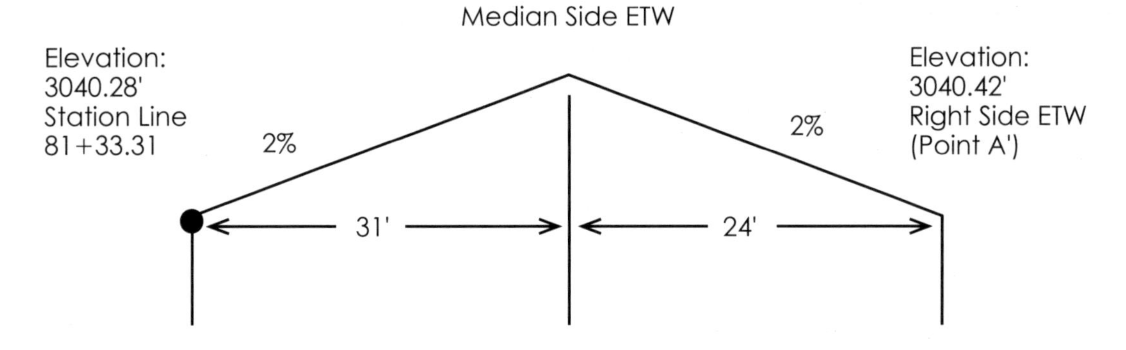

Figure 9-71 Elevation of the Westbound Off Ramp at Freeway Side

Station 81+33.31 is located in the sag curve. Its grade is to be approximately 0% from the view of stations (the West to East direction). To ensure grade consistency, the grade for the ramp entrance segment should also be 0%. However, the 0% grade may cause drainage problems. In this tutorial project we select 0.5%, the minimum value for the grade immediately coming out of Route 8.

The design of the freeway-side grade segment for the Westbound Off Ramp is shown in Figure 9-72. Note that the grade for the segment is 0.5% from the view of stations (the West to East direction).

a. Go to the ***MicroStation → Window Area*** command and zoom in to the Westbound Off Ramp Profile.

b. Go to the ***MicroStation → Move/Copy Parallel*** command to get a horizontal offset line at elevation 3,040.42 ft.

   Note the offset from 3,040 ft gridline is 2.10 ft. due to the vertical exaggeration of 5, 2.10 ft/5 = 0.42 ft.

c. Select the ***MicroStation → Place Line*** command to draw the 0.5% grade segment on the Westbound Off Ramp Profile (see Figure 9-72).

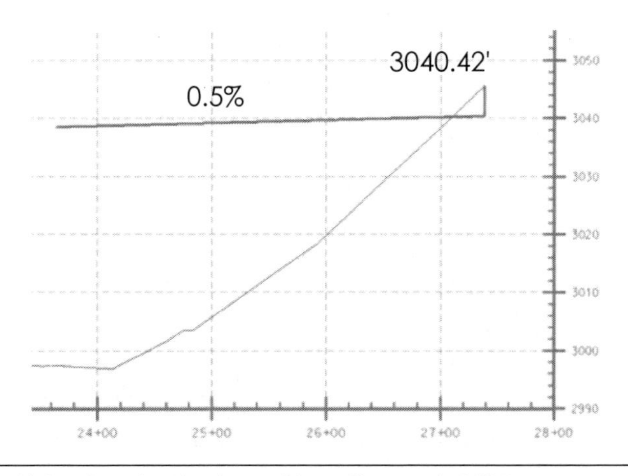

Figure 9-72 Freeway-Side Grade Design of Westbound Off Ramp

3. Location and design of grade line in the middle

The grade line design needs to consider the location and elevation of the culvert that passes through the ramp.

Go to the ***MicroStation → Measure Distance*** command to measure the culvert's location or the intersection point as shown in Figure 9-73. The culvert is located at Station 24+12.28, where the breakline 3 intersects the station line of the Westbound Off Ramp (see Figure 9-22).

The culvert elevation is 2,996.91 ft at Station 24+12.28. The culvert diameter is given to be 4 ft. The elevation of the culvert top side therefore is 3,000.91 ft. We assume that a 2-ft buffer is needed between the culvert and the ramp's pavement surface. The minimum elevation that the ramp grade line can be lowered is 3,002.91 ft.

From Figure 9-73, you see that the culvert is significantly below the grades at the two ends (or local-road-side and freeway-side grade segments). Therefore the culvert is not a controlling feature affecting the ramp design.

Go to the ***MicroStation → Extent Elements to Intersection*** command to extend the two grade segments and get the intersection point (see Figure 9-73).

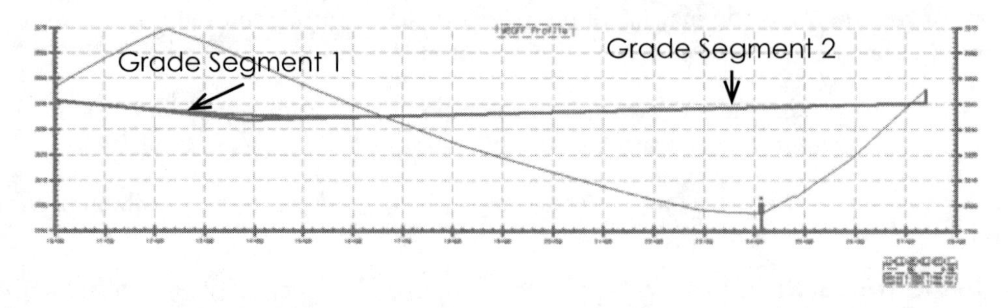

Figure 9-73 Vertical Alignment Design of Westbound Off Ramp

a. Go to ***MicroStation → Tools → Tool Boxes ....***

b. Check on the **Fillets** option in the **Tool Boxes** window.

The **Fillets** toolbox appears.

c. Select the **Construct Parabolic Fillet** tool in the **Fillets** toolbox.

The **Construct Parabolic Fillet** window appears.

d. Type the following in the **Construct Parabolic Fillet** window:

Distance: ***600.00***

Type: ***Horizontal***

Truncate: ***None***

e. Select the grade segment 1 and the grade segment 2 (see Figure 9-73).

The parabolic curve of 600 ft. is filleted into the grade segments.

4. Caltrans Standards Conformance Check

Table 9-5 shows the graphical elements that form the vertical alignment of Westbound Off Ramp.

24 + 12.05

Table 9-5 Conformance Check

Element Type	Length (L)	Conformance Check
Grade 1 Segment	67.83' –2%	–2% matches the cross slope of the ramp intersection. OK.
Parabolic Curve 1	600.00'	The algebraic grade difference is |0.5 – (–2.0%)| = 2.5%. As the curve is in the middle of the vertical alignment, the design speed on the curve is about 40 mph. Following the equation L_{min} = 10V, the minimum length of the vertical curve is 10 × 40 mph = 400 ft. The minimum length of the vertical curve is less than 200 (see Figure 9-6). Select 400 ft. 600 ft > 400 ft. OK.
Grade Segment 2	1,069.31' 0.5%	The grade matches closely the grade on Route 8. OK

5. Import the graphical elements into InRoads to create the vertical alignment of the Westbound Off Ramp
   a. Make sure the **WBOFF_VA** level is active in MicroStation. If **WBOFF_VA** is not available, create the level using MicroStation commands.
   b. Make sure the **tutorial** geometry project is active. If it is not active, set it to be active.
   c. Make sure the horizontal alignment **WBOFF** is active.
   d. In MicroStation, create a complex chain using the **Create Complex Chain** command.
   e. Select the **two tangent grades** for the Westbound Off Ramp and create a chain.
   f. Go to ***InRoads → Import → Geometry ...***.

      The **Import Geometry** window appears.
   g. Select the **From Graphics** tab and type the following:

Type:	**Vertical Alignment**
Name:	**WBOFF**
Description:	**WBOFF Vertical Alignment**
Style:	**Default**
Vertical	
Curve Definition:	**Parabolic**
Other fields:	**Default values**

   h. Select ***Apply***.

      InRoads prompts you to select the graphical elements.

i. Select the chain you just created.

j. Select ***Accept*** and ***Reset***.

k. Select ***Close***.

l. Check in the **Workspace Bar** window to make sure the vertical alignment **WBOFF** is created. You will see a Workspace Bar window similar to Figure 9-74.

m. Go to ***InRoads → Geometry → View Geometry → Active Vertical.***

n. Go to ***InRoads → Geometry → Vertical Curve Set → Define Curve ….***

The **Define Curve** window appears (see Figure 9-75).

o. In the **Length** field, type ***600.00.***

p. Select ***Apply*** and ***Close***.

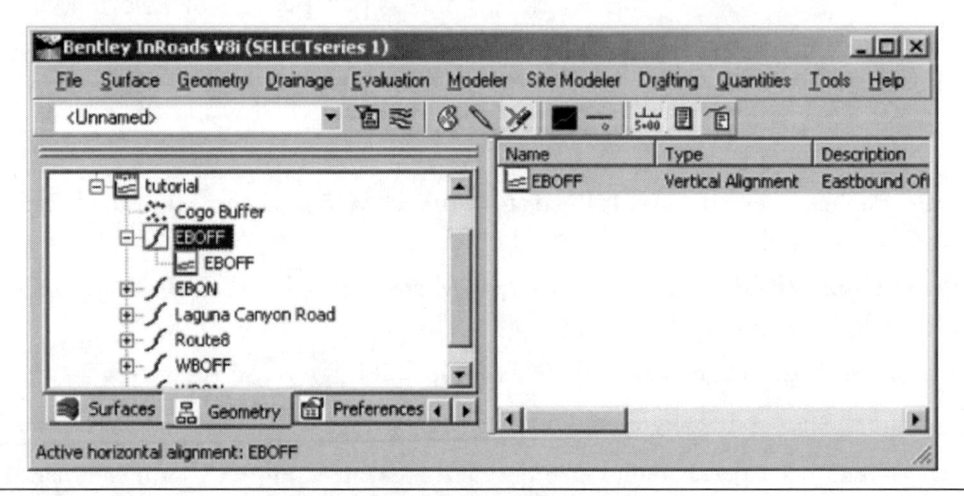

Figure 9-74 Vertical Alignment of Westbound Off Ramp

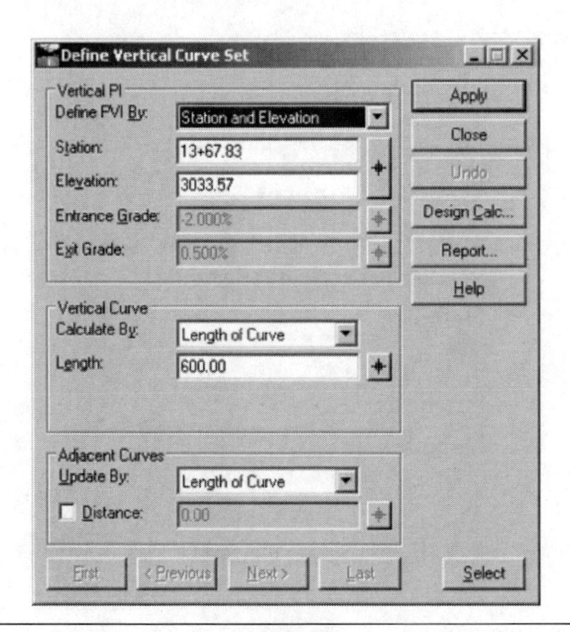

Figure 9-75 Define Curve Window

6. Save the **tutorial** geometry project

   a. Make sure the **tutorial** geometry project is set to be active.

   b. Go to ***InRoads → File → Save → Geometry Project.***

   c. Save the **tutorial** geometry project as ***tutorial.alg*** in the **C:/Temp/Tutorial/Chp9/Master Files** folder.

7. The **WBOFF** vertical alignment is created and the **tutorial** geometry project is saved. Now we can view the **WBOFF** vertical alignment using InRoads tools.

   a. Make sure the **tutorial** geometry project is set to be active.

   b. Make sure the **WBOFF** vertical alignment is set to be active.

   c. Go to ***InRoads → Geometry → View Geometry → Vertical Annotation …***.

      The **Vertical Annotation** window appears.

   d. Select the **Points** tab and make sure the following appears in the Symbology listbox:

PVC Text:	**Check on**
PVI Text:	**Check on**
PVT Text:	**Check on**
High Point Text:	**Check on**
Low Point Text:	**Check on**
PVC Point:	**Check on**
PVI Point:	**Check on**
PVT Point:	**Check on**
High Point:	**Check on**
Low Point:	**Check on**
All other fields:	**Check off**

   e. Select the **Curves** tab. Make sure all the fields are same as shown in Figure 9-76.

   f. Select the **Tangents** tab and do the following (see Figure 9-77):

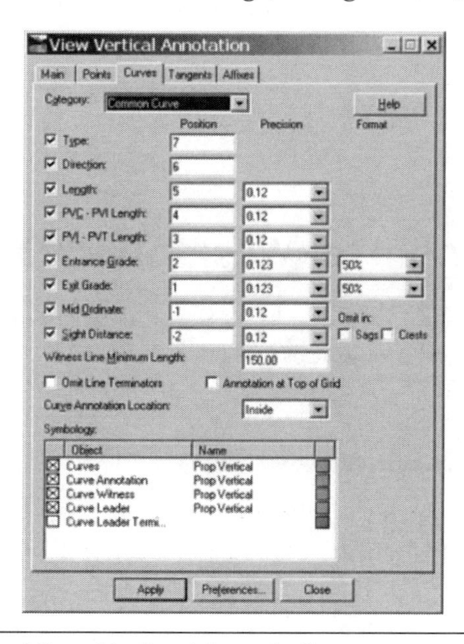

Figure 9-76 Curve Tab in the View Vertical Annotation

Grade: **Check on**

Other fields: **Same as shown in Figure 9-77**

g. In the **View Vertical Annotation** window, select ***Apply*** and ***Close***.

The vertical annotations are displayed along the vertical alignment of the Westbound Ramp (see Figure 9-78).

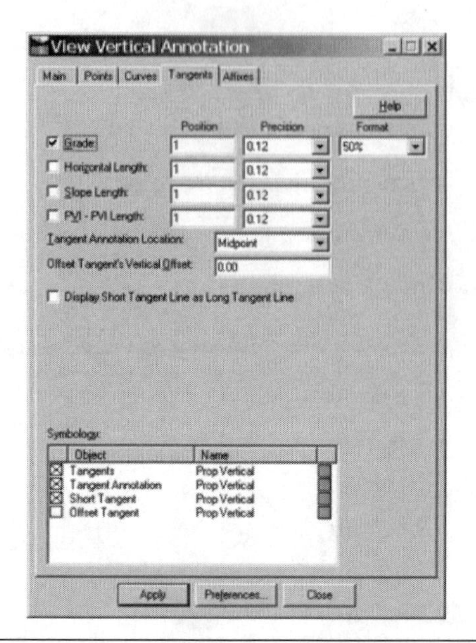

Figure 9-77 Tangents Tab of the View Vertical Annotation

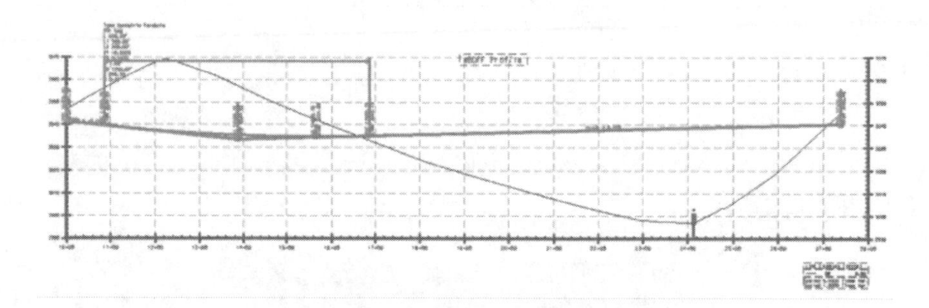

Figure 9-78 Vertical Alignment for Westbound Off Ramp

9.3.6 Vertical Alignment for Westbound On Ramp

The vertical alignment design for the Westbound On Ramp consists of three components: freeway-side grade, local-road-side grade, and a series of grades and curves that connect to Route 8 and Laguna Canyon Road.

1. Location and design of freeway-side grade

The ramp's freeway-side grade segment should have the same elevation and grade as those on Route 8 in the transition area (or the gore area) where drivers change direction from the ramp to Route 8.

Go to the ***MicroStation* → *Place Line*** command to create a line segment perpendicular to the Route 8 Station Line and connect to the intersection (Point A') between the Route 8 right side ETW line and the ramp station line (see Figure 9-79).

The intersection A' is 55' (or 31' + 12' + 12') away from the Route 8 station line at Station 42+29.50.

Use the ***MicroStation* → *Measure Distance*** command to measure the elevation of Route 8 at Station 42+29.50 in the Route 8 Profile.

The elevation is measured to be 3,070.43 ft. (see Figure 9-80). The elevation at the freeway side (or Point A') can be calculated (see Figure 9-81):

$= 3{,}070.43' + 2\% \times 31 - 2\% \times 24$

$= 3{,}070.57'$

Station 42+29.50 is located in the sag curve. Its grade is to be approximately −1% from the view of driving direction. To ensure the grade consistency, the grade for the ramp entrance segment is also to be −1%.

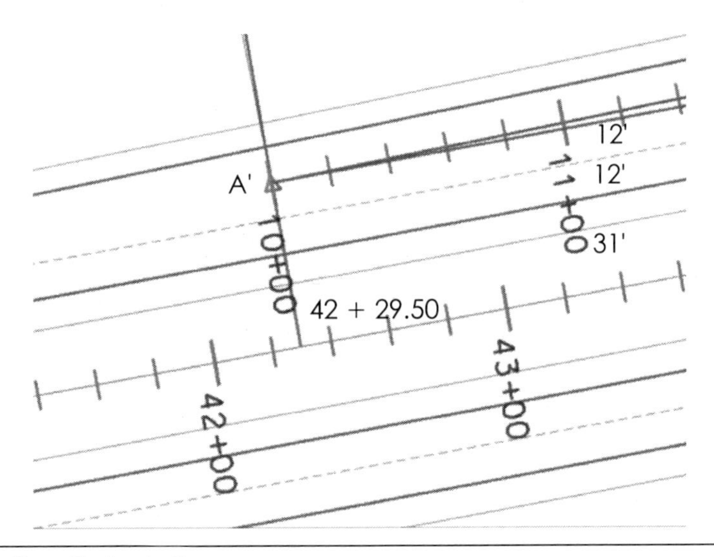

Figure 9-79 Elevation Calculation

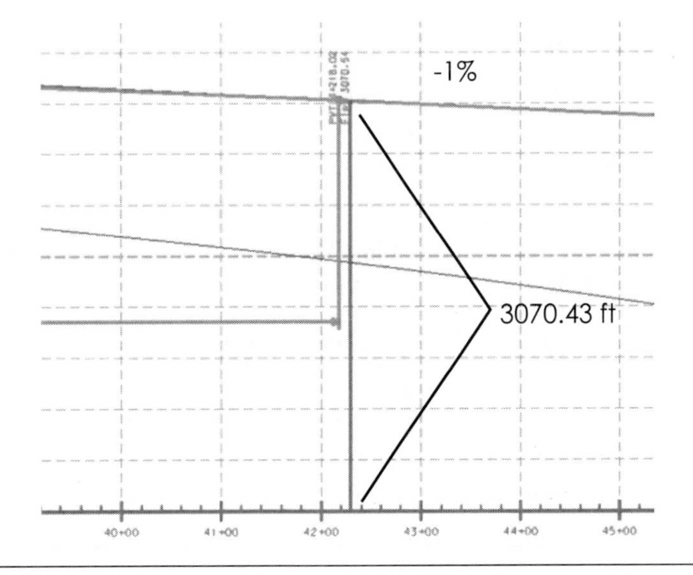

Figure 9-80 Elevation of Route 8 at Station 42+29.50

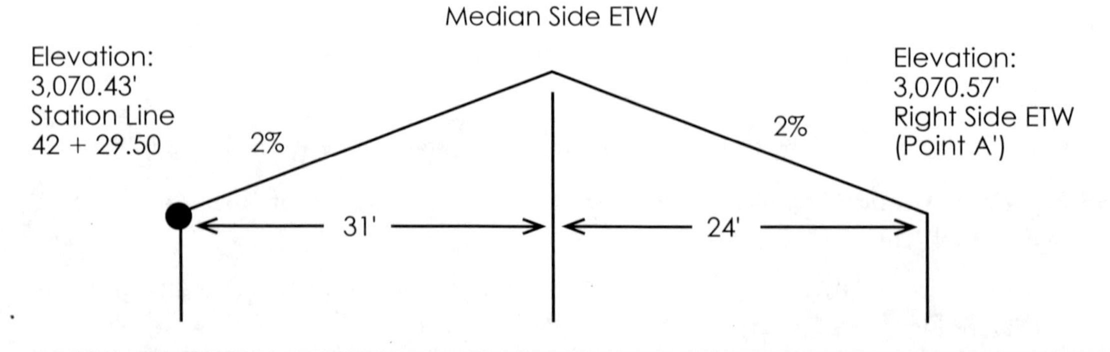

Figure 9-81 Elevation of the Westbound On Ramp at Freeway Side

a. Go to the ***MicroStation → Settings → Levels → Manager.***

   The **Level Manager** window appears.

b. Go to ***Levels → New*** in the **Level Manager** window.

c. Name the new level **WBON_VA.** Select the **Color, Line Style,** and **Line Weight** of your choice.

d. Set the ***WBON_VA*** level as the active level.

e. Close the **Level Manager** window.

f. Go to the ***MicroStation → Window Area*** command and zoom in to the Westbound On Ramp Profile.

g. Go to the ***MicroStation → Move/Copy Parallel*** command to get a horizontal offset line at elevation 3,070.57 ft.

   Note the offset from 3,070 ft gridline is 2.85 ft due to the vertical exaggeration of 5, 2.85 ft/5 = 0.57 ft.

h. Select the ***MicroStation → Place Line*** command to draw the –1% grade segment on the Westbound On Ramp Profile (see Figure 9-82).

   The freeway-side grade segment design for the Westbound On Ramp is shown in Figure 9-82.

2. Location and design of local-road-side grade

The local-road-side grade segment connects to Laguna Canyon Road. The ramp elevation should be the same as that at the intersection (Point A') between Laguna Canyon Road's ETW line and the ramp station line (see Figure 9-83). The intersection (Point A') is 24 ft away from the centerline or station line of Laguna Canyon Road at 28+06.19.

Use the ***MicroStation → Measure Distance*** command to measure the elevation of Point A' at Station 28+06.19. The elevation is measured to be 3,041.41 ft.

The local-road-side elevation of the ramp is then calculated (see Figure 9-84):

= 3,041.41 ft – 24 × 2%

= 3,040.93 ft

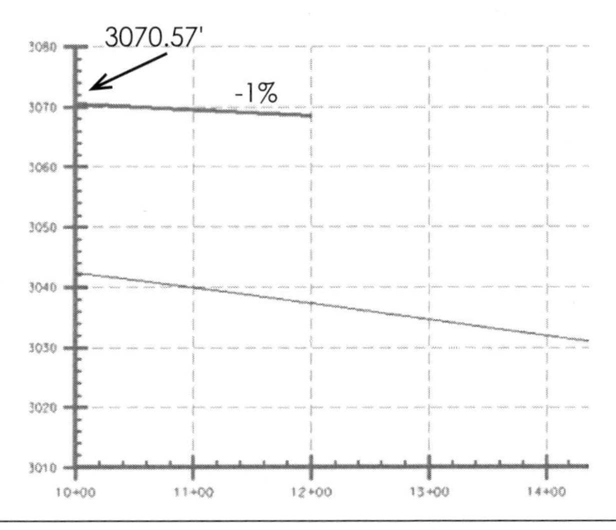

Figure 9-82 Freeway-Side Grade Design of Westbound On Ramp

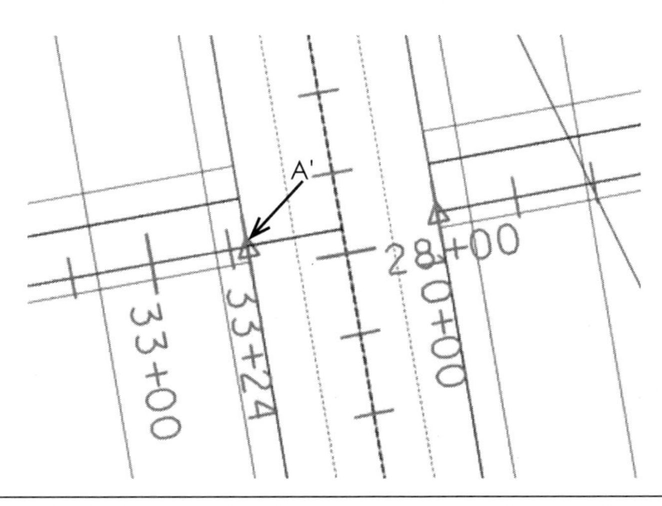

Figure 9-83 Location of Westbound On Ramp on Laguna Canyon Road

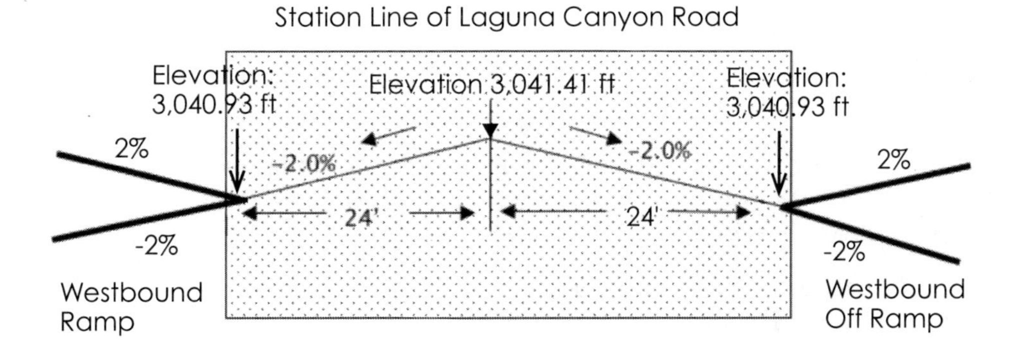

Figure 9-84 Local-Road-Side Elevation of the Westbound On Ramp

The local-road-side grade of the Westbound On Ramp is 2% (from West to East) for this tutorial project.

a. Go to the ***MicroStation → Window Area*** command and zoom in to the Westbound On Ramp Profile.

b. Go to the ***MicroStation → Move/Copy Parallel*** command to get a horizontal offset line at elevation 3,040.93 ft.

   Note the offset from 3,040 ft. gridline is 4.65 ft due to the vertical exaggeration of 5, 4.65 ft/5 = 0.93 ft.

c. Select the ***MicroStation → Place Line*** command to draw the 2% grade segment on the Westbound On Ramp Profile (see Figure 9-85).

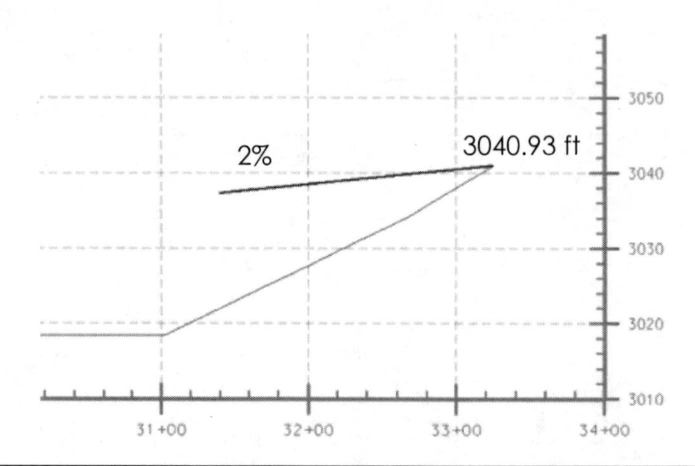

Figure 9-85 Local-Road-Side Grade Design of Westbound On Ramp

3. Location and design of grade line in the middle

The grade line design needs to consider the location and elevation of the culvert that passes through the ramp.

Go to the ***MicroStation → Measure Distance*** command to measure the location of the culvert or the intersection point as shown in Figure 9-86. The culvert is located at Station 28+59.33, where the breakline 1 intersects the station line of the Westbound On Ramp.

Go to the ***MicroStation → Measure Distance*** command to measure the elevation of the ground that supports the culvert at Station 28+59.33. The measurement should be done on the Westbound On Profile. The elevation is measured to be 3,012.86 ft.

The diameter of the culvert is given to be 4 ft. The elevation of the culvert top side therefore is 3,016.86 ft. We assume that a 2-ft buffer is needed between the culvert and the ramp's pavement surface. The minimum elevation that the ramp grade line can be lowered is 3,018.86 ft.

From Figure 9-86, you see that the culvert is significantly below the grades at the two ends (or local-road-side and freeway-side grade segments). Therefore the culvert is not a controlling feature affecting the design of the ramp.

Now we need to create the grade line that connects to the two end grade segments. Considering the maximum grade of 8% that can be used for ramps and after some trial and error, we use 3%.

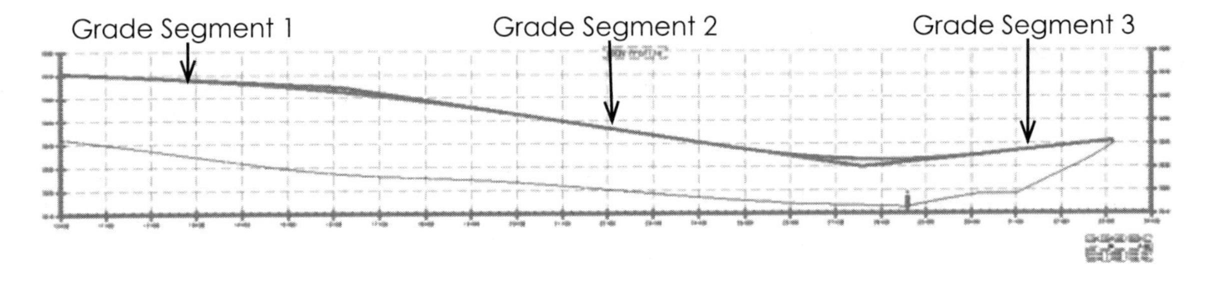

Figure 9-86 Vertical Alignment Design of Westbound Off Ramp

a. Go to the ***MicroStation* → *Place Line*** command to create a segment with a 3% grade.

b. Go to the ***MicroStation* → *Move*** command to move the grade segment to intersect the two end grade segments (see Figure 9-86).

The stations and elevations of the two intersecting points are measured as:

Station/Elevation:	16+14.00/3,064.43 ft.
Station/Elevation:	27+68.14/3,029.81 ft.

c. Go to ***MicroStation* → *Tools* → *Tool Boxes* ….**

d. Check on the **Fillets** option in the **Tool Boxes** window.

The **Fillets** toolbox appears.

e. Select the **Construct Parabolic Fillet** tool in the **Fillets** toolbox.

The **Construct Parabolic Fillet** window appears.

f. Type the following in the **Construct Parabolic Fillet** window:

Distance:	***700.0000***
Type:	***Horizontal***
Truncate:	***None***

g. Select the grade segment 1 and the grade segment 2.

The parabolic curve of 700 ft. is filleted into the grade segments.

h. Again select the **Construct Parabolic Fillet** tool in the **Fillets** toolbox.

i. Type the following in the **Construct Parabolic Fillet** window:

Distance:	***500.0000***
Type:	***Horizontal***
Truncate:	***None***

a. Select the grade segment 2 and the grade segment 3.

The parabolic curve of 500 ft. is filleted into the grade segments.

4. Caltrans Standards Conformance Check

Table 9-6 shows the graphical elements that form the vertical alignment of the Westbound On Ramp.

Table 9-6 Conformance Check

Element Type	Length (L)	Conformance Check
Grade 1 Segment	264.00' −1.0%	The grade of −1.0% matches the grade on Route 8. OK.
Parabolic Curve 1	700.00'	The algebraic grade difference is | −3.0% − (−1.0%) | = 2.0%. Following the equation $L_{min} = 10V$, the minimum length of the vertical curve is 10 × 50 = 500 ft. The minimum length of the vertical curve is about 600 ft (see Figure 9-5 or Figure 201.4 of Caltrans *HDM*). 600 ft > 500 ft. Select 600 ft. 700 ft > 600 ft. OK.
Grade Segment 2	554.14' -3.00%	−3% > − 8%. OK.
Parabolic Curve 2	500.00'	The algebraic grade difference is | 2.0% − (−3.0%) | = 5.0%. The curve is in the middle of the ramp. The design speed is assumed to be 40 mph. Following the equation $L_{min} = 10V$, the minimum length of the vertical curve is 10 × 40 = 400 ft. The minimum length of the vertical curve is about 300 ft (see Figure 9-6 or Figure 201.5 of Caltrans *HDM*). 400 ft > 300 ft. Select 400 ft. 500 ft > 400 ft. OK.
Grade Segment 3	306.20' 2.0%	2% matches the cross slope of the ramp intersection. OK.

5. Import the graphical elements into InRoads to create the vertical alignment of the Westbound On Ramp

   a. Make sure the **WBON_VA** level is active in MicroStation. If **WBON_VA** is unavailable, create the level using MicroStation commands.

   b. Make sure the **tutorial** geometry project is active. If it is not active, set it to be active.

   c. Make sure the horizontal alignment **WBON** is active.

   d. In MicroStation, create a complex chain using the **Create Complex Chain** command.

   e. Select the **three tangent grades** for the Westbound On Ramp and create a chain (see Figure 9-86).

   f. Go to ***InRoads → Import → Geometry ...***.

   The **Import Geometry** window appears.

   g. Select the **From Graphics** tab and type the following:

Type:	**Vertical Alignment**
Name:	**WBON**
Description:	**WBON Vertical Alignment**

Style:	**Default**
Vertical	
Curve Definition:	**Parabolic**
Other fields:	**Default values**

h. Select ***Apply.***

InRoads prompts you to select the graphical elements.

i. Select the chain you just created.

j. Select ***Accept*** and ***Reset.***

k. Select ***Close.***

l. Check in the **Workspace Bar** window to make sure the vertical alignment **WBON** is created.

You will see a **Workspace Bar** window similar to Figure 9-87.

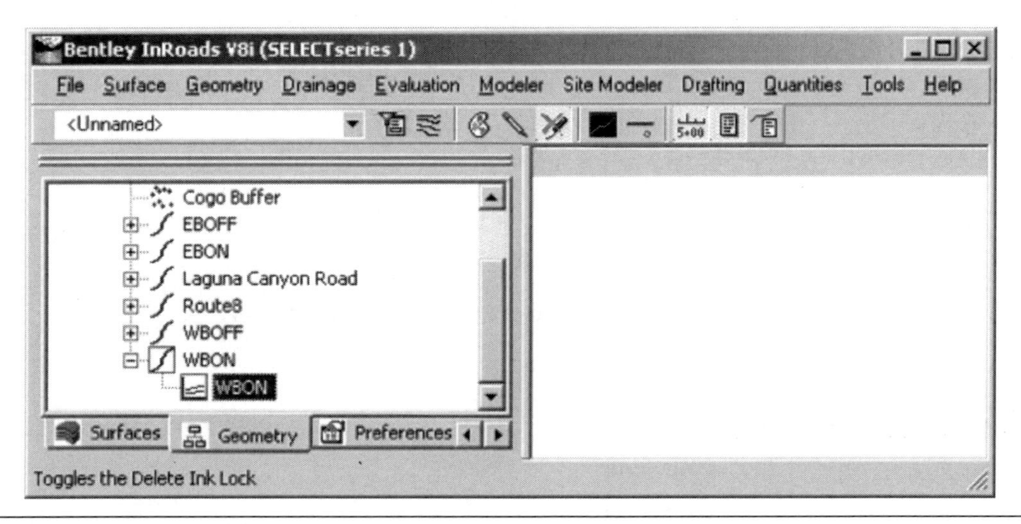

Figure 9-87 Vertical Alignment of Westbound On Ramp

m. Go to ***InRoads → Geometry → View Geometry → Active Vertical.***

n. Go to ***InRoads → Geometry → Vertical Curve Set → Define Curve ….***

The **Define Curve** window appears (see Figure 9-88).

o. In the **Length** field, type ***700.00.***

p. Select ***Apply.***

q. Select ***Next.***

r. In the **Length** field, type ***500.00.***

s. Select ***Apply*** and ***Close.***

6. Save the **tutorial** geometry project

a. Make sure the **tutorial** geometry project is set to be active.

b. Go to ***InRoads → File → Save → Geometry Project.***

c. Save the **tutorial** geometry project as ***tutorial.alg*** in the **C:/Temp/Tutorial/Chp9/Master Files** folder.

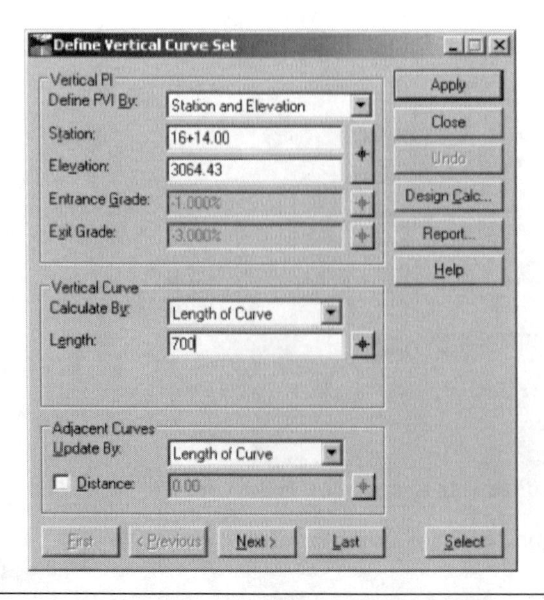

Figure 9-88 Define Curve Window

7. The **WBON** vertical alignment is created and the **tutorial** geometry project is saved. Now we can view the **WBON** vertical alignment using InRoads tools.
   a. Make sure the **tutorial** geometry project is set to be active.
   b. Make sure the **WBON** vertical alignment is set to be active.
   c. Go to ***InRoads → Geometry → View Geometry → Vertical Annotation …***.

      The **Vertical Annotation** window appears.
   d. Select the **Points** tab and make sure the following appears in the Symbology listbox:

      PVC Text: **Check on**
      PVI Text: **Check on**
      PVT Text: **Check on**
      High Point Text: **Check on**
      Low Point Text: **Check on**
      PVC Point: **Check on**
      PVI Point: **Check on**
      PVT Point: **Check on**
      High Point: **Check on**
      Low Point: **Check on**
      All other fields: **Check off**
   e. Select the **Curves** tab and make sure all the fields are same as shown in Figure 9-89.
   f. Select the **Tangents** tab and make sure all the fields are same as shown in Figure 9-90.
   g. In the **View Vertical Annotation** window, select ***Apply*** and ***Close***.

      Vertical annotations are displayed along with the vertical alignment of Westbound On Ramp (see Figure 9-91).
   h. Save the working **tutorial_2D_Profile.dgn** file as **C:/Temp/Tutorial/Chp8/Master Files/tutorial_2D_VA.dgn.**

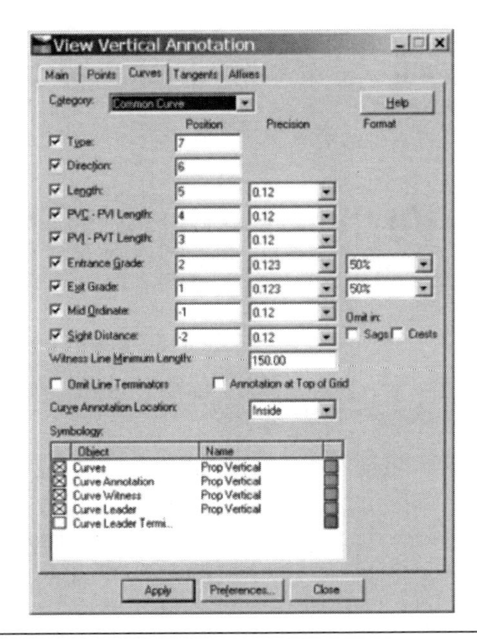

Figure 9-89 Curve Tab in the View Vertical Annotation

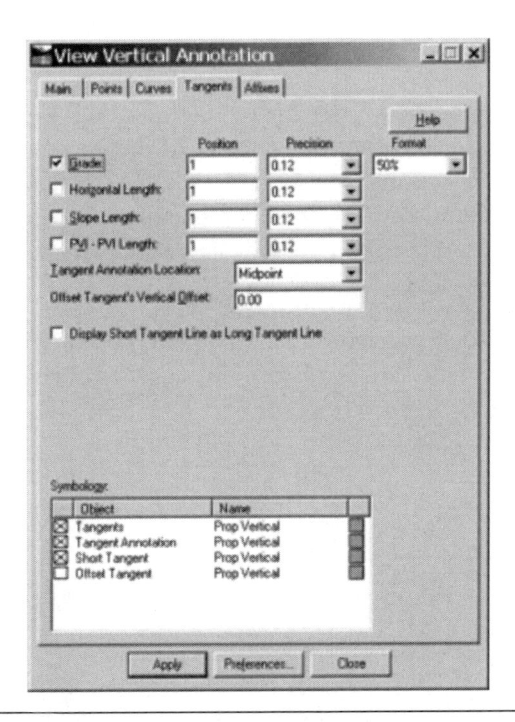

Figure 9-90 Tangents Tab of the View Vertical Annotation

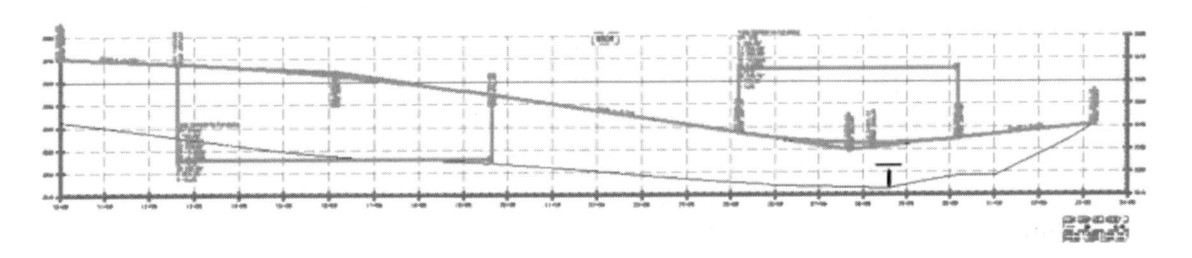

Figure 9-91 Vertical Alignment for Westbound On Ramp

The difference between the two files is that the **tutorial_2D_VA.dgn** has vertical alignments, while the **tutorial_2D_Profile.dgn** does not. The **tutorial_2D_VA.dgn** will be used in later chapters.

9.4 Summary of Vertical Alignment Design

This chapter describes the process of designing six vertical alignments for the tutorial project. The design process applies Caltrans standards (design speed, minimum and maximum grade, sight distance, minimum length of curve, falsework requirement, vertical clearance, etc.) to the tutorial project.

Through this design process, you should now understand that vertical alignments are the result of your "artful" work that balances all the constraints such as standards, project control agreements (A, B, and C points), and driver behaviors (variance of design speed along ramps). There are many alternatives when you design a vertical alignment. The vertical alignments we have created are one of the alternatives that meet the Caltrans standards.

The design process emphasizes the importance of safety (or Caltrans standards) in creating vertical alignments. Given all the factors that govern the shape of vertical alignments, the safety factors (such as stopping sight distance, falsework, and vertical clearance) must be considered first.

This chapter takes advantage of your solid knowledge of MicroStation. We first created graphical elements for a vertical alignment. We then checked the conformance of these elements to Caltrans standards. After all the graphical elements were checked for meeting Caltrans standards, we imported the grade segments into InRoads to form a vertical alignment. We then completed the vertical alignment by adding vertical curves. InRoads has a limitation of importing parabolic curves (created from the MicroStation **Parabolic Fillet** command) into vertical alignment.

InRoads uses a geometry project to store vertical alignments. It communicates with MicroStation to display vertical alignments and their associated attributes (such as PVCs, PVIs, and PVTs).

Name________________________________ Date______________

9.5 Questions

Q9.1 Design speed is a critical design consideration in vertical alignment. List Caltrans standards related to design speeds.

Q9.2 Is a profile a vertical alignment? To produce a vertical alignment, what are the steps used in this chapter?

Q9.3 Do Caltrans standards require grade lines to be the same for both divided and undivided highways?

Q9.4 What is the maximum grade for a rural highway segment in an area of rolling terrain?

Q9.5 What is the minimum grade for California highways?

Q9.6 Assume the grade is 3.5% for a segment on a freeway in the rolling area. The length of the segment is 1,000 ft.

Does this grade meet Caltrans standards?

Do we need a climbing lane?

Q9.7 Why do we use parabolic curves instead of circular curves for vertical alignments?

Q9.8 Assume the design speed of a freeway curve is 60 mph. G_1 = 3.0%, G_2 = -2.0%.

What is the algebraic grade difference or the total grade change?

What is the minimum required length of the vertical curve?

Q9.9 According to Caltrans standards, do we need to have a vertical curve if the algebraic grade difference is less than or equal to 0.5%?

Q9.10 An overcrossing bridge is designed for a new freeway. The freeway runs under the bridge. What is the required vertical clearance from the pavement surface of the freeway to the bottom of the bridge deck?

Q9.11 Given the bridge layout below, what are the vertical clearance and the falsework depth required for construction?

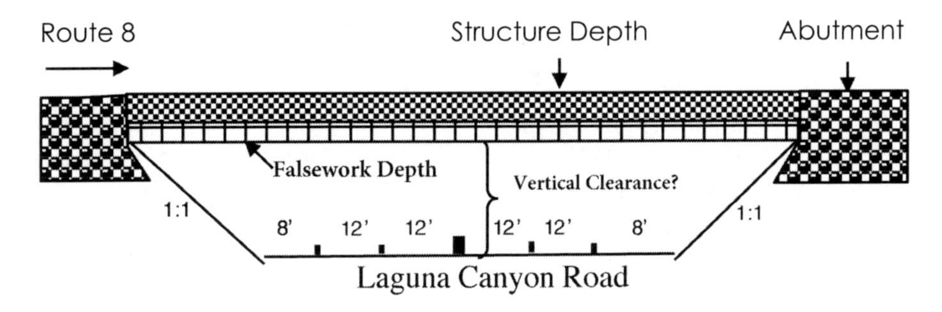

Q9.12 What are the differences between cast-in-place structure and precast structure as they relate to falsework?

Q9.13 How do MicroStation and InRoads work together to create a vertical alignment?

Q9.14 List the steps to create a new level in MicroStation.

Q9.15 Assume the design elevation is 3,095.67 ft. for Route 8. How do you use the **Move/Copy Parallel** command in MicroStation to locate the elevation at Station 10+00 in the Route 8 profile?

Q9.16 Assume the elevation exaggeration is 10 in a profile. Draw a slope of 4% for the profile.

Q9.17 Is the bridge layout in Figure 9-21 for one or for two Route 8 bridges?

Q9.18 List the steps to determine if a breakline is a valley line or a ridge line.

Q9.19 List the control features considered in the design of the Route 8 vertical alignment.

Chapter 10

Cross Section Templates

LEARNING OBJECTIVES

After completing this chapter you should know:

1. Caltrans standards of roadway cross sections related to the tutorial project;
2. How to create a cross section template library;
3. How to create template components including simple, constrained, and end conditions;
4. How to assemble template components to create new templates.

SECTIONS

As mentioned in the previous chapters, horizontal alignments, vertical alignments, and cross sections are the three main components for the geometric design of a roadway. Different from the former two components, which provide the plan and profile view of a roadway, the cross sections illustrate the transverse geometry of a roadway.

Generally, the transverse geometry includes number of lanes, lane width, shoulder width, median width and slope, side slope, and so on (see Figure 10-1).

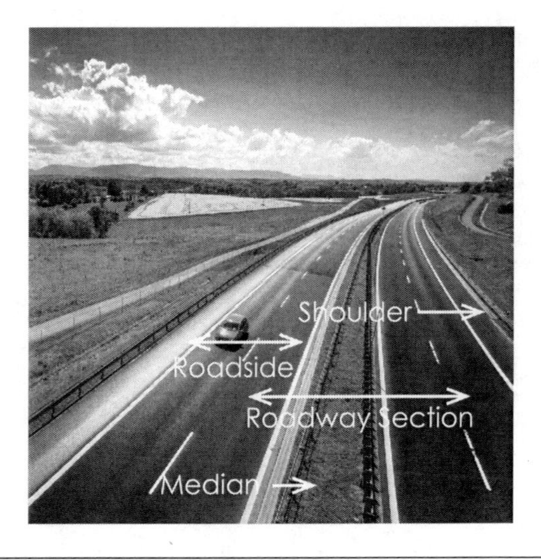

Figure 10-1 Transverse Geometry of a Roadway

A cross section is made of two parts: roadway section and roadside section. The *roadway section* is a paved section that supports vehicular travel. It is divided into lanes. The dimension of the roadway section is controlled by the number of lanes and the lane width. The *roadside section* consists of shoulder and side slope in the median area or outside the roadway section (to the left side of the driving direction). The *geometric design* of the roadside section involves the determination of side slope and shoulder.

10.1 Cross Section Elements of the Tutorial Project

It is described in Chapter 4 that there are two lanes in each direction for both Route 8 and Laguna Canyon Road, whereas the number of lanes for each ramp is only one. The widths of median, inside, and outside shoulders vary significantly between Route 8 and the ramps, ranging from 4 ft. to 62 ft. (Note that the median is not designed for Laguna Canyon Road.)

Figure 10-2 shows the cross section elements for Route 8. There are two roadway sections. Each of them is bounded by the edge of traveled way (ETW). The median is depressed with the slope down to the centerline. The roadside is the range between the outside ETW and the right-of-way (R/W) line. The side slope is intersected with the existing ground. The intersection point is often referenced as to the catch point in geometric design.

Figure 10-3 shows the cross section elements for Laguna Canyon Road. The cross section centerline is located at the top of the crown section. Figure 10-4 shows the cross section elements of ramps. Note that the cross section has only one slope between the two ETWs. The left ETW is on the left side of the traveling direction.

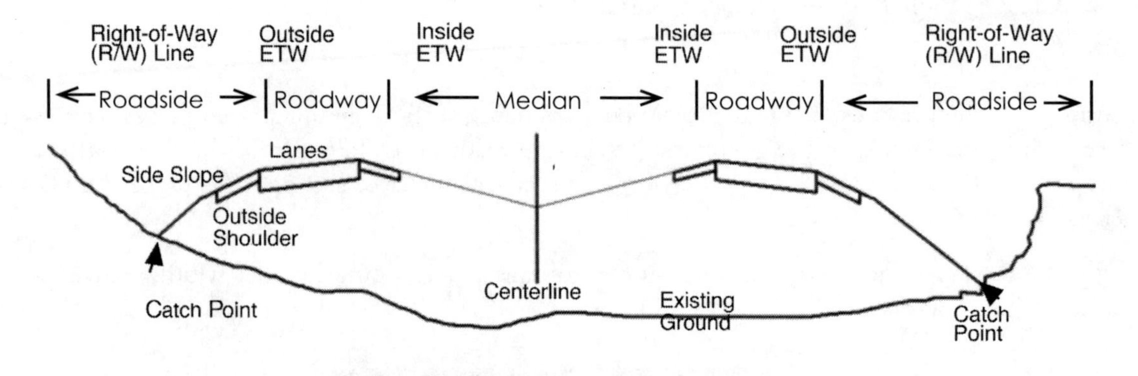

Figure 10-2 Cross Section Elements of Route 8

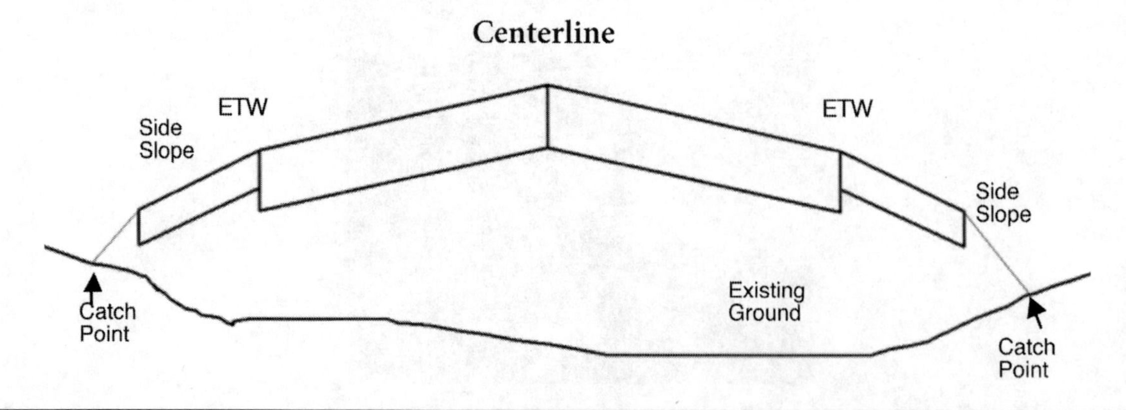

Figure 10-3 Cross Section Elements of Laguna Canyon Road

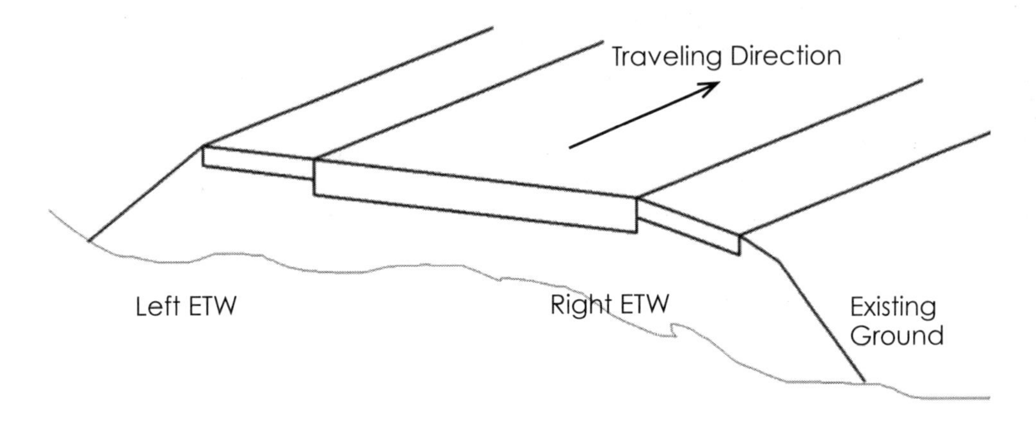

Figure 10-4 Cross Section Elements of Ramps

10.2 Design Standards for Cross Sections

Similar to the horizontal and vertical alignments, *Caltrans' Highway Design Manual* (*HDM*) governs the design of cross section elements for the tutorial project. The detailed description of the standards for traveled way width, shoulder width, median width, cross slope, and dike type is presented in sequence.

10.2.1 Traveled way width

Traveled way width depends on several factors including number of lanes, available land, predicted traffic volume, median type, etc. The traveled way width includes only the width of the traveled lanes, excluding shoulders, curbs, dikes, gutters, or gutter pans. According to Caltrans *HDM* Index 301.1, the basic lane width for new construction on two-lane and multilane highways, ramps, collector roads, and other appurtenant roadways shall be 12 ft. For roads with curve radii of 300 ft. or less, widening due to offtracking should be considered.

For the tutorial project, since all radii of the horizontal curves as designed previously are greater than 300 ft., widening is not required for offtracking. Hence, the traveled way width of Route 8 with four lanes is 24 ft. for each direction. The widths for Laguna Canyon Road and ramps are 24 ft. (one direction) and 12 ft., respectively.

10.2.2 Shoulder width

For freeway and conventional highway, adequate shoulder width is essential for the safety of parked vehicles, maintenance activities, and emergency responses. Table 302.1 of Caltrans *HDM* provides detailed requirements for minimum paved shoulder width. The requirements related to the tutorial projects are outlined in Table 10-1.

Table 10-1 Minimum Paved Shoulder Width

Facility Type	Min. Paved Shoulder Width (Ft.)	
	Left	Right
Freeway and Expressway		
4 lanes	5	10
Single-lane ramps	4	8
Conventional Highways		
Multilane undivided	N/A	8

In the tutorial project, we follow the recommended minimum widths for most of the shoulders, especially the left and right shoulder widths chosen for Route 8 are 10 ft. The Laguna Canyon Road is designed to be 8 ft. for both left and right shoulders. The ramp has a right shoulder of 8 ft. and a left shoulder of 4 ft.

10.2.3 Median width

Median width is the dimension between the inside edges of traveled way, including the inside shoulder. The minimum median width for freeways and expressways in suburban areas should be 62 ft. according to the Caltrans *HDM*.

Since Route 8 is located within a suburban area, the median width selected for Route 8 is 62 ft. There are no medians for Laguna Canyon Road and the ramps.

10.2.4 Cross slope

A cross slope is required to drain storm water away from roadway surface. The cross slope varies widely across roadway sections including traveled way, shoulder, and median. Caltrans requirements on cross slopes are shown below in sequence.

1. Cross slope of traveled way

Caltrans *HDM* indicates that the standard cross slope to be used for new construction of traveled ways for all surface types shall be 2%.

Therefore, the cross slopes of 2% are selected for all traveled ways in the tutorial project.

2. Shoulder cross slope

Caltrans *HDM* requires that shoulders in the depressed median sections shall be sloped at 2% away from the roadway's left ETW. In paved median sections, shoulders to the left of traffic shall be designed in the plane of the traveled way. In normal tangent sections, shoulders to the right of traffic shall be sloped at 2% to 5% away from the traveled way.

Based on the above criteria, the cross slope of the left shoulder for Route 8 is selected to be 2% away from the traveled way, while the cross slope for the right shoulder is 5%. For Laguna Canyon Road, there is no left shoulder, but the right shoulder employs the same cross slope of 5%. For ramps, the right shoulder has the same cross slope of 5%. However, the left shoulder is designed in the plane of the traveled way, using the same cross slope of traveled way, or 2%.

3. Median cross slope

Caltrans standards require that unsurfaced medians up to 65 ft. wide should be sloped downward from the adjoining shoulders to form a shallow valley in the center. Cross slopes should be 10:1 or flatter; 20:1 being preferred. For four-lane freeways, medians 22 ft. or less in width should be paved.

In the tutorial project, since median width for Route 8 (62 ft.) is greater than 22 ft., the unsurfaced median criteria are selected. Additionally, the relatively flat slope, 2% (i.e., 50:1), is used. There are no medians for Laguna Canyon Road and the ramps so median cross slope criteria are not applicable.

10.2.5 Dike uses

Generally, a roadway cross section includes dike, curb, or side gutters, whose main purposes include, but are not limited to, protecting pedestrians from vehicles, establishing access control to properties, and preventing the erosion of pavement and side slopes from storm water runoff. Caltrans *HDM* Topic 303 provides detailed guidelines on the selection of dike facilities and indicates the Type E dike is the preferred type under most conditions.

Based on Caltrans guidelines and recommendations, the project selects Dike E as the facility to control storm water runoff. Note that other types of facilities might be also appropriate.

Per Caltrans Standard Plan A87B, Dike (Type E) dimensions are illustrated in Figure 10-5. Note 1 indicates that the dimensions apply to HMA (hot mix asphalt) shoulders only.

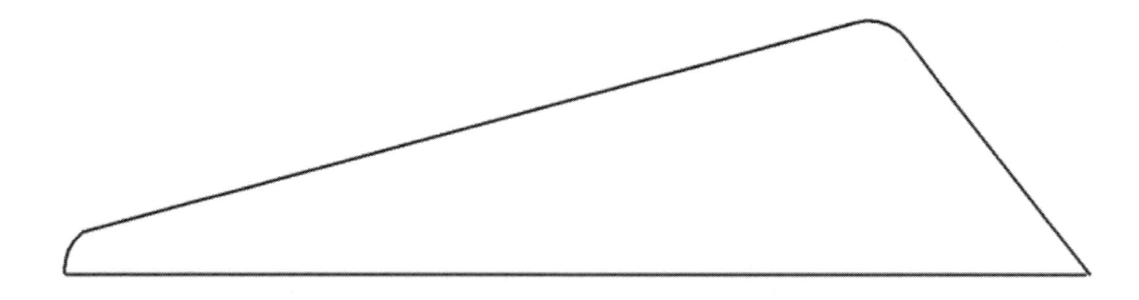

Figure 10-5 Dike Type E Layout

10.2.6 Side slopes

A variety of factors affect side slope design, including safety, economics, aesthetics, etc. Usually embankment slopes can be divided into three types: recoverable slopes (4:1 or flatter), traversable but not recoverable slopes (between 3:1 and 4:1), and nonrecoverable and nontraversable slopes (steeper than 3:1). Caltrans *HDM* Topic 304 presents detailed side slope standards.

Considering the topography varies widely along a roadway, six alternative side slopes are provided in the tutorial project to cover various road conditions: 4:1 (H:V) cut, (4:1) fill, 3:1 cut, 3:1 fill, 2:1 cut, and 2:1 fill. Based on Caltrans standards, the lower the side slope, the better the design. Therefore, 4:1 cut/fill is assigned higher priority values compared with other side slopes.

10.2.7 Geometric cross section

Based on the aforementioned discussions, the final geometric cross section designs for Route 8, Laguna Canyon Road, and the ramps are illustrated in Figures 10-6a, 10-6b, and 10-6c.

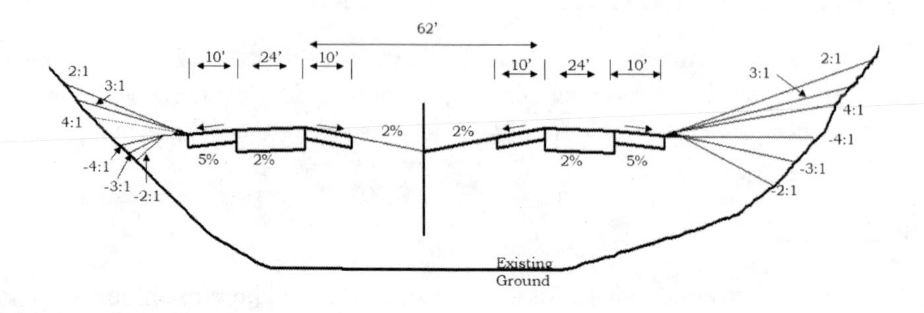

Figure 10-6a Cross Section Design of Route 8

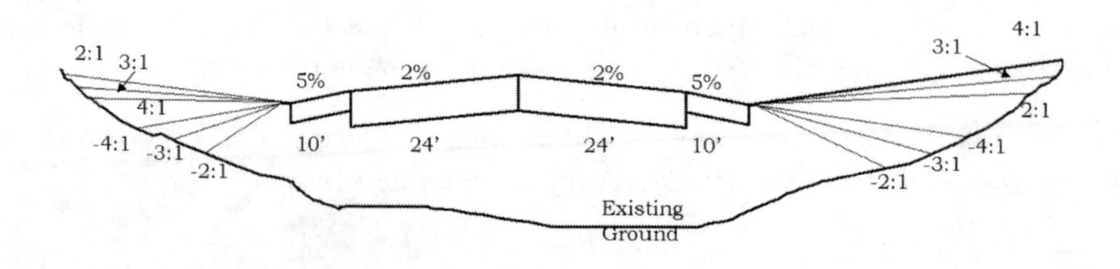

Figure 10-6b Cross Section Design of Laguna Canyon Road

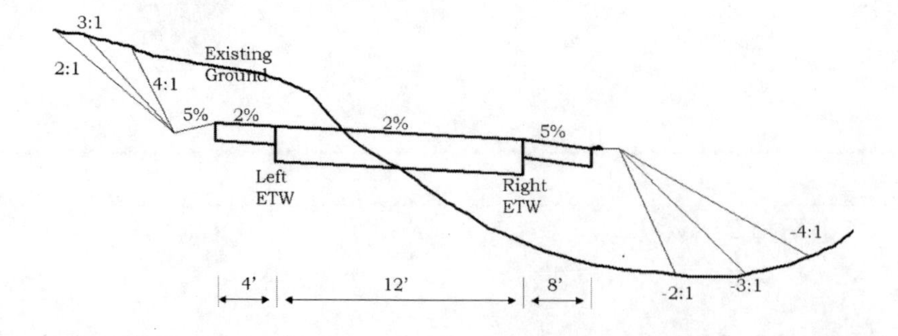

Figure 10-6c Cross Section Design of Ramps

10.3 Cross Section Template Design Using InRoads

Cross section element design is done through InRoads templates, which are stored in template library files (.itl). Even though only one template library file may be open in InRoads at a given time, the library file may contain multiple templates related to a roadway project. For example, three templates should be designed and stored in a library for a roadway that has three different segments: a two-lane segment, a three-lane segment, and a transition segment (see Figure 10-7).

InRoads templates consist of template points and template components. The former are used to define the longitudinal breakline features in a roadway surface model, while the latter are a series of connected points (in either an open or closed form) that represent cross section elements such as dike, median barrier, concrete pavement, and so on.

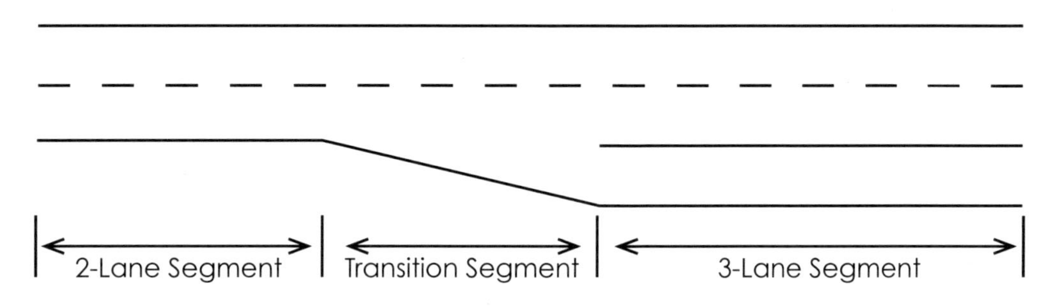

Figure 10-7 Three Different Segments in a Roadway

InRoads components fall into one of five types: simple, constrained, unconstrained, null point, and end condition. In this chapter, only the first three cross section components (applicable to the tutorial project) are described at length.

Cross section template design in InRoads starts with the creation of a template library file, which stores all the templates relevant to a roadway project. A template can be established by developing various components or by assembling other templates. The following subsections cover the detailed procedures of creating the template library and templates for Route 8, Laguna Canyon Road, and the four ramps.

10.3.1 Create the template library file for the tutorial project

Before we create the cross section templates for the tutorial project, you need to have MicroStation and InRoads installed on your computer. You also need to copy the **tutorial_2D_VA.dgn** file from the **C:/Temp/Book/Chp10/Master Files** folder to the **C:/Temp/Tutorial/Chp10/Master Files/** folder or you can download the files from the book website (www.csupomona.edu/~xjia/GeometricDesign/Chp10/Master Files/).

You need to create a template library in which to store the templates. Follow these steps to create a new template library file:

a. Start MicroStation and InRoads (if they are not open).

b. Load the **tutorial_2D_VA.dgn** file from the **C:/Temp/Tutorial/Chp10/Master Files/** folder.

c. Select the ***InRoads → Template*** tab from the workspace bar.

d. In the workspace bar window, right click on the **Template Library entry** (see Figure 10-8).

e. Select the **New...** option.

In the **Save As** dialog box that appears, browse to your project location (or the **C:/Temp/Tutorial/Chp10/Master Files/** folder) and save the file with the name **Tutorial.itl**. Now you have a typical template library created for the tutorial project (see Figure 10-8). The library is empty at this time.

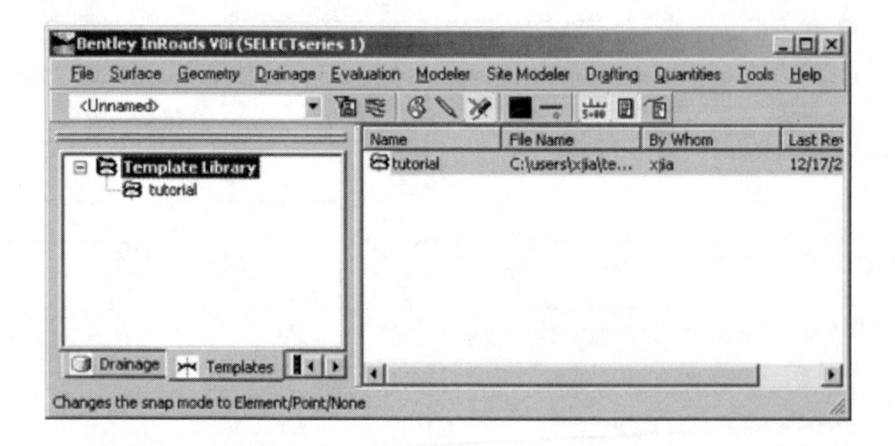

Figure 10-8 Create Template Library

10.3.2 Create the templates for Route 8

Once the template library file is created, you can create the Route 8 templates and store them in the template library.

1. Create a template for Route 8 backbone

   a. Go to ***InRoads → Modeler → Create Template.***

   The **Create Template** dialog box appears as shown in Figure 10-9.

Figure 10-9 Create Template Window

b. Right click over the file name as shown in the **Template Library** list window (see the top left corner of Figure 10-9). Go to ***New* → *Template*** and name the template **Route 8.**

   Now you have an empty template for Route 8 (Figure 10-10).

c. From the Create Template window, go to ***Tools* → *Options....***

   The **Template Options** window opens (Figure 10-11).

d. Set the **Pointes Seed Name** to **Pavement**.

e. Click on **Apply Affixes.** In the **Prefix** field, type **L-** and **R-.**

f. Click **OK**.

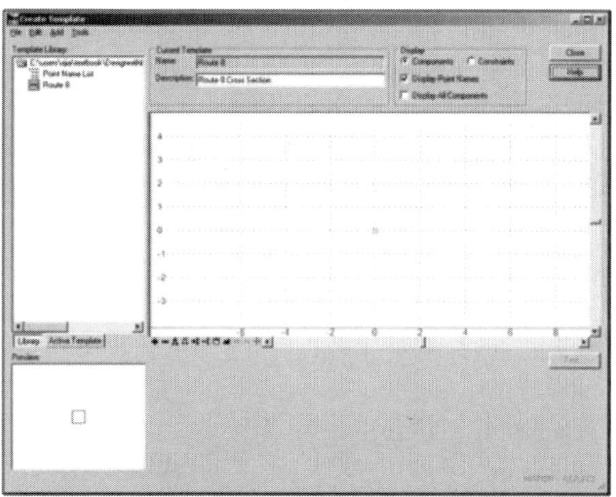

Figure 10-10 Route 8 Template Created

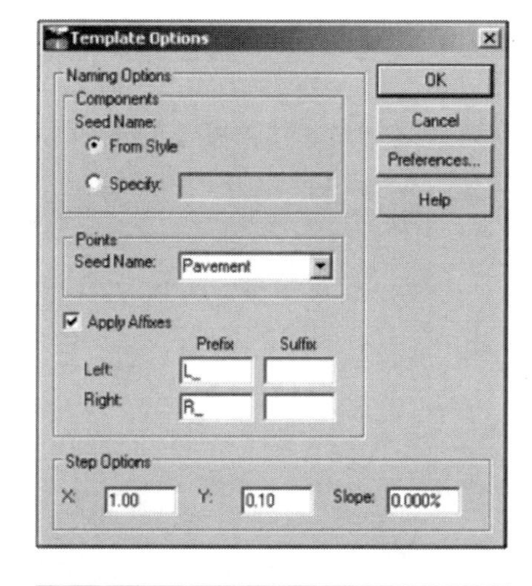

Figure 10-11 Template Option Window

g. Go to ***Create Template* → *Add* → *Simple*** or right click in the **Template Editor** window. Go to ***Add New Component* → *Simple.***

h. At the bottom of the **Create Template** dialog box, type the information into the **Current Component** section as shown in Figure 10-12.

   Note: The **Default** style provided by InRoads is selected as you do not have the preferred style. Assume the median slope thickness to be 0 ft. You should select the specific style and the thickness as required by your project needs.

   Current Component Name: ***Median Slope***
   Slope: ***2%***
   Thickness: ***0.00***
   Width: ***21***

   You should see a **Median Slope** component attached to your cursor when you navigate the Editor window.

i. Right click in the window and go to ***Context Menu* → *Mirror*** (see Figure 10-13).

   Both left and right side components are now attached to your cursor.

j. Zoom to the center of the Editor window. Place your cursor at the **0,0 origin** (magenta square) and click to place the components.

k. Use the ***Create Template → Fit*** tool at the bottom left corner of the Editor window.

   The components should look like what is shown in Figure 10-14.

l. Right click in the **Template Editor** window. Go to ***Add New Component → Simple.***

m. At the bottom of the **Create Template** dialog box, type the information shown in Figure 10-15 into the **Current Component** section.

   Note that the thickness of the paved shoulder is 0.40 ft.

n. Place the cursor over the Point **R_Pavement** and click to place the **ShoulderinMedian components.** Use the ***Create Template → Fit*** tool at the bottom left corner of the Editor window to exhibit the whole view.

   The components should look like what is shown in Figure 10-16.

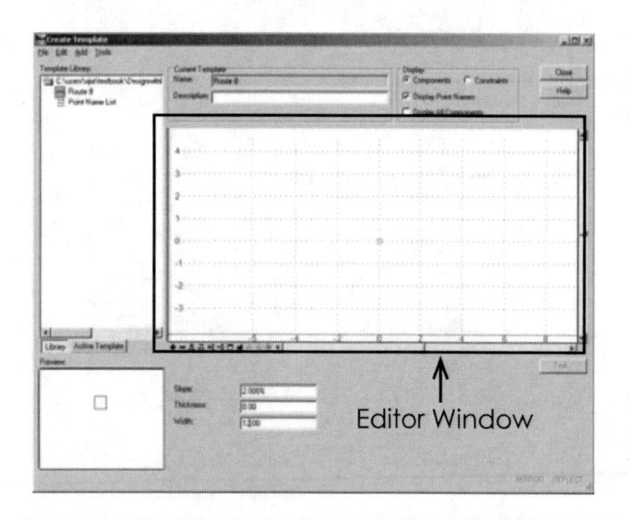

Figure 10-12 Create Median Slope

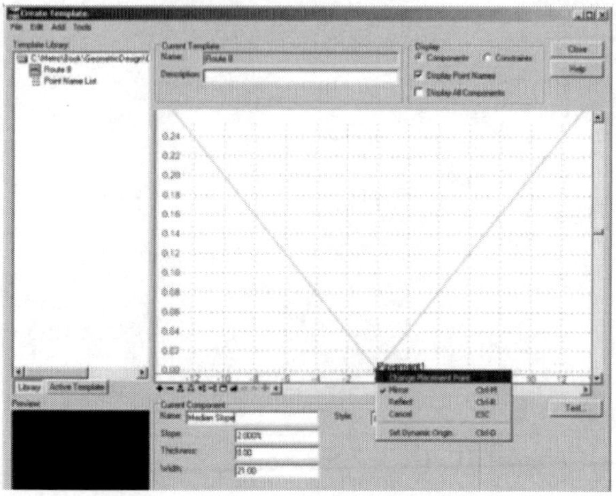

Figure 10-13 Context Menu

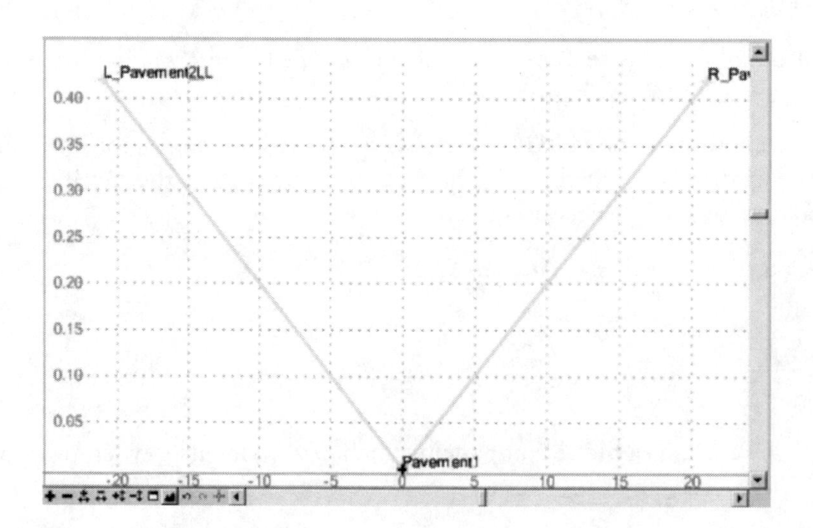

Figure 10-14 Median and Inside Shoulder Component

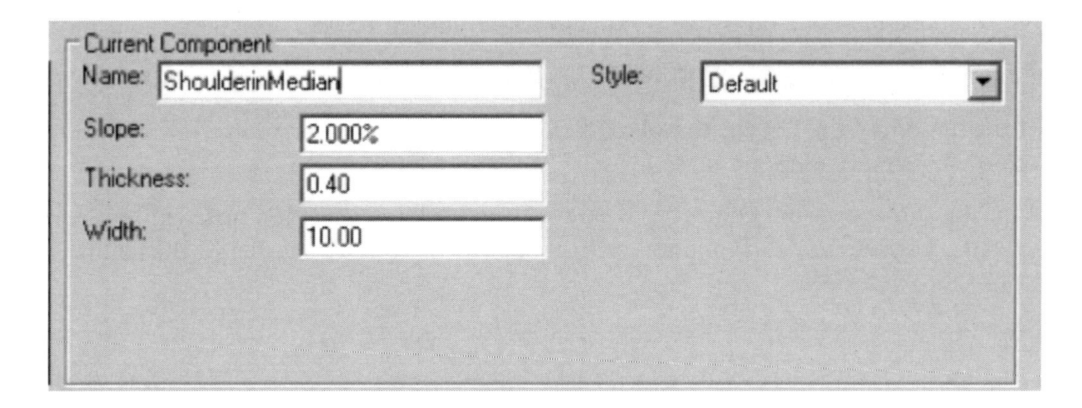

Figure 10-15 Information for Shoulder in Median Components

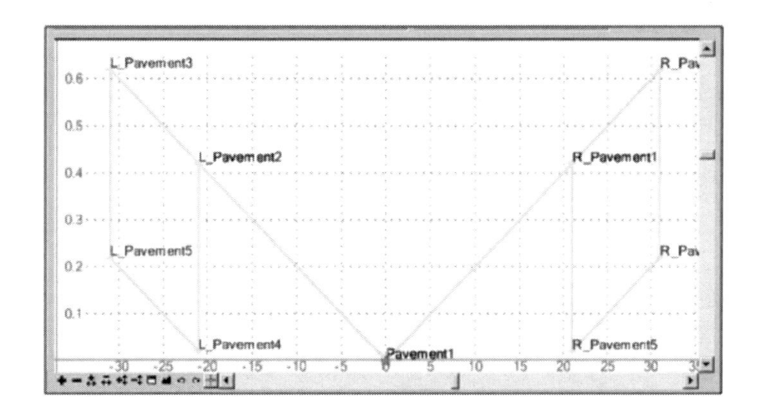

Figure 10-16 Shoulder in Median Components

Current Component
Name: Lanes
Style: Default
Slope: -2.000%
Thickness: 1.00
Width: 24.00

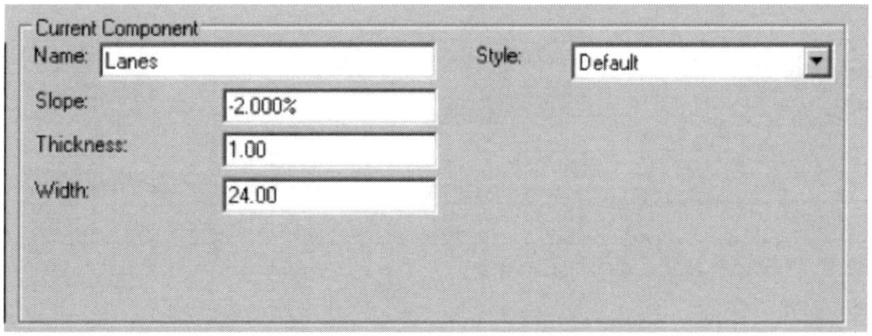

Figure 10-17 Lane Components

o. Right click in the **Template Editor** window. Go to ***Add New Component → Simple.***

p. At the bottom of the **Create Template** dialog box, type the information shown in Figure 10-17 into the **Current Component** section.

   Note that the pavement thickness of the lanes is 1.0 ft.

q. Place the cursor over the Point **R_Pavement3** and click to place the **Lanes components**. Use the ***Create Template → Fit*** tool at the bottom left corner of the Editor window to exhibit the whole view.

The components should look like what is shown in Figure 10-18.

r. Right click in the **Template Editor** window. Go to ***Add New Component → Simple.***

s. At the bottom of the **Create Template** dialog box, type the information shown in Figure 10-19 into the **Current Component** section.

t. Place the cursor over the Point **R_pavement6** and click to place the **Outside Shoulders components.** Use the **Create Template → Fit** tool at the bottom left corner of the editor window to exhibit the whole view.

The components should look like what is shown in Figure 10-20.

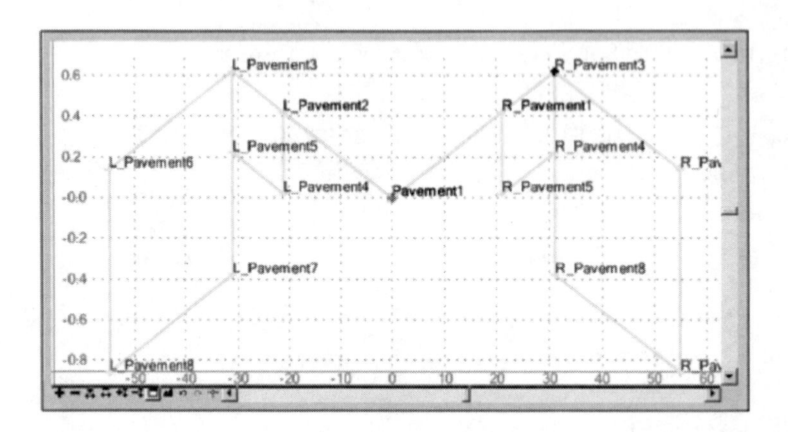

Figure 10-18 Lane Components

Figure 10-19 Outside Shoulder Components

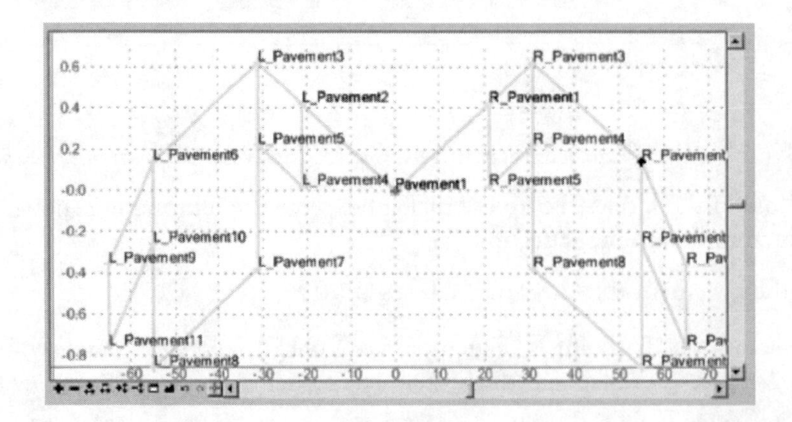

Figure 10-20 Outside Shoulder Components

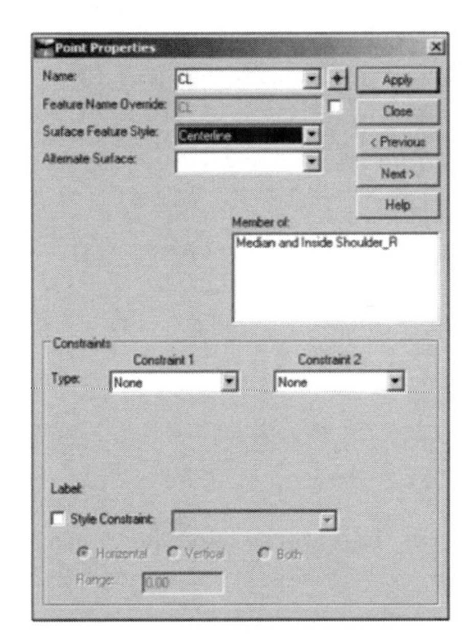

Figure 10-21 Change Point Name

The components representing the shoulders and traveled lanes are now complete. All the names are automatically assigned by InRoads based on the point seed name as shown in Figure 10-11.

To easily identify the different points, you must manually edit the points by reassigning the names and selecting the appropriate surface feature styles.

Point-editing can be initiated by double clicking on each point. The **Point Properties** window appears. Figure 10-21 shows an example of editing for point **Pavement**.

Rename it **CL** and select the surface feature style of **Centerline** (select the feature style as required by your project needs). Follow the same procedure to edit the other points. After editing is complete, the final view should look like what is shown in Figure 10-22. Note that some points may have more than one name. For example, the **CL** point is also called the **CL_Base** point.

The last step is to save what you have done so far. Go to ***File → Save*** in the **Create Template** window.

Now you have created the template for the Route 8 backbone. In the following subsection you will learn how to create a template including constrained components.

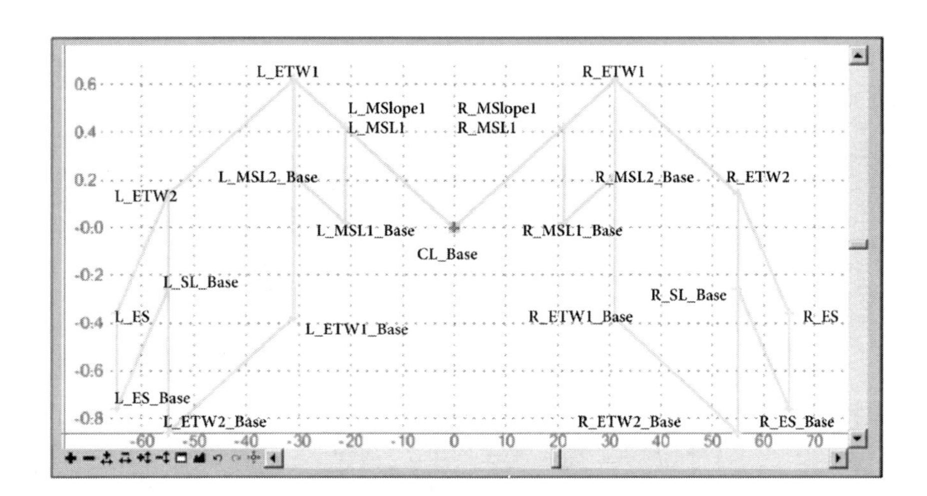

Figure 10-22 Point Names for Route 8 Backbone Template

2. Create a template for constrained components

A *constrained component* consists of points that are all restricted to the previous point placed. When a point (parent) is moved, any constrained point (child) also moves. This type of component is useful for establishing median barriers, curbs and gutters, sidewalks, dikes, and so on.

The geometric layout of the Dike E is shown in Figure 10-23. Note that the layout has two curves. However, InRoads cannot create these two curves using the Template Editor. Use two line segments to represent the two curves (see the dashed lines in Figure 10-23).

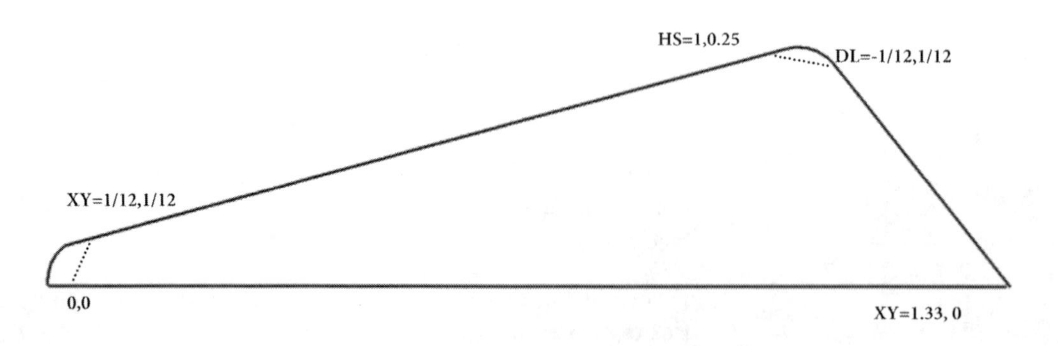

Figure 10-23 Dike E Layout

Follow these steps to create the Dike E template:

a. In the **Template Library** list window of InRoads, right click over the **Tutorial** folder previously created.

b. Go to ***New* → *Template*** and name the template **Dike E.**

Now you have two templates under **Tutorial** folder, **Route 8** and **Dike E** (see Figure 10-24).

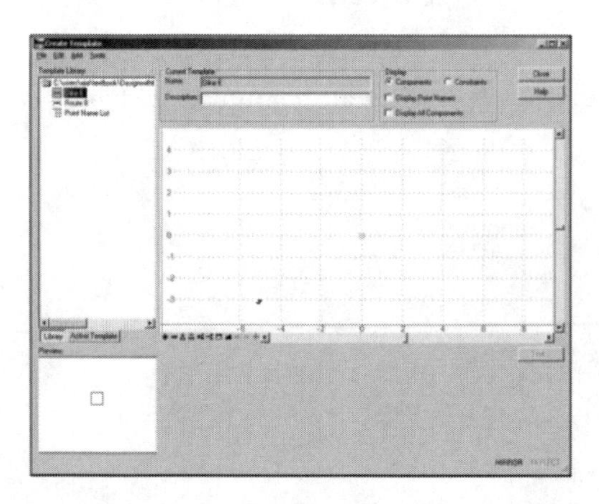

Figure 10-24 Dike Type E (Empty) Template

c. From the **Create Template** window, go to ***Tools* → *Dynamic Settings.***

The **Dynamic Settings** dialog box opens (see Figure 10-25).

» Set the **Point Name** to **Dike_1.**
» Set the **Point Style** to **Curb.**
» Click **Apply Affixes** to OFF.

Note the following terminologies:

XY	Type absolute coordinates
DL	Key in delta coordinates from last point placed (defaults to the dynamic origin if it is the first point of a component)
HS	Key in horizontal delta distance and slope from last point placed
VS	Key in vertical delta distance and slope from last point placed
OL	Key in delta coordinates from dynamic origin
OS	Key in horizontal delta distance and slope from dynamic origin

Without closing this box, go to the next step.

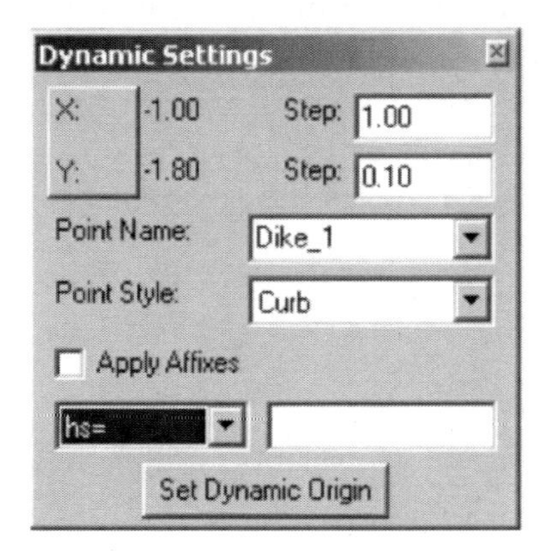

Figure 10-25 Dynamic Settings Window

d. Right click in the **Template Editor** window. Go to ***Add New Component → Constrained.***

e. At the bottom of the **Create Template** dialog box, type the information shown in Figure 10-26 into the **Current Component** section.

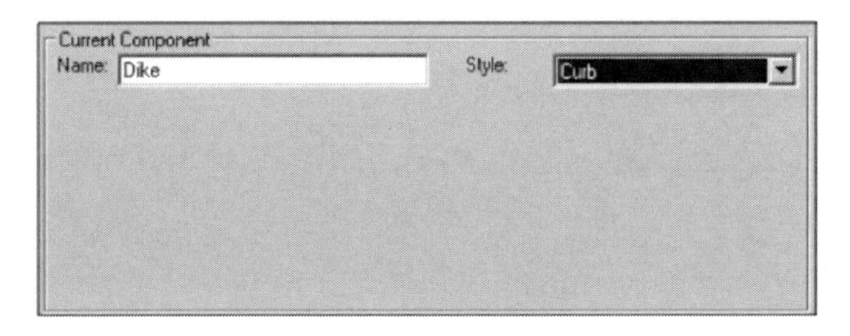

Figure 10-26 Information for Dike E

f. Zoom to the center of the Editor window. Place your cursor at the **0,0 origin** and click to place the point.

g. Return to the **Dynamic Settings** dialog box. Select **xy**= from the drop-down list. In the adjacent field, type **1/12, 1/12.** Press **Enter** at keyboard.

h. If you see two lines in the Template Editor window, right click in the window and toggle OFF **Mirror** in the **Context** menu.

   Note: The mirror function remains effective since you created the template for the Route 8 backbone. You don't need this feature when creating the template for constrained component. That is why you turn off this function before placing the component points here.

   The *second* point, **Dike_11,** has been placed in the window.

i. Return to the **Dynamic Settings** dialog box. Select **hs**= from the drop-down list and in the adjacent field type **1.00, 0.25**. Press **Enter** at keyboard.

   The *third* point, **Dike_12,** has been placed in the window.

j. Return to the **Dynamic Settings** dialog box. Select **dl**= from the drop-down list and in the adjacent field type **1/12, –1/12**. Press **Enter** in the keyboard.

The *fourth* point, **Dike_13,** has been placed in the window.

k. Return to the **Dynamic Settings** dialog box. Select **xy**= from the drop-down list and in the adjacent field type **1.33, 0.** Click **Enter** in the keyboard.

The *fifth* point, **Dike_14,** has been placed in the window.

l. Return to the **Dynamic Settings** dialog box. Select **xy**= from the drop-down list and in the adjacent field type **0, 0.** Click **Enter** in the keyboard.

The *sixth* point, **Dike_15,** has been tied back to the first point, **Dike_1.**

m. Right click on the **Template Editor** window and select **Finish.**

All the names are automatically assigned by InRoads based on the **Point Name** set as in Figure 10-25.

To easily identify the points, rename them by double clicking each point.

a. Using the ***Create Template → Fit*** tool at the bottom left corner of the Editor window to exhibit the whole view, you see the new **Dike E** template as shown in Figure 10-27.

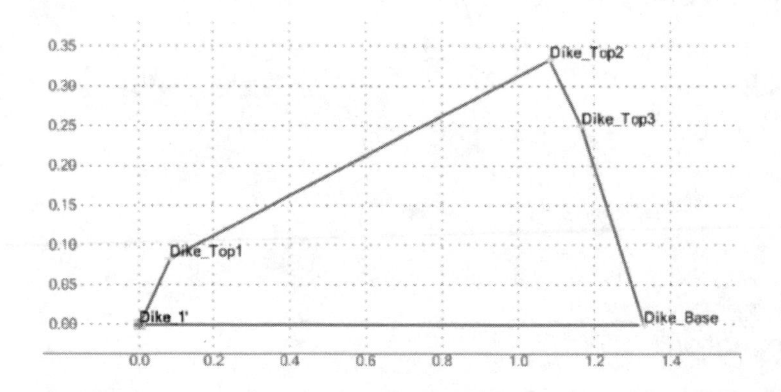

Figure 10-27 Point Names for Dike E Template

b. Go to the **Create Template** window and select ***File → Save.***

3. Create a template for end condition components

An *end condition component* is a special open-shaped component that targets a surface, an elevation, or an alignment. An end condition component is used to create cut/fill slopes of a roadway cross section. Multiple end conditions can be created with one template. However, different end conditions are assigned different priority values that specify your preferred order in which the end conditions are tried by InRoads.

The following procedures show you how to develop the conditions you want InRoads to try. Specifically, you will create six end conditions, three representing the cut situation and three representing the fill condition.

a. In the **Template Library** list window of InRoads, right click over the **Tutorial** folder previously created. Go to ***New → Template***, and name the template **Cut/Fill**.

   Now you have three templates under the **Tutorial** folder, **Route 8**, **Dike E**, and **Cut/Fill.**

b. From the **Create Template** window, go to ***Tools → Dynamic Settings.***

   The **Dynamic Settings** dialog box opens up.

   » Set the **Point Name** to **Origin**.
   » Set the **Point Style** to **Cut**.
   » Change **Apply Affixes** to OFF.

   Without closing this box, go to the next step.

c. Right click in the **Template Editor** window. Go to ***Add New Component → End Condition.***

d. At the bottom of the **Create Template** dialog box, type the information into the **Current Component** section as shown in Figure 10-28.

   Priority value is set as 1. This means InRoads will test this end condition first.

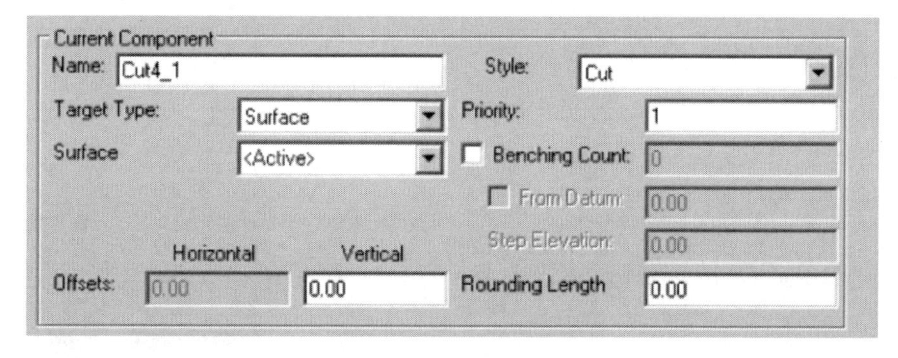

Figure 10-28 Information for Cut/Fill Template

e. Zoom to the center of the Editor window. Place your cursor at the **0,0 origin**. Click to place the point.

f. Return to the **Dynamic Settings** dialog box. Develop the settings as shown in Figure 10-29. After you Key-in **hs = 30, 0.25** (30 ft. to the right at 4:1 (H:V) slope), press **Enter** at keyboard.

   With this setting, InRoads will try to connect a 4:1 cut slope with the target surface with the horizontal constraint distance of 30 ft.

g. Right click again in the **Template Editor** window and select **Finish**.

   Now you have a component line as shown in Figure 10-30.

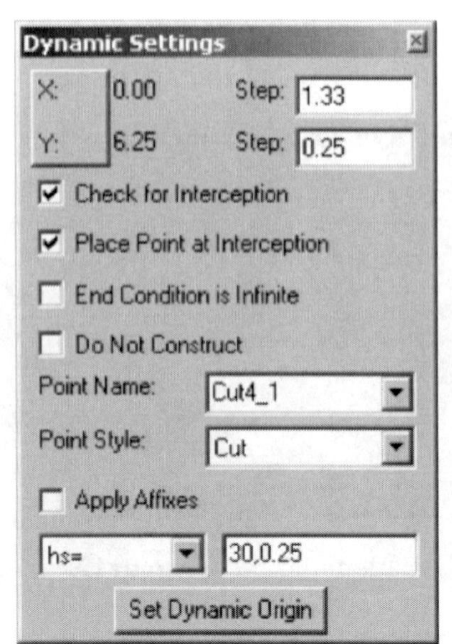

Figure 10-29 Information for 4:1 Cut Slope

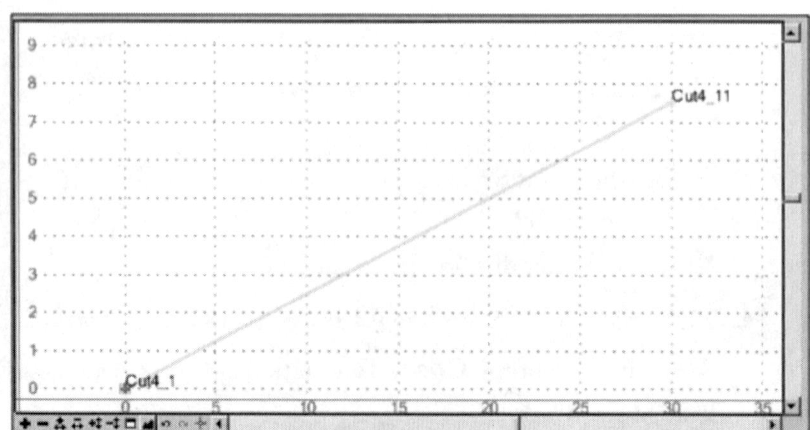

Figure 10-30 4:1 Cut Slope

h. Return to the **Dynamic Settings** dialog box and do the following:

» Set the **Point Name** to **Origin**.
» Set the **Point Style** to **Fill**.
» Change **Apply Affixes** to OFF.

Without closing this box, go to the next step.

i. Right click in the **Template Editor** window. Go to ***Add New Component → End Condition.***

j. At the bottom of the **Create Template** dialog box, type the information shown in Figure 10-31a into the **Current Component** section.

Priority value is set as 2. This means InRoads will test the previously created end condition of Cut4_1 first. If that end condition fails to connect with the target surface, the current end condition will be tested next by InRoads.

k. Zoom to the center of the Editor window. Place your cursor at the **0,0 origin** and click to place the point.

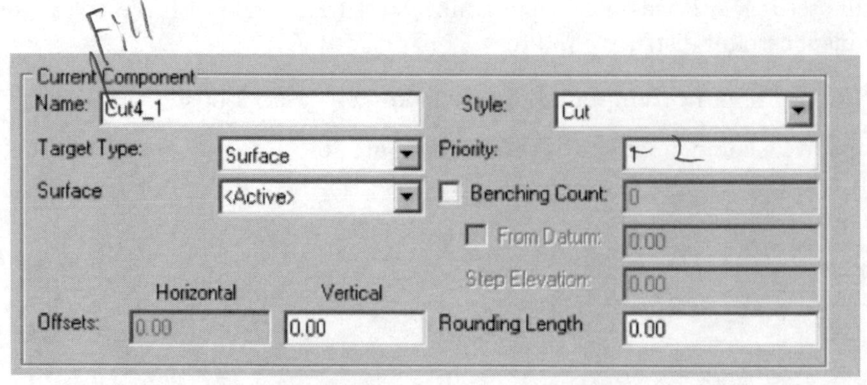

Figure 10-31a 4:1 Fill Slope

l. Return to the **Dynamic Settings** dialog box. Develop the settings as shown in Figure 10-31b. After you Key-in **hs = 1.67, 0.05** (1.67 ft. to the right at 5% slope down), press **Enter** at keyboard.

   1.67 ft. is selected to make sure the buffer zone off the shoulder point is 3 ft. (1.67 here + 1.33 from Dike E) as required by Caltrans.

m. Return to **Dynamic Settings** and do the following:

   » Set the **Point Name** to **Fill4_1.**
   » Set the **Point Style** to **Fill.**
   » Change **Apply Affixes** to OFF.
   » Set **hs = 28.33, -0.25.** Press **Enter** at keyboard.

   The value of 28.33 ft. is selected to make sure the total horizontal distance off the point **Dike_Base** of Dike E is 30 ft. (28.33 ft. + 1.67 ft.).

n. Use the ***Create Template → Fit*** tool at the bottom left corner of the Editor window to exhibit the whole view.

   The components should look like what is shown in Figure 10-32.

o. Repeat the procedure to create other end conditions including 3:1 cut, 3:1 fill, 2:1 cut, and 2:1 fill. The associated priority values are 3, 4, 5, and 6, respectively.

   The final **Cut/Fill** template should look like what is illustrated in Figure 10-33.

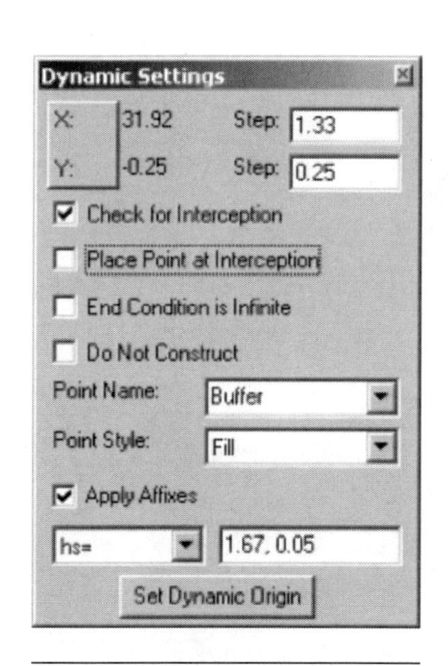

Figure 10-31b Settings for 4:1 Fill Slope

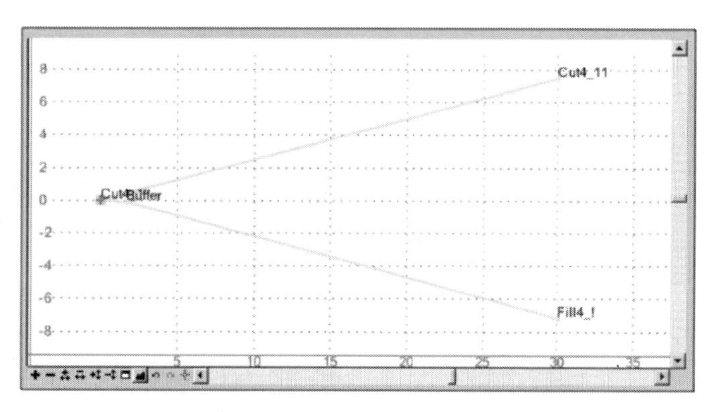

Figure 10-32 4:1 Fill Slope

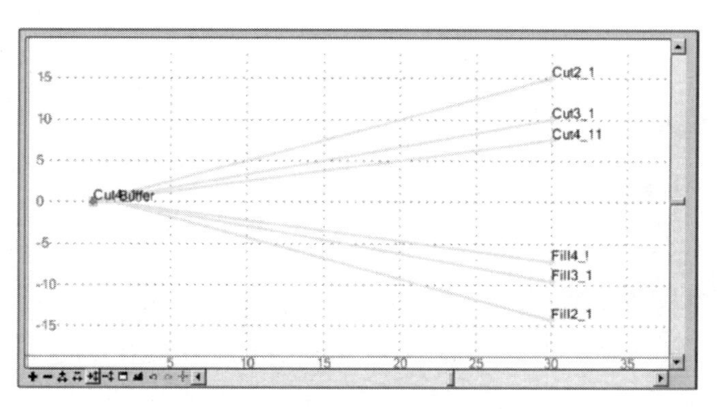

Figure 10-33 Cut/Fill Template

In your end condition components template, the created end conditions will try to extend a maximum horizontal distance of 30 ft. to catch the target surface (remember the previous key-ins: **hs = 30, 0.25; hs = 28.33, –0.25**, etc.). If one condition fails to catch the target surface within the distance, InRoads moves to next alternative end conditions with larger priority values.

Sometimes InRoads fails to connect the slopes with the target surface after all the end conditions have been tested. In these situations, you must usually define the end condition point as having an infinite horizontal reach, which allows the points to intercept the target surface regardless of their distances.

Generally, the last priority conditions for Cut and Fill (in our case, 2:1 Cut and 2:1 Fill) will be selected for infinite end conditions. This can be done through the following steps.

a. Double click the Point **Cut2_1.**

   The **Point Properties** dialog box appears (see Figure 10-34).

b. Check on **End Condition is Infinite**. Click **Apply**.

c. Click the **Next** button going to Point **Fill2_1.**

d. Check on **End Condition is Infinite.** Click **Apply**.

e. To close this dialog box, click **Close.**

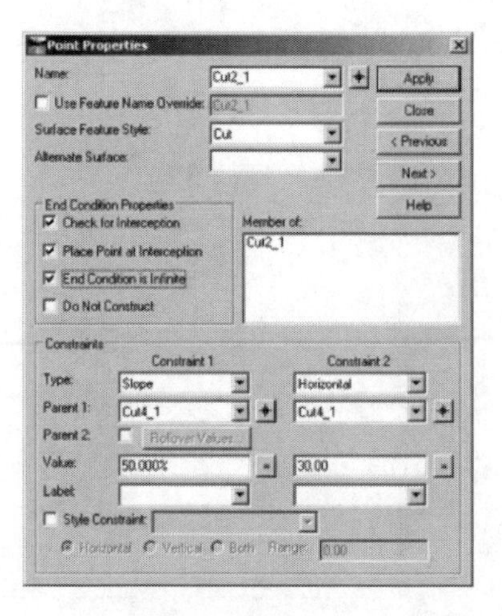

Figure 10-34 Set Cut2_1 Point to Be with End Condition Infinite

Now it is time to test the end conditions.

a. Click the **Test...** button below the **Template Editor** window.

   The **Test End Condition** window opens.

b. Click **Draw**.

c. Test the various end conditions by moving the surface slope line with your cursor.

   Since you clicked on **End Condition is Infinite** for 2:1 Cut and 2:1 Fill, you can see (from Figure 10-35) the slopes do not terminate at the horizontal distance of 30 ft. away from the **Origin** point for the two end conditions. It actually extends with the movement of the cursor.

d. When you are done checking the end conditions, click **Close**.

e. Go to the **Create Template** window and select ***File → Save.***

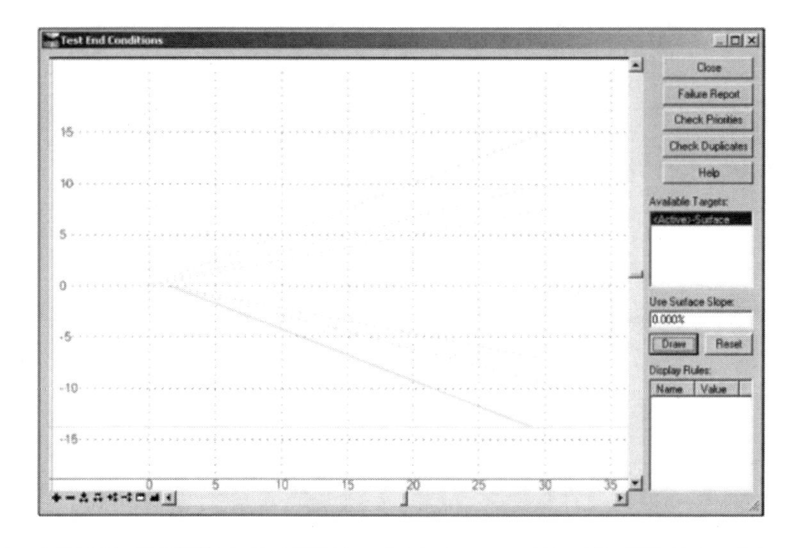

Figure 10-35 Test the Cut/Fill Template

Now you are done creating the template for end condition components.

In the following subsection you will learn how to assemble the three templates, **Route 8, Dike E,** and **Cut/Fill**, to create a complete template for Route 8.

4. Create a complete template including Route 8 backbone, dike, and end condition components

The previously created template, **Route 8,** can be updated by adding the **Dike E** and **Cut/Fill** templates to form a complete template including Route 8 backbone, dike, and end condition.

a. In the left window listing of the **Create Template** dialog box, set the template **Route 8** to active.

b. Click and drag the **Dike E** template to the **Template Editor** window. You should see two dikes in the Editor window.

   » If you see only one dike, press your left mouse button, and then right click to activate the context-sensitive shortcut menu. Turn ON the **mirror** option.

   » If you did it wrong, right click in the Editor window → **Delete Components.** Then click and drag over the components you want to delete.

c. Move your cursor over the point **R_ES**. Click to place the two dikes on either side of the roadway backbone.

d. Do the same for the **Cut/Fill** template. Click and drag the **Cut/Fill** template to the **Template Editor** window.

e. Move your cursor over the point **Dike_Top_3** and click to place the end conditions on either side of the roadway backbone.

f. From the bottom left corner of the Editor window, use the ***Create Template → Fit*** tool to exhibit the whole view.

   You should see new **Route 8** template as shown in Figure 10-36.

g. In the **Create Template** dialog box, go to ***Create Template → File → Save.***

   You have created a complete template for Route 8, including backbone, dikes, and end conditions.

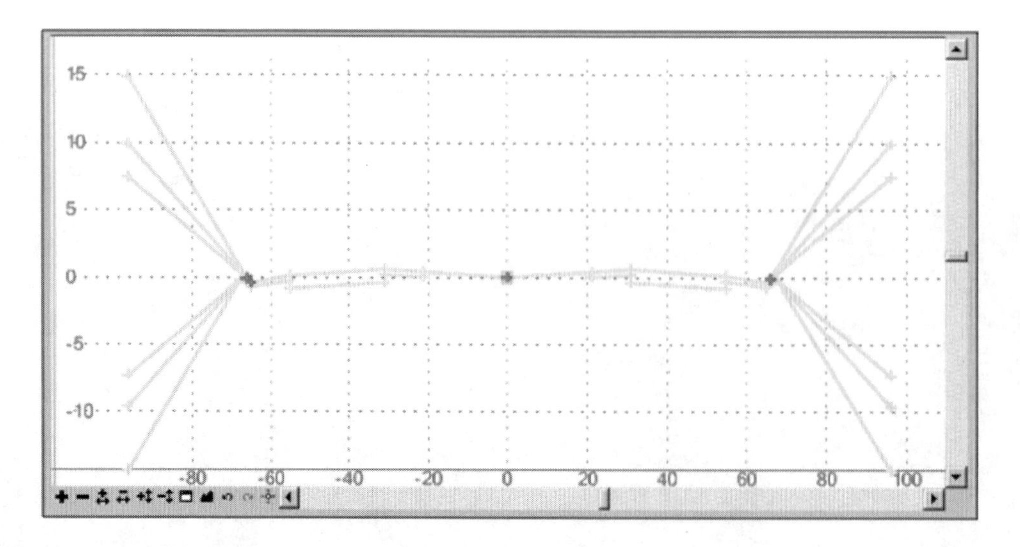

Figure 10-36 Route 8 Template

10.3.3 Create templates for Laguna Canyon Road

In this section you will focus on creating the template for Laguna Canyon Road. The procedures are similar to those for Route 8. The major difference lies in the median design. The median width for Route 8 is 62 ft., whereas there is no median for Laguna Canyon Road. First, you will design the backbone for Laguna Canyon Road. Then you will assemble it with other templates to create a complete template. The detailed procedures follow:

1. Create a template for Laguna Canyon backbone

   a. In the **Template Library** list window of the **Create Template** dialog box, right click over the Tutorial folder previously created. Go to *New* → *Template*, and name the template **Laguna Canyon.**

      Now you have a project roadway folder and one empty template for Laguna Canyon Road.

   b. From the Create Template window, go to ***Tools*** → ***Options....***

      The **Template Options** window opens.

      » Set the **Pointes Seed Name** to **Pavement.**
      » Click on **Apply Affixes.** Type **L_** and **R_** in the **Prefix** field.
      » Click **OK.**

   c. Right click in the **Template Editor** window. Go to ***Add New Component*** → ***Simple.***

   d. At the bottom of the **Create Template** dialog box, type the information shown in Figure 10-37 into the **Current Component** section.

      You should see two **Traveled Way** components attached to your cursor when you navigate around the editor window.

      » If you only see one component, right click in the window to activate the **Shortcut Menu** and select **Mirror.**
      » Zoom to the center of the editor window.
      » Place your cursor at the **0,0 origin** and click to place the components.
      » To merge the components, right click on the vertical line adjacent to the point **pavement.** Go to ***Shortcut Menu*** → ***Merge Components.***

The components have merged.

e. Right click in the **Template Editor** window. Go to ***Add New Component → Simple.***

f. At the bottom of the **Create Template** dialog box, type the information shown in Figure 10-38 into the **Current Component** section.

g. Place the cursor over the Point **R_Pavement** and click to place the **Shoulders** components. Use the ***Create Template → Fit*** tool at the bottom left corner of the editor window to exhibit the whole view.

The components should look like what is shown in Figure 10-39.

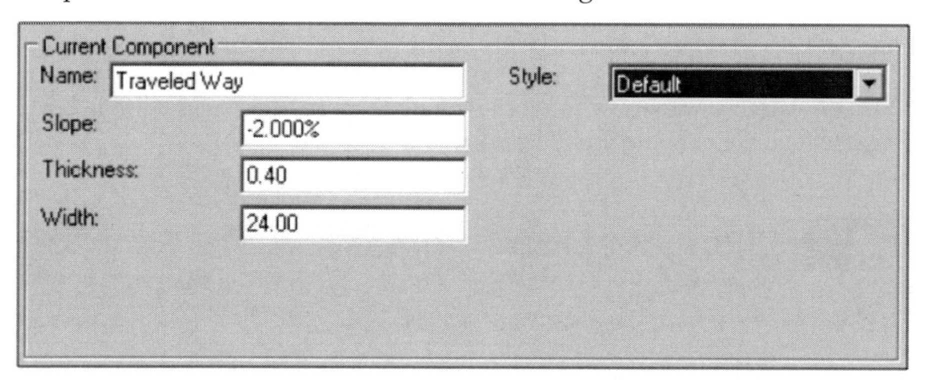

Figure 10-37 Traveled Way Component

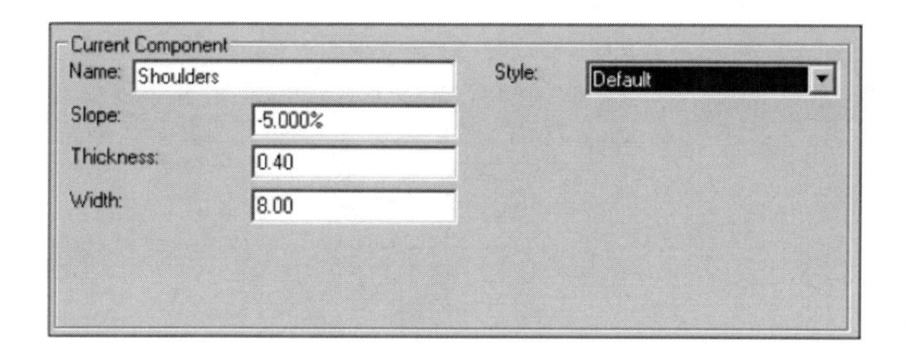

Figure 10-38 Traveled Way Component

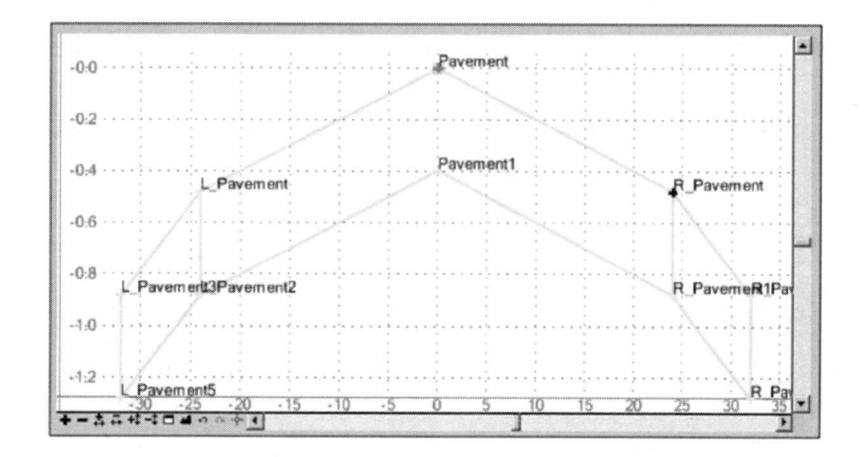

Figure 10-39 Laguna Canyon Road Backbone

The components representing the shoulders and traveled lanes are complete. However, all the names were automatically assigned by InRoads based on the point seed name of "Pavement." To easily identify the points, you must manually edit them by reassigning the names and selecting the appropriate surface feature styles.

a. Initiate point-editing by double clicking on every point.

   The **Point Properties** window appears. Figure 10-40 shows an example of editing for point **Pavement**.

b. Rename the point **CL** and select the surface feature style of **Centerline** (select the feature style as required by your project needs).

c. Follow the same procedures to edit the other points.

   Table 10-2 presents a summary of properties of all the points you have edited. After the editing is complete, the final view should look like what is shown in Figure 10-41a.

Table 10-2 Changes of Point Names

Original Point Name	New Point Name
L_Pavement3	L_ES
L_Pavement	L_ETW
Pavement	CL
R_ Pavement	R_ETW
R_ Pavement 3	R_ES
L_ Pavement 4	L_ES_BASE
L_ Pavement 2	L_ETW_BASE
Pavement	CL_BASE
R_ Pavement 2	R_ETW_BASE
R_ Pavement 4	R_ES_BASE

2. Create a complete template including Laguna Canyon Road backbone, dike, and end condition components

The previously created template **Laguna Canyon** can be updated by adding the **Dike E** and **Cut/Fill** templates to form a complete template including Laguna Canyon backbone, dike and end condition components.

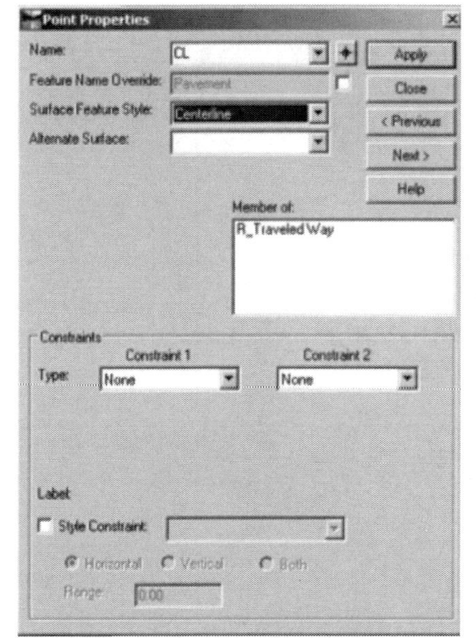

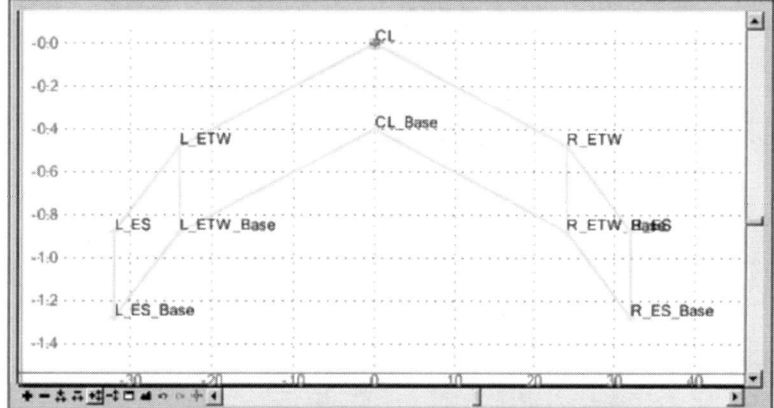

Figure 10-40 Example of Changing Point Name to CL

Figure 10-41a Laguna Canyon Road Backbone with New Point Names

a. In the left window listing of the **Create Template** dialog box, set the template **Laguna Canyon** to active.

b. Click and drag the **Dike E** template to the **Template Editor** window.

   You should see two dikes in the editor window.

   » If you can see only one dike, press your left mouse button. Then right click to activate the context-sensitive **shortcut menu** and turn the **mirror** option to ON.
   » If you did it wrong, right click in the **Editor Window → Delete Components.** Click and drag over the components you want to delete.

c. Move your cursor over the point **R_ES** and click to place the two dikes on either side of the roadway backbone.

d. Do the same for the **Cut/Fill** template.

e. Click and drag the **Cut/Fill** template to the **Template Editor** window.

f. Move your cursor over the point **Dike_Top_3**. Click to place the end conditions on either side of the roadway backbone.

g. Using the ***Create Template → Fit*** tool at the bottom left corner of the Editor window to exhibit the whole view , you should now have the new **Laguna Canyon** template as shown in Figure 10-41b.

h. In the Create Template dialog, go to ***File → Save.***

   Now you have created a complete template for Laguna Canyon including Laguna Canyon backbone, dikes, and end conditions.

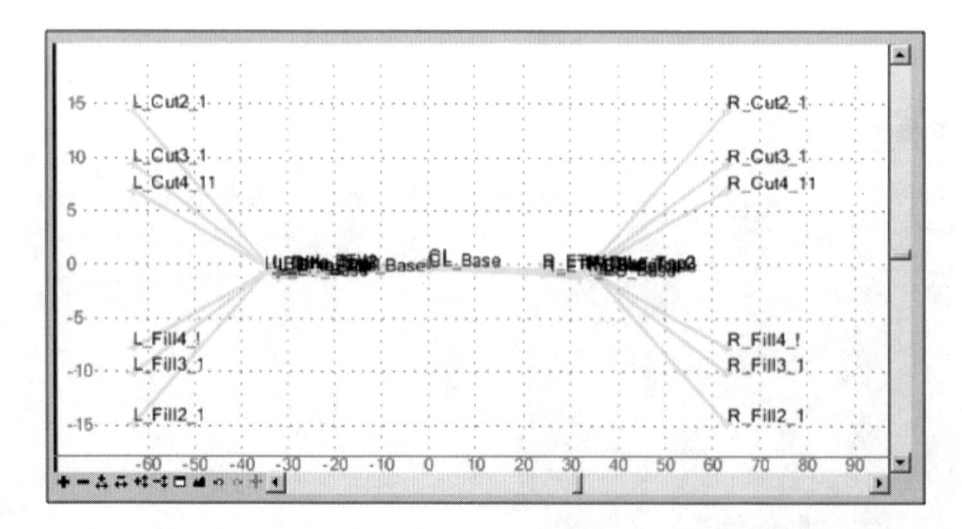

Figure 10-41b Laguna Canyon Road Template

10.3.4 Create templates for ramps

In this section we will focus on creating the template for ramps. The procedures are similar to those for Route 8 and Laguna Canyon Road. The major difference lies in the shape of the cross sections. The templates for Route 8 and Laguna Canyon Road are symmetric, while the ramp template is asymmetric. First, you will design the backbone for the ramps. Then you will assemble it with other templates to create a complete template. Detailed procedures follow.

1. Create a template for ramp backbone
   a. Go to ***InRoads → Modeler → Create Template.***

      The **Create Template** dialog box appears.

   b. In the **Create Template** dialog box, go to ***New → Template,*** and name the template **Ramps**.

   c. From the **Create Template** dialog box, go to ***Tools → Options....***

      The **Template Options** window opens.

      » Set the **Pointes Seed Name** to **Pavement**.
      » Change **Apply Affixes** to ON. Click **OK**.

   d. Right click in the **Template Editor** window. Go to ***Add New Component → Simple.***

   e. At the bottom of the **Create Template** dialog box, type the information shown in Figure 10-42 into the **Current Component** section.

      Select the **Default** style provided by InRoads since you do not have the preferred style. Assume the pavement thickness is 0.40 ft. Select the specific style and thickness required by your project needs.

      You should see a **Left Shoulder** component attached to your cursor when you navigate around the Editor window.

      If you see two symmetrical components, follow these steps:

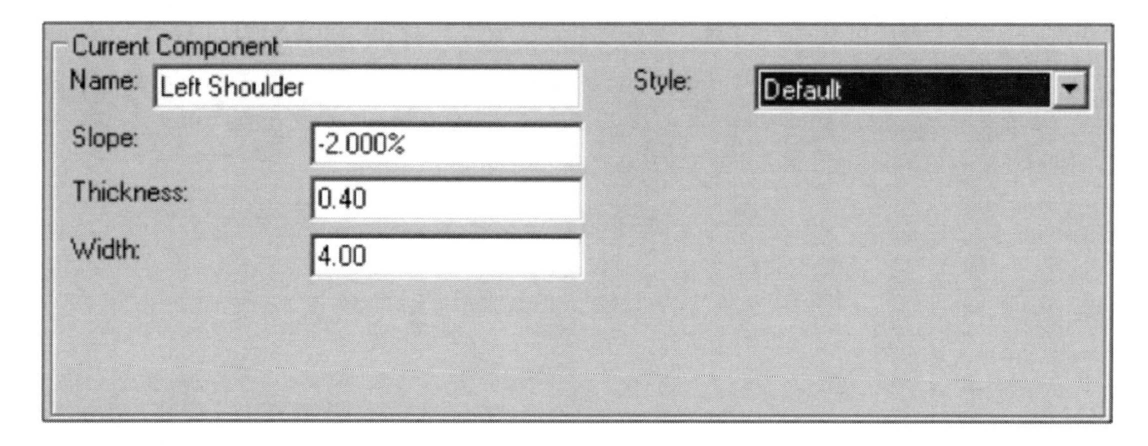

Figure 10-42 Information for Left Shoulder Component

» Right click in the window. Go to the **Shortcut Menu** and toggle the **Mirror** to OFF.
» Zoom to the center of the Editor window. Place your cursor at the **0,0 origin** and click to place the components.
» Use the ***Create Template → Fit*** tool at the bottom left corner of the Editor window to exhibit the whole view.

The components should look like what is shown in Figure 10-43a.

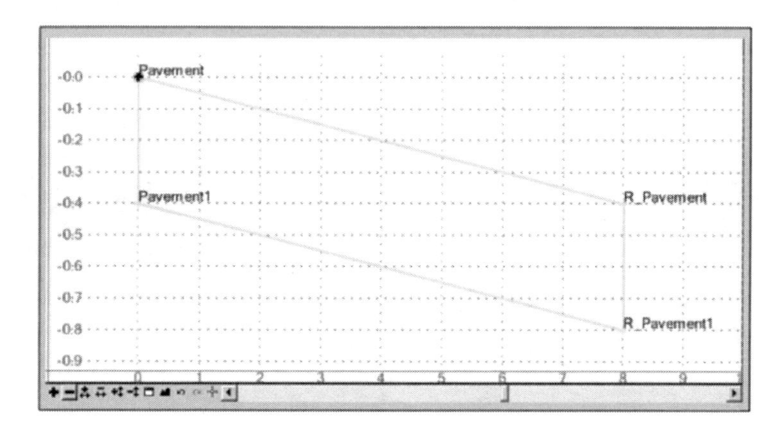

Figure 10-43a Left Shoulder Component

f. Right click in the **Template Editor** window, then go to ***Add New Component → Simple.***

g. At the bottom of the **Create Template** dialog box, type the information shown in Figure 10-43b into the **Current Component** section.

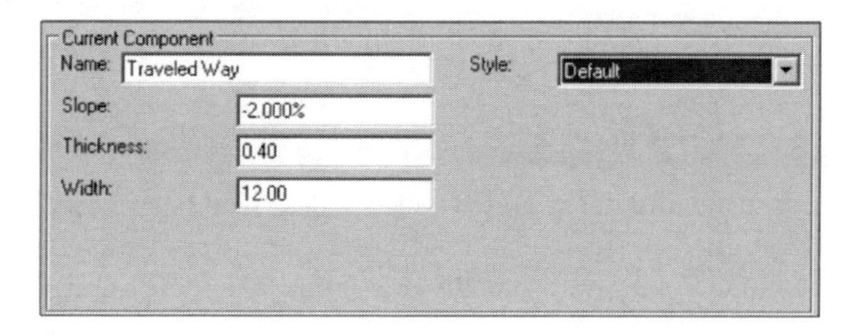

Figure 10-43b Information for Traveled Way Component

h. Place the cursor over the Point **R_Pavement** and click to place the **Traveled Way** component.

i. Use the ***Create Template*** → ***Fit*** tool at the bottom left corner of the editor window to exhibit the whole view.

The components should look like what is shown in Figure 10-44.

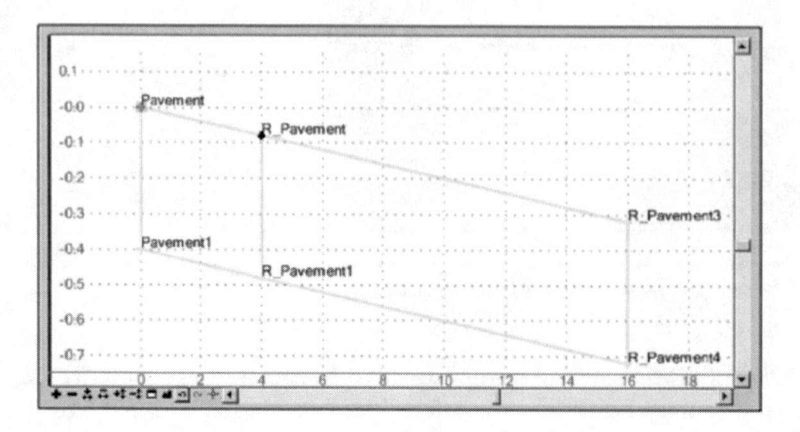

Figure 10-44 Traveled Way Component for Ramps Template

j. Right click in the **Template Editor** window, then go to ***Add New Component*** → ***Simple.***

k. At the bottom of the **Create Template** dialog box, type the information shown in Figure 10-45 into the **Current Component** section.

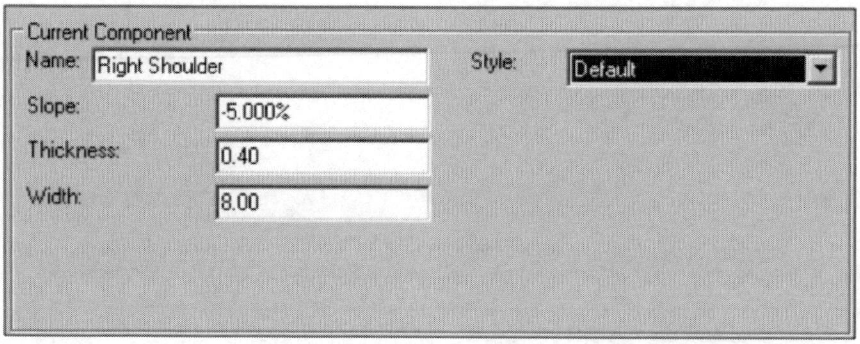

Current Component
Name: Right Shoulder
Style: Default
Slope: -5.000%
Thickness: 0.40
Width: 8.00

Figure 10-45 Information for Right Shoulder Component

l. Place the cursor over the Point **R_Pavement3** and click to place the **Right Shoulder** component.

m. Use the ***Create Template*** → ***Fit*** tool at the bottom left corner of the Editor window to exhibit the whole view.

The components should look like what is shown in Figure 10-46.

The components representing the shoulders and traveled lanes have been completed. However, all the names were automatically assigned by InRoads based on the point seed name as **Pavement**. To easily identify the points, you need to manually edit them by reassigning the names and selecting the appropriate surface feature styles.

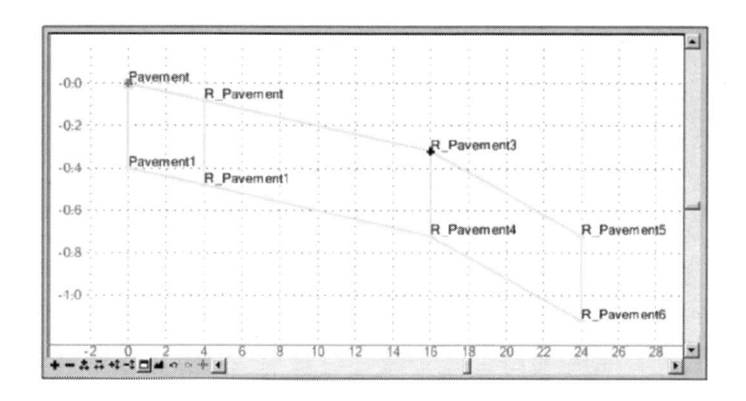

Figure 10-46 Backbone of Ramp Template

a. Initiate point editing by double clicking on every point.

   The **Point Properties** window appears. Figure 10-47 shows an example of editing for point **Pavement**.

b. Rename it **L_ES**, and select the surface feature style of **Shoulder** (select the feature style as required by your project needs).

c. Follow the same procedure to edit all the other points.

   Table 10-3 presents the summary of properties of all the points you have edited. After the editing is complete, the final view should look like what is shown in Figure 10-48.

Table 10-3 List of Point Names

Original Point Name	New Point Name
Pavement	L_ES
R_Pavement	L_ETW
R_Pavement 3	R_ETW
R_Pavement 5	R_ES
R_Pavement 1	L_ETW_BASE
R_Pavement 4	R_ETW_BASE
R_Pavement 6	R_ES_BASE
Pavement 1	L_ES_BASE

d. In the **Create Template** window, go to ***File*** → ***Save.***

   Now you have created the template for the ramp backbone. The constrained component template for the ramps is the same as the one for Route 8 and Laguna Canyon Road, consisting of Type E dike. However, the end conditions for the ramps are different from those for the other road sections. In the following subsection you will learn how to create the end condition components for the ramps.

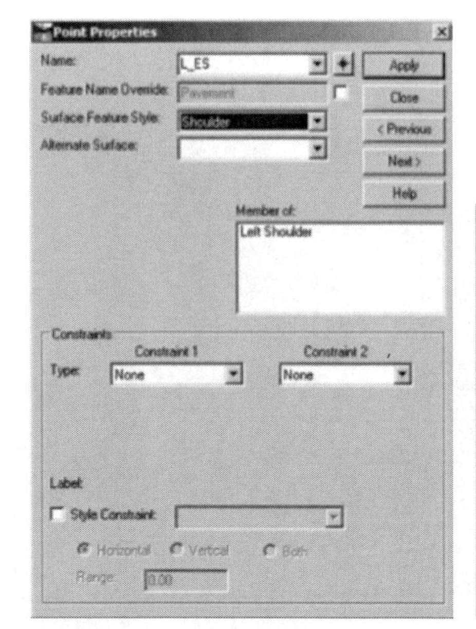

Figure 10-47 Change Pavement to L_ES

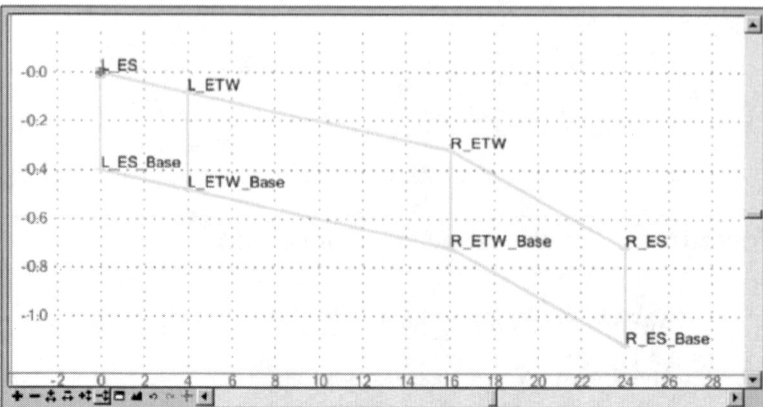

Figure 10-48 Shoulders and Traveled Way for the Ramp Template

2. Create the template of end condition for the ramps

The end conditions on the right side of the ramps are exactly the same as those of Route 8 and Laguna Canyon Road. However, the end conditions on the left side are different. A description of left end conditions is covered in detail as follows.

a. In the **Template Library** list window of InRoads, right click over the **Tutorial** folder previously created. Go to ***New → Template*** and name the template **Left Cut/Fill for Ramps.**

   You now have six templates under **Tutorial** library, **Route 8, Dike E, Cut/Fill, Laguna Canyon, Ramps,** and **Left Cut/Fill for Ramps.**

b. From the **Create Template** window, go to ***Tools → Dynamic Settings.***

   The **Dynamic Settings** dialog box opens.

   » Set the **Point Name** to **Origin.**
   » Set the **Point Style** to **Default.**
   » Change **Apply Affixes** to OFF.

   Without closing this box, go to the next step.

c. Right click in the **Template Editor** window. Go to ***Add New Component → End Condition.***

d. At the bottom of the **Create Template** dialog box, type the information shown in Figure 10-49 into the **Current Component** section.

   Priority value is set as 1. This means InRoads will test this end condition first.

e. Zoom to the center of the Editor window. Place your cursor at the **0,0 origin** and click to place the point.

f. Return to the **Dynamic Settings** dialog box and develop the settings as shown in Figure 10-50. After you Key-in **hs = −3, 0.05** (3 ft. to the left at 5% slope), press **Enter** at keyboard.

   InRoads has created a buffer zone of 3 ft. to the left of the left shoulder.

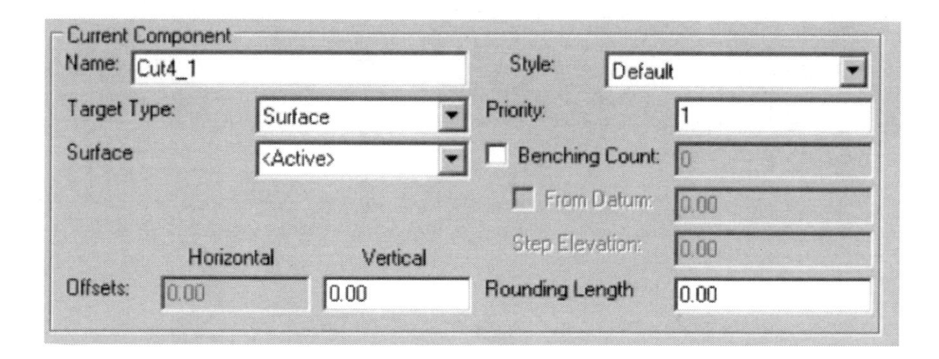

Figure 10-49 4:1 Cut Slope Information

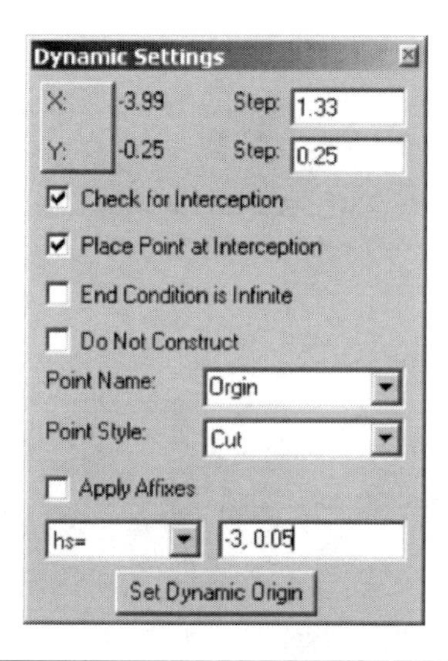

Figure 10-50 Information for Side Slope Buffer Component

g. Return to **Dynamic Settings** and follow these steps:

» Set the **Point Name** to **Cut4_1.**
» Set the **Point Style** to **Default.**
» Check off **Apply Affixes.**
» Set **hs** = **−30, −0.25** and press **Enter** at keyboard.

InRoads will try to connect the 4:1 cut with the target surface within 30 ft. off the buffer zone.

h. Right click again in the **Template Editor** window and select **Finish.**

i. Use the ***Create Template → Fit*** tool at the bottom left corner of the Editor window to exhibit the whole view.

The components should look like what is shown in Figure 10-51.

j. Return to **Dynamic Settings** dialog box and follow these steps:

» Set the **Point Name** to **Origin**.
» Set the **Point Style** to **Default**.
» Change **Apply Affixes** to OFF.

Without closing this box, go to the next step.

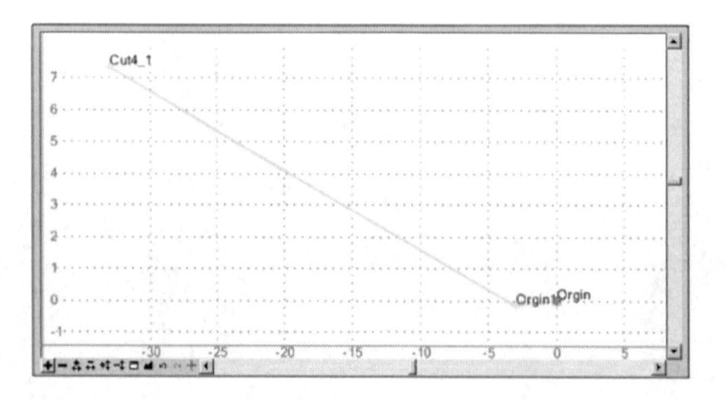

Figure 10-51 4:1 Cut Slope

k. Right click in the **Template Editor** window. Go to ***Add New Component* → *End Condition.***

l. At the bottom of the **Create Template** dialogue, type the information shown in Figure 10-52 into the **Current Component** section.

Priority value is set as 2. InRoads will test the previously created end condition of Cut4_1 first. If that end condition fails to connect with the target surface, InRoads will test the current end condition.

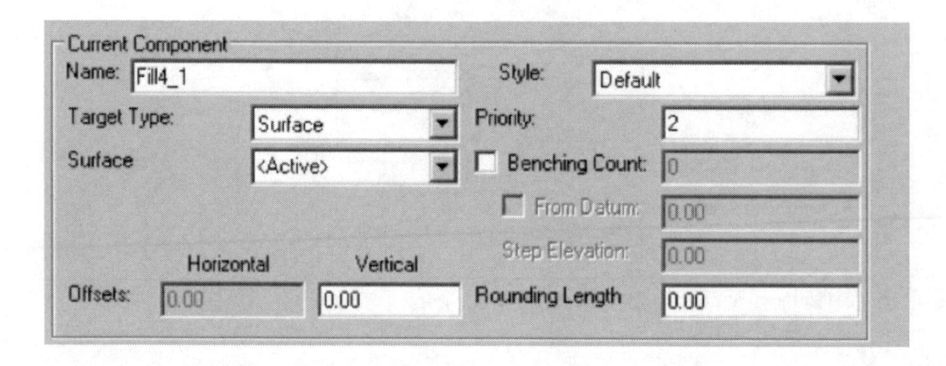

Figure 10-52 Information for 4:1 Fill Slope

m. Zoom to the center of the Editor window. Place your cursor at the **0,0 origin** and click to place the point.

n. Return to the **Dynamic Settings** dialog box and develop the settings as shown in Figure 10-53.

o. After you Key-in **hs** = **−3, 0.05** (3 ft. to the left at 5% slope), press **Enter** at keyboard.

3 ft. is selected to make sure the buffer zone off the shoulder point is 3 ft. as required by Caltrans.

p. Return to **Dynamic Settings** and do the following:

» Set the **Point Name** to **Fill4_1.**
» Set the **Point Style** to **Default.**
» Change **Apply Affixes** to OFF.
» Set **hs** = **−30, 0.25** and press **Enter** at keyboard.

Note that the value of 30 ft. is selected to make sure the total horizontal distance off the buffer zone is 30 ft.

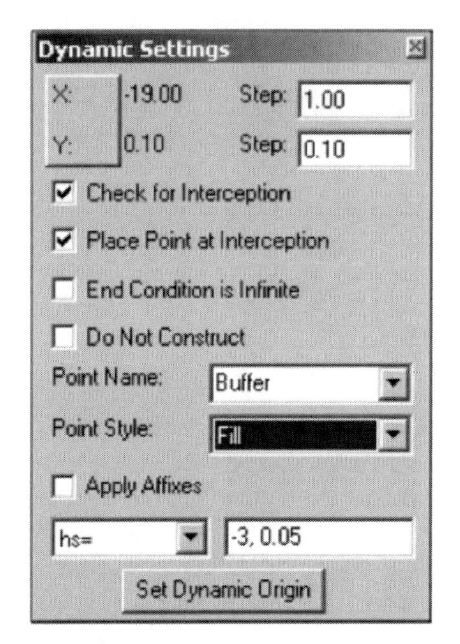

Figure 10-53 Information for 4:1 Fill Slope Buffer

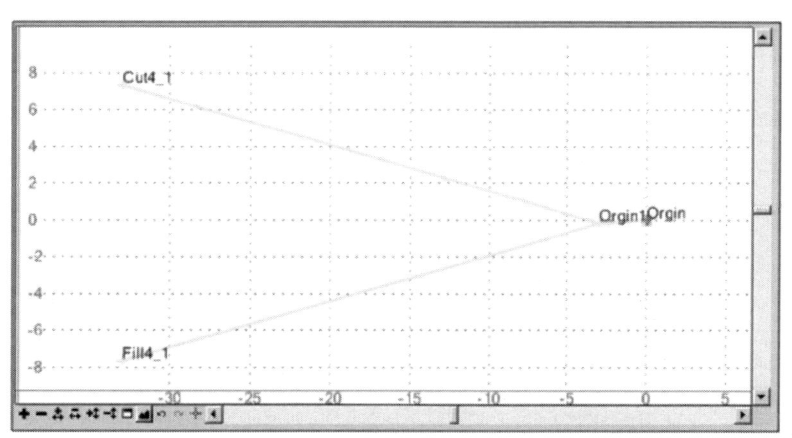

Figure 10-54 4:1 Cut and Fill Slopes

q. Use the ***Create Template → Fit*** tool at the bottom left corner of the editor window to exhibit the whole view.

The components should look like what is shown in Figure 10-54.

r. Repeat the procedures to create other end conditions including 3:1 cut, 3:1 fill, 2:1 cut, and 2:1 fill. The associated priority values are 3, 4, 5, and 6, respectively.

The final **Left Cut/Fill for Ramps** template should look like what is illustrated in Figure 10-55.

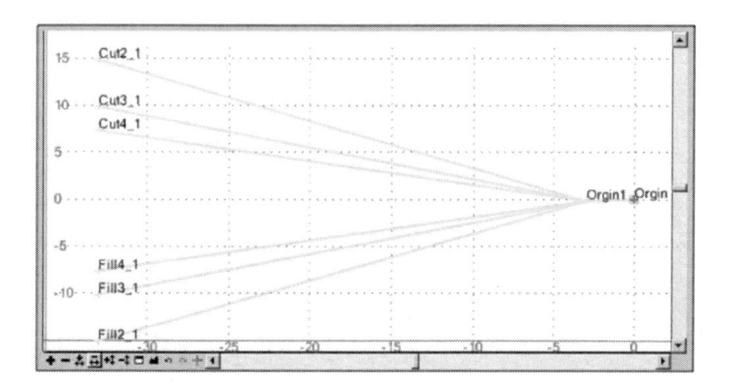

Figure 10-55 Cut/Fill Slopes for Ramps

Like the end conditions for Route 8 and Laguna Canyon Road, you also must define the infinite end conditions in case all the alternative slopes fail to intercept the target surface within the given horizontal distance of 30 ft. The last priority conditions for Cut and Fill (in this case, 2:1 cut and 2:1 fill) will be selected for infinite end conditions. Do that by following these steps.

a. Double click the Point **Cut2_1.**

The **Point Properties** dialog box appears (see Figure 10-56).

b. Check on **End Condition is Infinite** and click **Apply**.

c. Click ***Next*** → go to Point **Fill2_1.**

d. Click on **End Condition is Infinite.**

e. Click **Apply**. Click **Close**.

The dialog box closes.

Test the end conditions by following these steps:

a. Click the **Test...** button below the **Template Editor** window.

The **Test End Condition** window opens.

b. After clicking the **Draw** button, test the end conditions by moving the surface slope line with your cursor.

Since you checked **End Condition is Infinite** for 2:1 Cut and 2:1 Fill, you can see the slopes do not terminate at the horizontal distance of 30 ft. away from the **Origin** point for the two end conditions. It actually extends with the movement of the cursor.

c. Once you are done checking the end conditions, click **Close**.

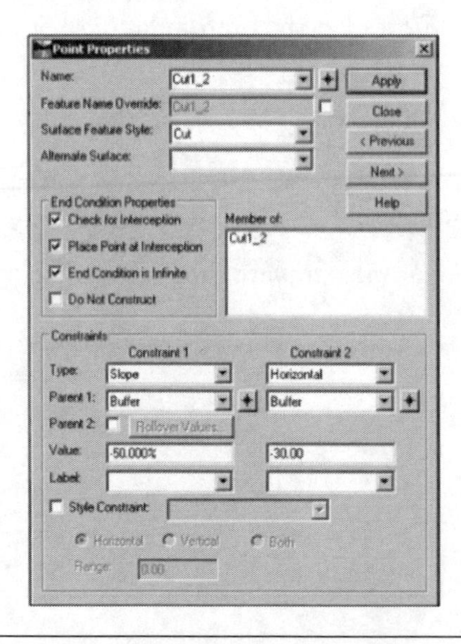

Figure 10-56 Set Cut2_1 Point to Be with End Condition Infinite

d. Save what you have done so far. Go to the **Create Template** window and select ***File* → *Save.***

Now you have created the template for the left end conditions of ramps.

In the following subsection you will learn how to assemble the four templates, **Ramps, Dike E, Cut/Fill,** and **Left Cut/Fill for Ramps**, to create a complete template for the ramps.

3. Create a complete template including ramp backbone, dike, and end condition components

The previously created template, **Ramps**, can be updated by adding the **Dike E** and **Cut/Fill** (both left and right sides) templates to form a complete template including ramp backbone, dike, and end condition components. Follow these steps:

a. In the left window listing of the **Create Template** dialog box, set the template **Ramps** to active.

b. Click and drag the **Dike E** template to the **Template Editor** window.

   You should see one dike in the Editor window.

   » If you see two dikes, press your left mouse button. Then right click to activate the context sensitive shortcut menu and turn OFF the **mirror** option.
   » If you did it wrong, right click in the ***Editor window*** → ***Delete Components.*** Click and drag over the components you want to delete.

c. Move your cursor over the point **R_ES** and click to place the dike on the right side of the ramp backbone.

d. Click and drag the **Cut/Fill** template to the **Template Editor** window.

e. Move your cursor over the point **Dike_Top_R3** and click to place the end conditions on the right side of the roadway backbone.

f. Do the same for the **Left Cut/Fill for Ramps** template. Click and drag the **Left Cut/Fill** for **Ramps** template to the **Template Editor** window.

g. Move your cursor over the point **L_ES** and click to place the end conditions on the left side of the roadway backbone.

   The final template for the ramps is shown in Figure 10-57.

h. In the **Create Template** dialog, go to ***File*** → ***Save.***

   You have created a complete template for the ramps including ramp backbone, dikes, and end conditions.

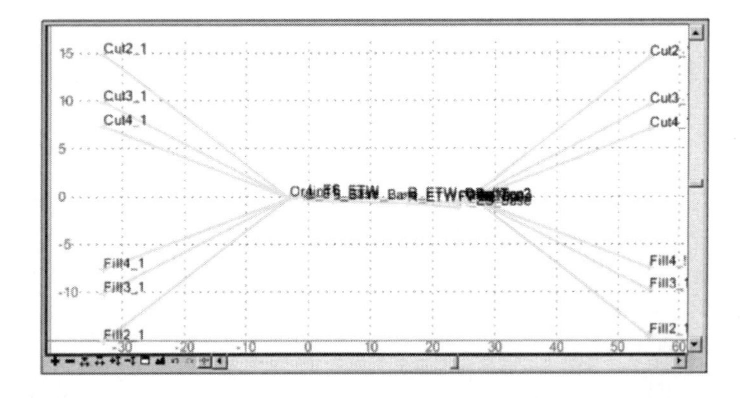

Figure 10-57 Ramp Template

i. To save the **tutorial** library, do the following:

   » Select the **tutorial** library.
   » Go to ***InRoads*** → ***File*** → ***Save*** → ***Template Library.***
   » Save the **tutorial** library as **tutorial.itl** in the **C:/Temp/Tutorial/Chp10/Master Files/** folder.

Name____ Date______________

10.4 Questions

Q10.1 Draw the cross section template for Route 8.

Q10.2 Draw the cross section template for the four ramps.

Q10.3 The term **Left** in the left side of a ramp refers to the driving direction. Is this correct? According to the Caltrans standards, what is the width of the left shoulder of a ramp? What is the width of the right shoulder of a ramp?

Q10.4 How many templates can be allowed in a highway project? Are they independent from horizontal alignments? Are they a part of a geometry project?

Q10.5 What is a backbone?

Q10.6 Is the typical template the same for all of the four ramps?

Q10.7 List the steps to create a template with a median curb. Hint: Check the InRoads Online help and see how to create a vertical line for a cross section template.

Q10.8 How do you save a cross section template?

Q10.9 What is a typical section library? How many cross section templates are allowed in a typical cross section library?

Q10.10 Is a typical section library independent from a geometry project?

Chapter 11

Roadway Designer and Superelevation

LEARNING OBJECTIVES

After completing this chapter you should know:

1. How to create a roadway designer (.ird) file and define a roadway corridor;
2. How to define superelevation in accordance with Caltrans standards;
3. How to create design surfaces.

SECTIONS

This chapter discusses how to create roadways or 3-D design surfaces for Route 8, Laguna Canyon Road, and four ramps.

3-D design surface modeling involves the use of horizontal and vertical alignments, original surface terrain, and cross section templates to create a three-dimensional design surface along a roadway. The horizontal and vertical alignments are used to define a framework for the roadway. The original surface terrain acts as a foundation to compute the catch points. The cross section templates are used to define the transverse geometry for the roadway.

Similar to the existing surface as discussed previously, the 3-D design surface is also a digital terrain model (DTM). To create the design surface, InRoads requires you to define a roadway corridor that applies the cross section templates to the horizontal and vertical alignments. The points in each template at different stations are connected with the corresponding points in adjacent templates. For example, the centerline point, left edge of shoulder, left edge of traveled way, right edge of shoulder, and right edge of traveled way on each template of a station are linked with those on the template of adjacent stations. All the connected lines between template points are considered as breaklines. InRoads uses these breaklines to create a DTM surface.

The roadway designer is a tool in InRoads that creates 3-D design surfaces. It requires an interactive process that integrates horizontal and vertical alignments with cross section elements. The process consists of defining roadway corridors, adding template drops, addressing template transitions, defining superelevation, etc. The rest of the chapter describes the Caltrans design standards associated with the superelevation design for the tutorial project. Then detailed procedures are used in the tutorial project to create roadway designer files and define corridors. The end products of this chapter are a series of design surfaces created for Route 8, Laguna Canyon Road, and ramps.

11.1 Design Standards for the Tutorial Project

As mentioned in the previous chapter related with horizontal alignment design, superelevation (along with side friction) is used to counteract the centrifugal force generated while traveling along a horizontal curve. Caltrans *HDM* governs the superelevation design for the tutorial project, which includes the superelevation rate and the superelevation transition length.

11.1.1 Superelevation rate

The selection of appropriate superelevation rates depends on a series of factors, including type of highway facility, traveling speed, curve radius, and side friction factor. Index 202.2 of Caltrans *HDM* provides the details of superelevation rate design under various conditions. Table 202.2 presents the recommended standard superelevation rates for different types of highway and curve radii. Refer to Caltrans *HDM* for details.

11.1.2 Superelevation transition

The superelevation transition is essential for the roadway's pleasing appearance and driver safety and comfort. In general, the transition consists of two components: crown runoff and superelevation runoff. The detailed standards of a superelevation transition can be found in the diagram and tabular data shown in Figure 202.5A of the Caltrans *HDM*.

11.1.3 Superelevation table (.sup)

Using the Caltrans standards related to superelevation, we can create a superelevation table (in a text file) to specify the transition lengths and the corresponding superelevation rates while defining superelevation in InRoads. The file has a file extension **.sup** and contains specific formatting and headings.

For our tutorial project, two example .sup files (see Tables 11-1 and 11-2) are created. These two files are named **CaltransRamp.sup** and **CaltransMultilane.sup.** You can access these two files from www.csupomona.edu/~xjia/GeometricDesign/Chp11/Master Files/. (You can modify them or create your own .sup files for your specific project needs). One is for single-lane ramps and the other is for Route 8 and multilane highways. For greater details, readers can refer to InRoads Help.

11.2 Create Roadway Designer File (.ird) of the Project

Before we work on the corridors for Route 8, ramps, and Laguna Canyon Road, you need to have MicroStation and InRoads installed on your computer. You also need to copy the **tutorial.dgn** file from the

C:/Temp/Book/Chp11/Master Files folder to the **C:/Temp/Tutorial/Chp11/Master Files/** folder or you can download the files from the book website (www.csupomona.edu/~xjia/GeometricDesign/Chp11/Master Files/). The steps to create a new roadway design file (.ird) follow:

1. Start MicroStation and InRoads (if they are not open).
2. Load the original project file **Tutorial.dgn** from the **C:/Temp/Tutorial/Chp11/Master Files/** folder.

   You cannot use **Tutorial_2D.dgn** for this chapter. You must create design surfaces for Route 8, ramps, and Laguna Canyon Road in a 3-D environment.
3. Select the **Corridors Tab** from the ***InRoads → Workspace Bar.***
4. Right click on the **Roadway Designer entry** in the workspace bar window.
5. Select the **New...** option.
6. In the **Save As** dialog box that appears, browse to your project location and save the file with the name of **Tutorial.ird**.

CaltransRamps.sup - Notepad

```
*  this file has not been checked; USE AT YOUR OWN RISK
*  values are not appropriate for Urban Roads, use the correct tables
*
* Design Speed in mph
*
*                                                        2-lane       4-lane
*                                                        tran         transition
*     Placeholder        Radius         Super Rate       length       lenght
*                                                                     (not used for CalTrans)
*
      -----------        --------       ----------       --------     --------
        Curve            20000.00           nc              0.0          0.0
        Curve            4500.000          .02            150.0        150.0
        Curve            3500.000          .03            150.0        150.0
        Curve            2700.000          .04            150.0        150.0
        Curve            2200.000          .05            150.0        150.0
        Curve            1900.000          .06            150.0        150.0
        Curve            1600.000          .07            150.0        150.0
        Curve            1350.000          .08            150.0        150.0
        Curve            1100.000          .09            180.0        180.0
        Curve             850.000          .10            210.0        210.0
        Curve             625.000          .11            210.0        210.0
        Curve             624.000          .12            240.0        240.0
        Curve               0.000          .12            240.0        240.0
```

Table 11-1 Superelevation Table for Single-Lane Ramp (CaltransRamps.sup)

Note: 'nc' in the Super Rate Field represents normal crown

CaltransMultiLane.sup - Notepad

```
* BENTLEY SYSTEMS INROADS SUPERELEVATION RATE TABLE
* From Caltrans HDM, July 24, 2009, Table 202.2 and Figure 202.5A
*  this file has not been checked; USE AT YOUR OWN RISK

* Design Speed in mph
*
*                                                        2-lane       4-lane
*                                                        tran         transition
*     Placeholder        Radius         Super Rate       length       lenght
*                                                                     (not used for CalTrans)
*
      -----------        --------       ----------       --------     --------
        Curve            20000.00           nc              0.0          0.0
        Curve            4500.000          .02            240.0        240.0
        Curve            3500.000          .03            330.0        330.0
        Curve            2700.000          .04            450.0        450.0
        Curve            2200.000          .05            510.0        510.0
        Curve            1900.000          .06            150.0        510.0
        Curve            1600.000          .07            510.0        510.0
        Curve            1350.000          .08            510.0        510.0
        Curve            1100.000          .09            510.0        510.0
        Curve             850.000          .10            510.0        510.0
        Curve             625.000          .11            510.0        510.0
        Curve             624.000          .12            510.0        510.0
        Curve               0.000          .12            510.0        510.0
```

Table 11-2 Superelevation Table for Route 8 (CaltransMultilane.sup)

Note: 'nc' in the Super Rate Field represents normal crown

You have now created a typical roadway designer file for the tutorial project (see Figure 11-1). The file is empty at this time.

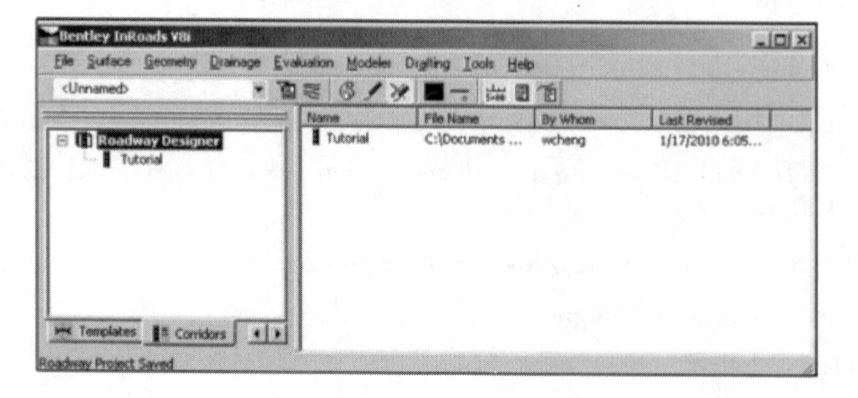

Figure 11-1 Creating the Tutorial Roadway Designer.

11.3 Roadway Corridor and Superelevation for Route 8

Once the roadway designer file is created, you can create the roadway corridor and associated templates and superelevation for Route 8 and store them in the roadway designer file. The procedures are shown below:

1. Define a Roadway Corridor for Route 8

Load **OG.dtm, Tutorial.alg**, and **Tutorial_Chapter11.itl** into InRoads. These files are available at **C:/Temp/Book/Chp11/Master Files/** or on the website (www.csupomona.edu/~xjia/GeometricDesign/Chp11/Master Files/). Copy these three files into **C:\Temp\Tutorial\Chp11\Master Files** folder before you load them into InRoads. The **Tutorial_Chapter11.itl** template library is different from the **Tutorial.itl** template library created in Chapter 10. It has two additional templates (Bridge and Bridge Railing).

a. Set the surface **OG**, horizontal alignment **Route 8** and vertical alignment **Route 8,** and template **Route 8** to active.

b. To make template Route 8 active, go to ***Modeler → Create Template....***

c. Select ***InRoads → Modeler → Roadway Designer.***

   The **Roadway Designer** dialog box with three empty view windows opens as shown in Figure 11-2.

d. From the ***InRoads → Roadway Designer*** dialog main menu, select ***Corridor → Corridor Management....***

e. Type the information shown in Figure 11-3 into the **Manage Corridors** dialog box.

f. Click **Add** to create the corridor **Route 8.** Click **Close.**

   You have now definend a new corridor for Route 8. The corridor is empty at this time.

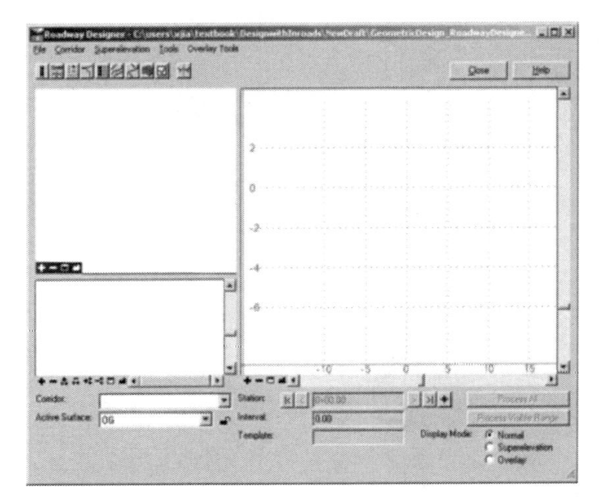

Figure 11-2 Roadway Designer Interface

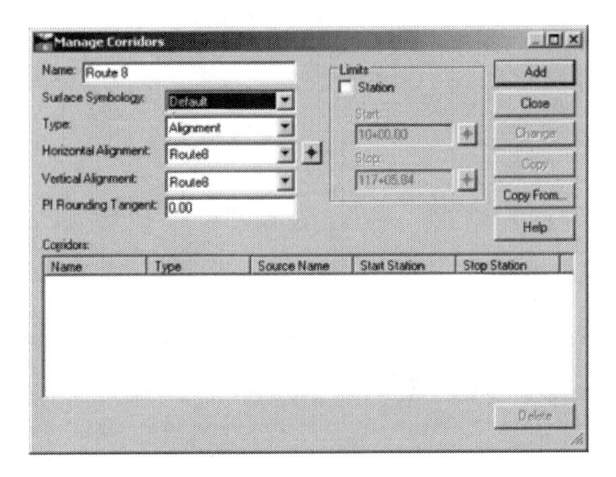

Figure 11-3 Manage Corridor Window

2. Add templates (or template drops) to Route 8 corridor
   a. From the ***InRoads → Roadway Designer*** main menu, select ***Corridor → Template Drops....***
   b. Type the information shown in Figure 11-4 into the **Template Drops** dialog box.

      In the **Library Templates** window, expand the library folders and select your design template, **Route 8**.
   c. Click **Add.**
   d. Type the information shown in Figure 11-5 into the **Template Drops** dialog box.

      Note that an additional template titled **Bridge** is available from the **Tutorial_Chapter11.itl** template library. To use the Tutorial.itl you created from the last chapter, insert the **Bridge** and **Bridge Railing** templates into your library.

      The bridge begins at station 64+30.67 (see Chapter 9). The interval of 10 ft. is selected due to the relatively short length of the bridge. The **Bridge** template is applied to the Route 8 corridor between stations 64+30.67 and 65+54.67.
   e. Click **Add.**

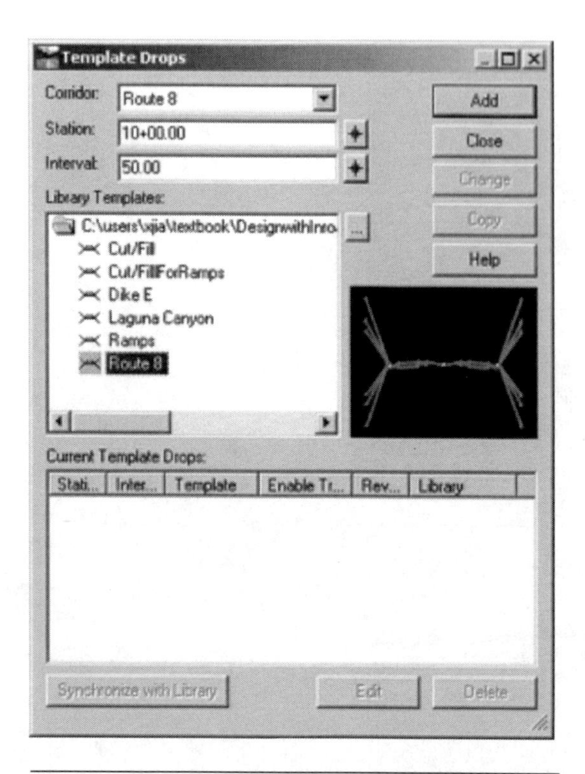

Figure 11-4 Select Route 8 Template

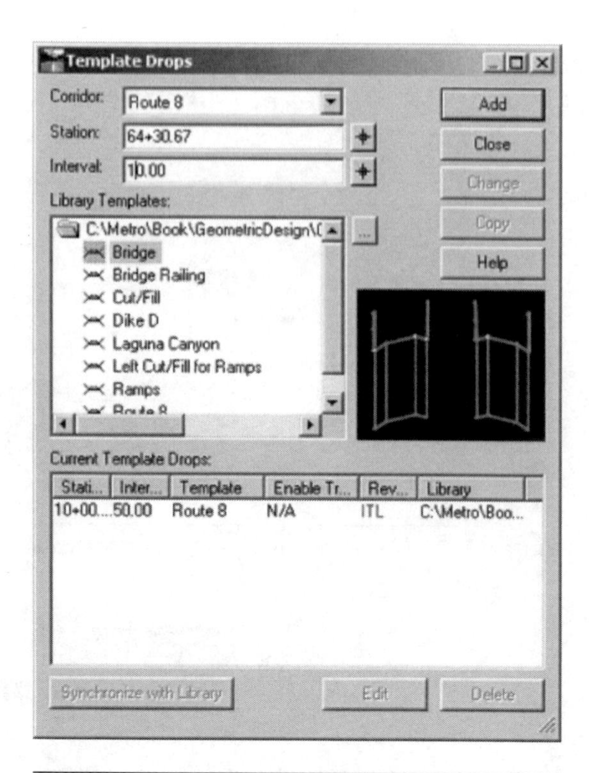

Figure 11-5 Adding the Bridge Template to the Route 8 Corridor

f. Type the information shown in Figure 11-6 into the **Template Drops** dialog box.

   The bridge ends at station 65+54.67. The interval of 50 ft. is selected due to the relatively long length of Route 8. The Route 8 template is re-inserted here, starting at Station 65+54.67.

g. Click **Add**.

   Now all three template drops (Route 8 template, Bridge template, and Bridge Railing template) have been added to the corridor **Route 8**. Note that two yellow sections are illustrated in the plan view of the **Roadway Designer** main dialog box (see Figure 11-7). It indicates there is a transition. A transition occurs when the points in adjacent templates do not match (e.g., different point names, different number of points, etc.). Template transitions are useful in situations like lane number change, median to no median, sidewalk to no sidewalk, etc. In this case, you will not transit the Route 8 mainline to bridge. Hence, you can delete the transition by right clicking over the two yellow sections and select **Delete**. The yellow highlights will disappear.

h. From the **Roadway Designer** main menu, select ***File → Save***.

   The corridor **Route 8** has been defined, which consists of three templates.

   In the next subsection you will learn how to define superelevation.

3. Define superelevation in the Route 8 corridor

Before you define the superelevation for the Route 8 corridor, it is important to review some crucial information associated with superelevation so you have a better understanding of the corresponding procedures.

Note 1 InRoads provides several ways to define the superelevation, including the AASHTO, the fixed length, and the table methods. For this tutorial project, use the table method, which requires you to load the previously created superelevation rate tables based on Caltrans standards.

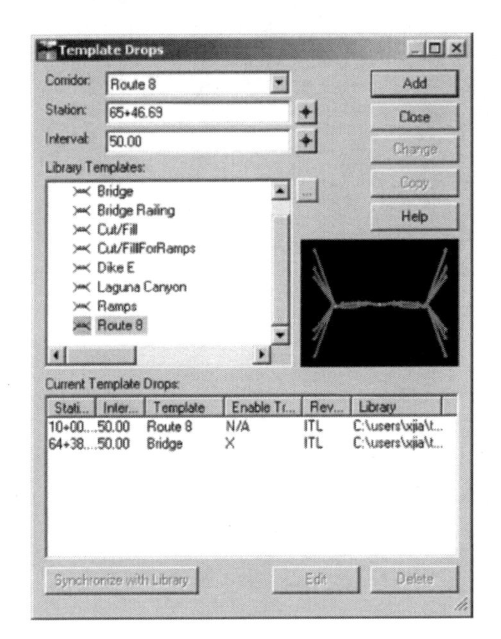

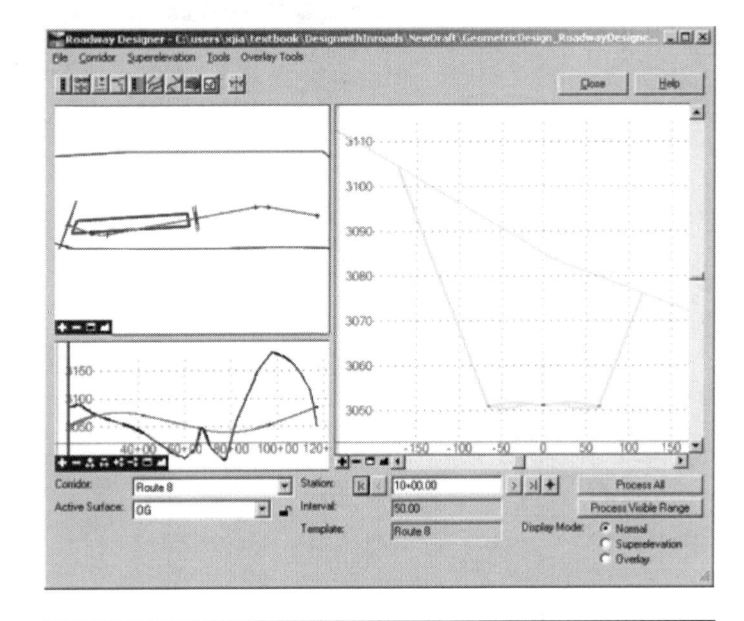

Figure 11-6 Adding the Route 8 Template to the Route 8 Corridor

Figure 11-7 Three Template Drops and Their Transitions

Note 2 Superelevation is only required at horizontal curves. That said, the definition of superelevation is not needed for Laguna Canyon Road since the whole roadway section follows inside the tangent line.

Note 3 The number of superelevation sections depends on the template shape. In general, if a template is symmetric and represents a two-way roadway, two superelevation sections must be defined. Conversely, if a template is asymmetric and represents a one-way roadway, only one superelevation section is required. Each superelevation section has its own Superelevation Range Points and Pivot Point. The *range points* specify the beginning and ending point of a cross section line on the pavement where the superelevation should be applied. The *pivot point* specifies the point on a cross section line on which the superelevation is rotated. Therefore, in our project, two superelevation sections are required for Route 8 and only one superelevation section for all the ramps.

Note 4 For the superelevation to work properly over the entire corridor the point names in different templates should be consistent. In our project, the Bridge template for Route 8 is created based on the Route 8 mainline template, and the corresponding points are named in a consistent way. Hence, we do not need to modify the point names for the templates to ensure consistency. Otherwise, the template drops in each corridor require editing and synchronization before defining the superelevation.

The process for setting up the superelevation is as follows.

a. From the ***InRoads → Roadway Designer*** main menu, select ***Superelevation → Create Superelevation Wizard → Table....***

   The **Table Wizard** dialog box appears as illustrated in Figure 11-8.

b. To browse to **C:/Temp/Book/Chp11/Master Files/**, click [...] to the right of the **Table** field. Select one of the previously created superelevation rate tables for Route 8 (CaltransMultilane.sup). You can also download this rate table from the textbook website (www.csupomona.edu/~xjia/GeometricDesign/Chp11/Master Files/).

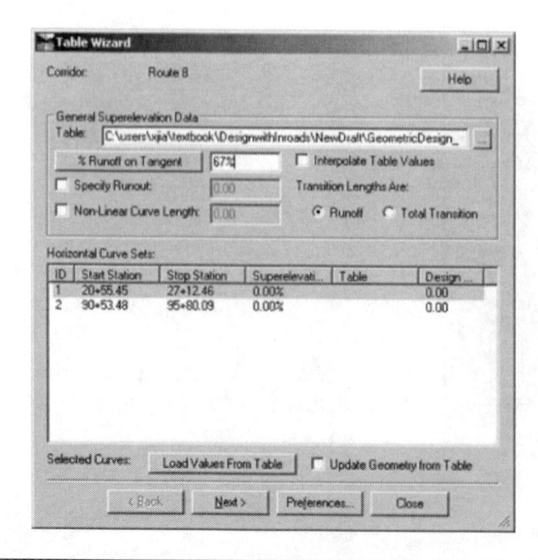

Figure 11-8 Select Superelevation Rate Table

The % Runoff on Tangent is set to 67%, meeting the Caltrans requirements.

c. From the **Horizontal Curve Sets** list, select curve 1. Click **Load Values From Table** (see Figure 11-9).

   The superelevation rate for curve 1 is updated from 0% to 9%, which corresponds to its radius.

d. From the **Horizontal Curve Sets** list, select curve 2. Click **Load Values From Table** (see Figure 11-9).

   The superelevation rate for curve 2 is updated from 0% to 8%, which corresponds to its radius.

e. In the **Table Wizard** dialog, click **Next>**.

   The **Superelevation Section Definitions** dialog box appears as shown in Figure 11-10.

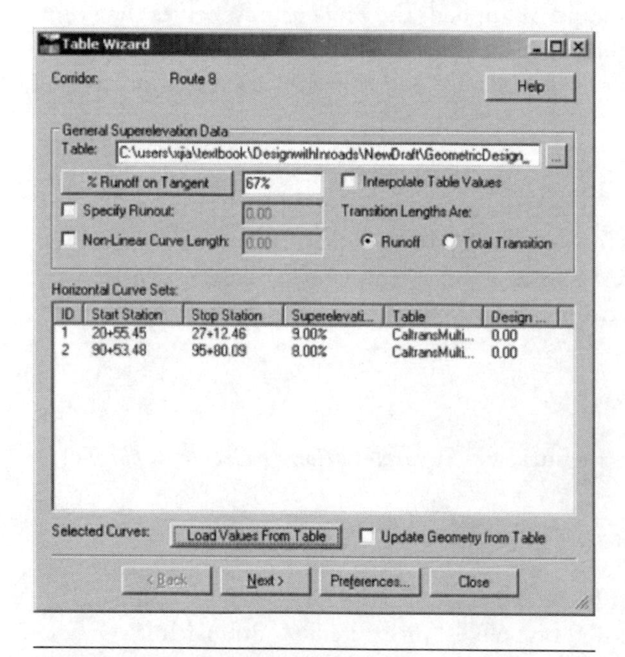

Figure 11-9 Apply Superelevation Rate Table to Curves

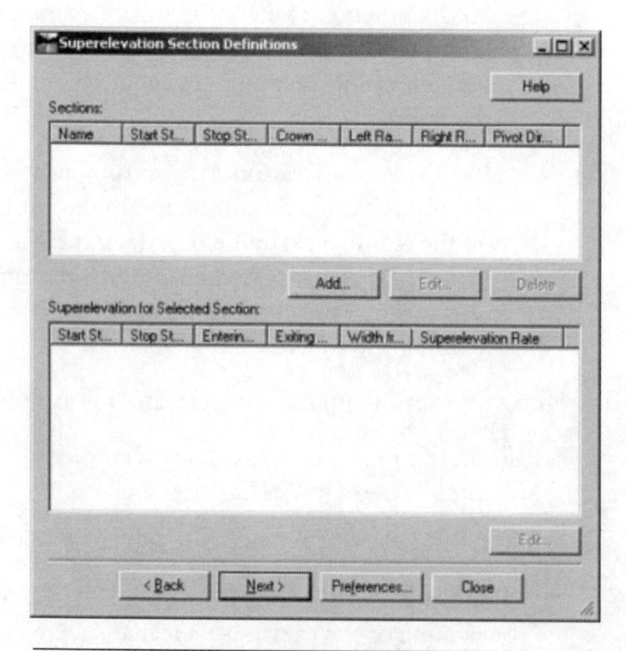

Figure 11-10 Superelevation Section Definitions

f. In the **Superelevation Section Definitions** dialog box, click **Add....**

   The **Add Superelevation Section** dialog box appears. Note that this refers to the superelevation section on the westbound direction of the Route 8 corridor.

Before we work on this step, it is critical to understand the location of the control points used for the superelevation sections (see Figure 11-11).

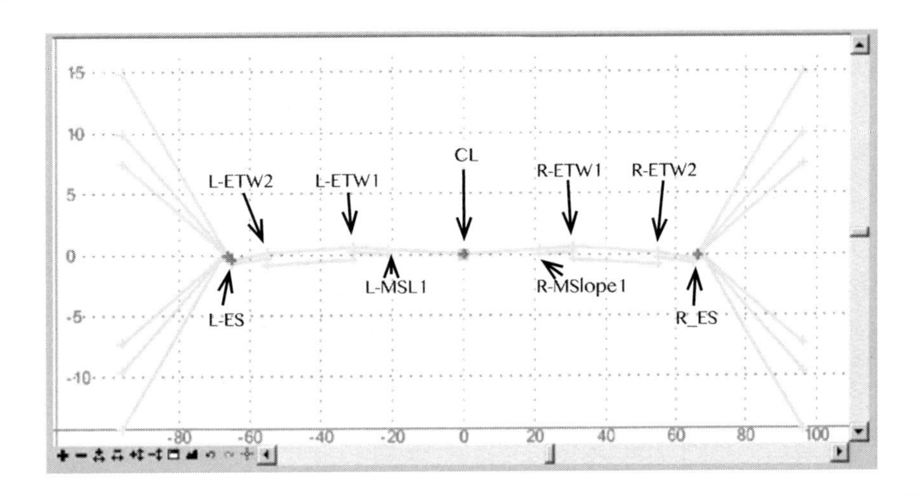

Figure 11-11 Control Points Used for Superelevation Sections

Your controls for the Route 8 template might not be the same as those shown in Figure 11-11. However, you can use yours (based on your template design from the previous chapter) to replace the corresponding control names used in the steps below.

a. In the **Add Superelevation Section** dialog box, type the information shown in Figure 11-12. Click **OK**.

   The first superelevation section for Route 8 is developed.

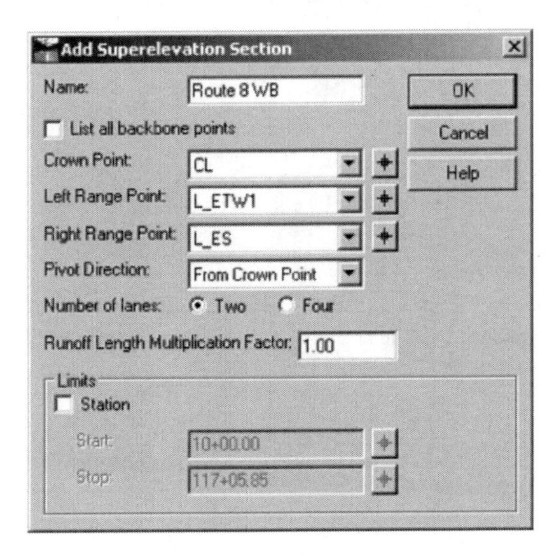

Figure 11-12 Add Superelevation Section 1

b. In the **Superelevation Section Definitions** dialog box (shown in Figure 11-13), click **Add…**.

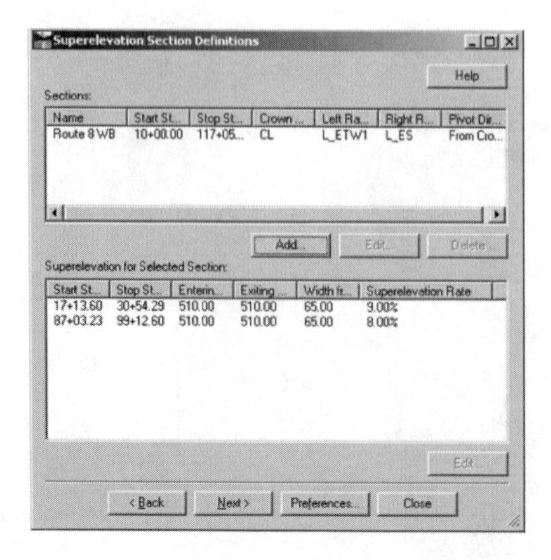

Figure 11-13 Prepare for Superelevation Section 2

c. In the **Add Superelevation Section** dialog box, type the information shown in Figure 11-14. Click **OK**.

The second superelevation section for Route 8 is developed. Note that this section refers to the superelevation section on the eastbound direction of the Route 8 corridor.

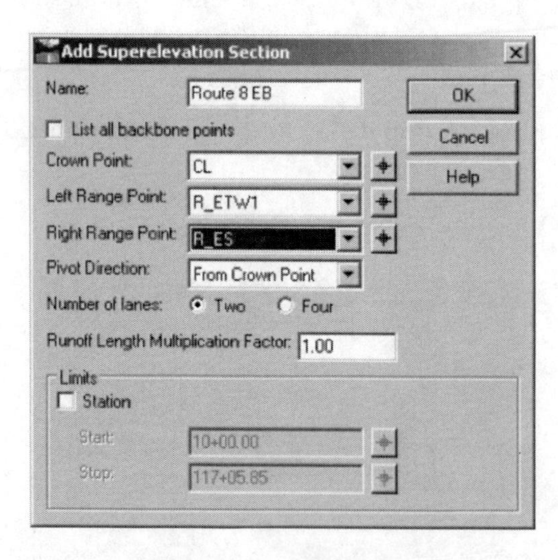

Figure 11-14 Add Superelevation Section 2

d. In the **Superelevation Section Definitions** dialog box (shown in Figure 11-15), click **Next >**.

e. In the **Superelevation Controls** dialog box, click **Finish**.

You will see the horizontal alignment view, the vertical alignment view, and the cross section view in Figure 11-16. You can use the > **Icon** to go through the Route 8 corridor at intervals of 50 ft. (see Figure 11-16).

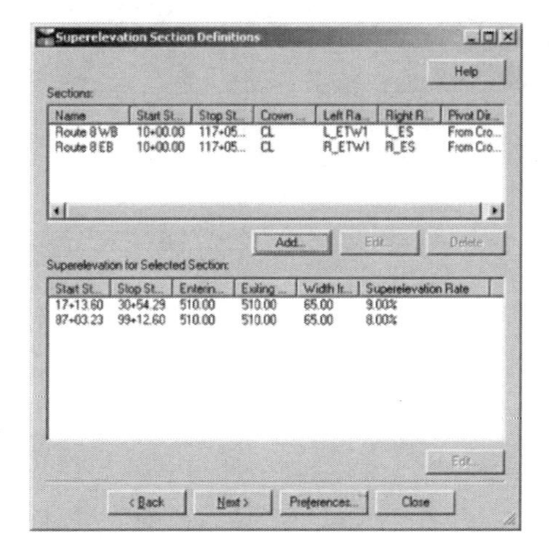

Figure 11-15 Two Superelevation Sections

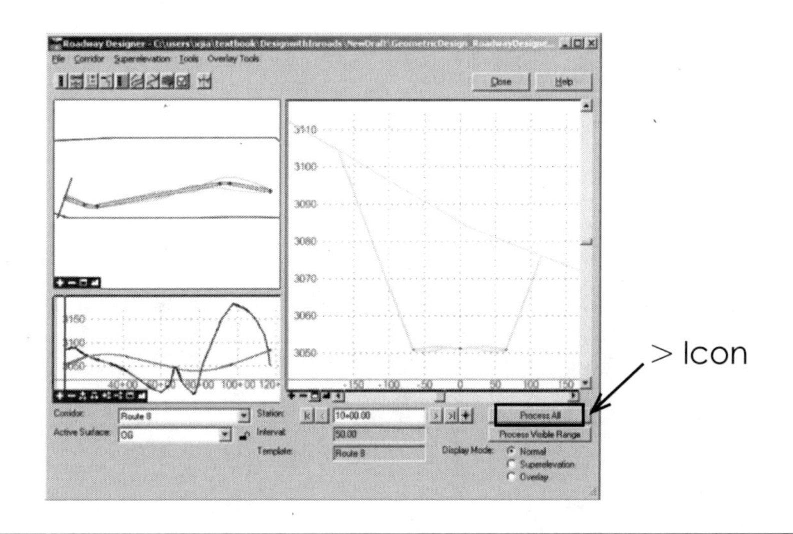

Figure 11-16 Three Views in the Roadway Designer Interface

f. In the **Roadway Designer** main menu, select ***File*** → ***Save***.

You have now defined the superelevation sections for the corridor **Route 8.**

» To see the four views associated with superelevation across the different stations, check the **Overlay** option in the **Roadway Designer** main dialog box (see Figure 11-17).

» To return to the normal roadway views, check the **Normal** option to ON as shown in Figure 11-17.

g. In the **Roadway Designer** main menu, select ***File*** → ***Save***.

4. Create the design surface for Route 8

a. Set the surface **OG.dtm** to active.

b. From the ***InRoads*** → ***Roadway Designer*** main menu, select ***Corridor*** → ***Create Surface....***

c. In the **Create Surface** dialog box, type the information as shown in Figure 11-18. Click **Apply**.

The **Results** dialog appears.

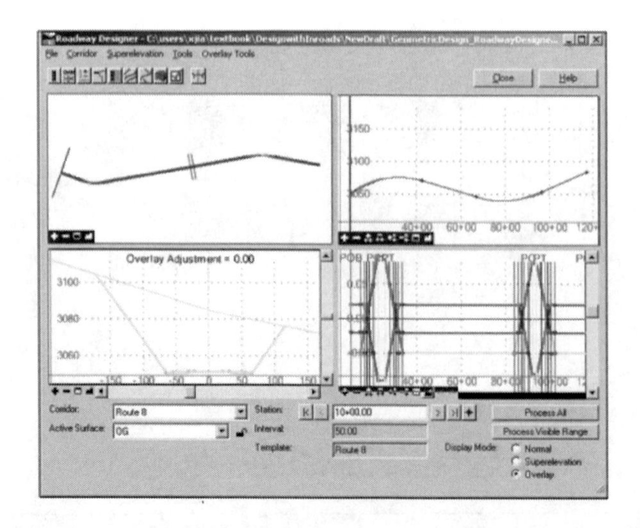

Figure 11-17 Four Views for the Route 8 Corridor

d. Review the results and click **Close**.

e. From the ***InRoads → Workspace Bar***, select the **Surface Tab**.

You will see a new surface named **Route 8.dtm** (see Figure 11-19).

f. Set this surface to active.

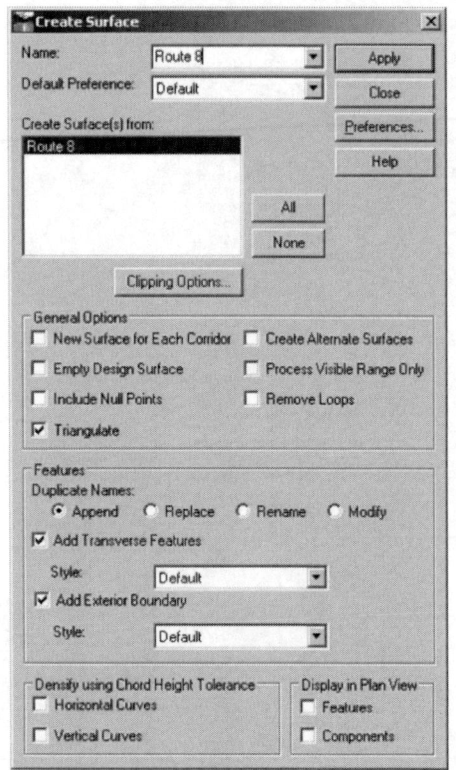

Figure 11-18 Create Route 8 Surface

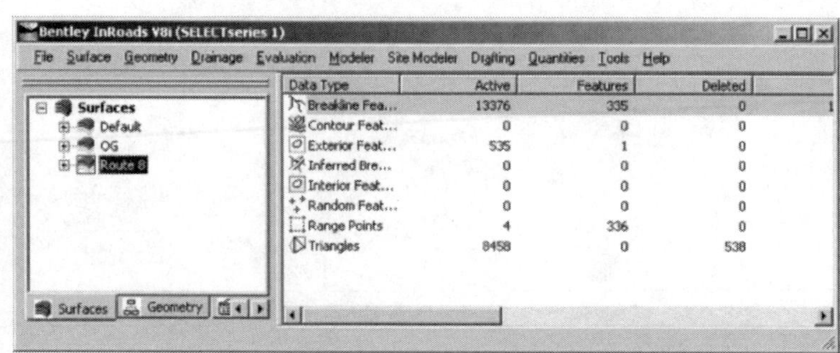

Figure 11-19 Route Surface Created

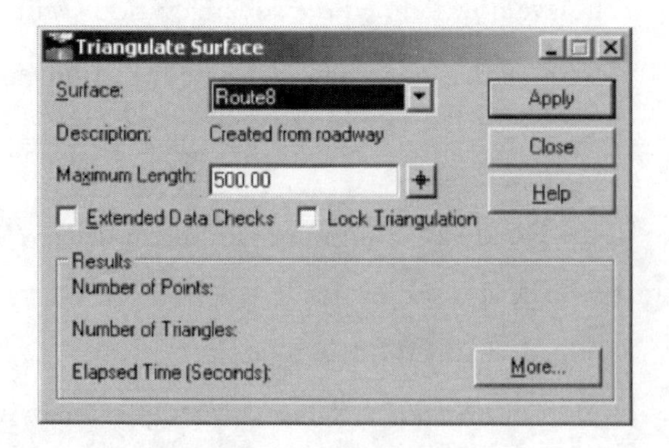

Figure 11-20 Triangulation for Route 8

g. Right click **Route 8.dtm**. Select **Triangulate.** Type **500** for in **Maximum Length** (see Figure 11-20).

h. Click ***Apply*** and ***Close***.

i. Go to ***InRoads → Surface → View Surface → Triangles....***

   You will see the triangles of Route 8 in the **Tutorial.dgn** file (similar to Figure 11-21).

j. To delete the triangulations, click ***MicroStation → Undo***.

   To view the Route 8 design surface in 3-D mode, follow the procedure in Chapter 6.

   Figure 11-22 shows a portion of the 3-D view of the Route 8 design surface.

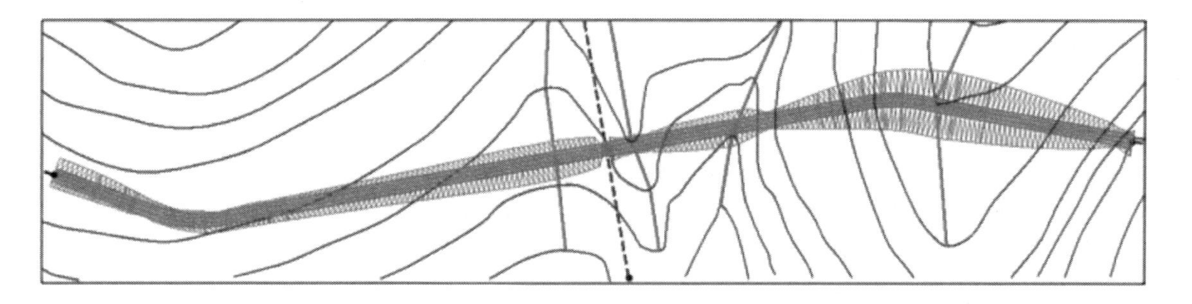

Figure 11-21 View of Route 8 Triangles

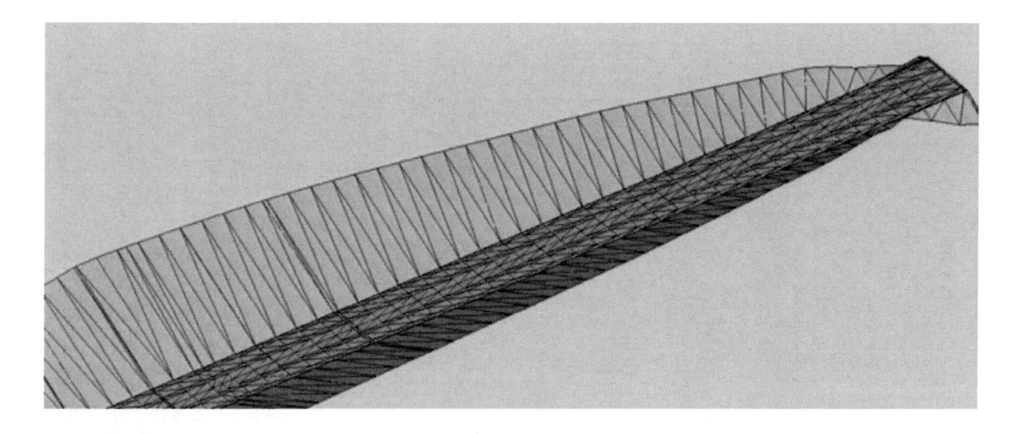

Figure 11-22 3D View of Route 8 Design Surface

k. Go to ***InRoads → File → Save → Surface.***

   A **Save As** dialog box appears.

l. Browse to the **C:/Temp/Tutorial/Chp11/MasterFiles/** folder. Save the file with the name of **Route 8.dtm**.

   You have now created the design surface for Route 8.

In the next subsection we will focus on creating the corridor and design surface for Laguna Canyon Road.

11.4 Roadway Corridor for Laguna Canyon Road

As discussed previously, since the Laguna Canyon Road section in the tutorial project consists only of a tangent line, the superelevation is not needed for the roadway section. Therefore, the corresponding procedures only include defining a roadway corridor, adding a template drop, and creating a design surface.

1. Define a roadway corridor for Laguna Canyon Road
   a. From the **C:/Temp/Tutorial/Chp11/Master Files/** folder, load **OG.dtm, Tutorial.alg**, and **Tutorial_Chapter11.itl** into InRoads.
   b. Set the surface **OG**, horizontal alignment **Laguna Canyon Road** and vertical alignment **Laguna Canyon Road**, and template **Laguna Canyon** to active.

      Follow these steps to set the template **Laguna Canyon** to active.

      » Go to ***InRoads → Modeler → Create Template....***
      » Select ***InRoads → Modeler → Roadway Designer.***
      » Select ***InRoads → Corridor → Corridor Management....***
      » Type the information shown in Figure 11-23 into the **Manage Corridors** dialog box.

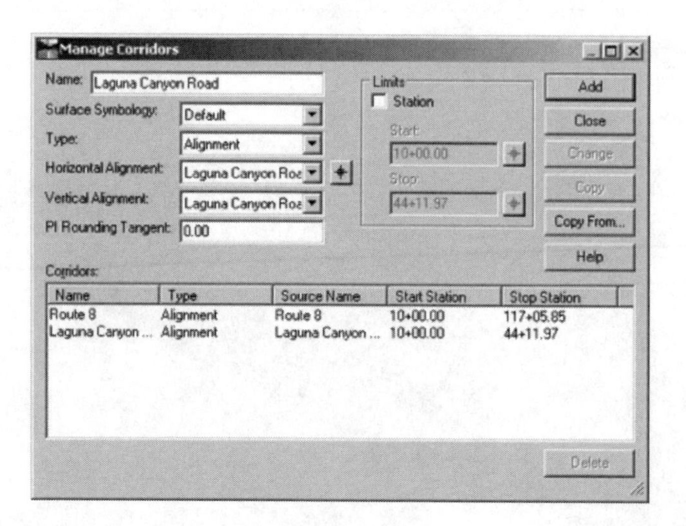

Figure 11-23 Create Laguna Canyon Road Corridor

   c. To create the corridor **Laguna Canyon Road**, click **Add.** Then click **Close.**

      You have now defined a new corridor for Laguna Canyon Road. The corridor is empty at this time.

2. Add template drop to the Laguna Canyon Road corridor
   a. From the **Roadway Designer** main menu, select ***Corridor → Template Drops....***
   b. Type the information shown in Figure 11-24 into the **Template Drops** dialog box.

      Note: In the **Library Templates** window you need to expand the library folders and select your design template, **Laguna Canyon**.
   c. Click **Add.**

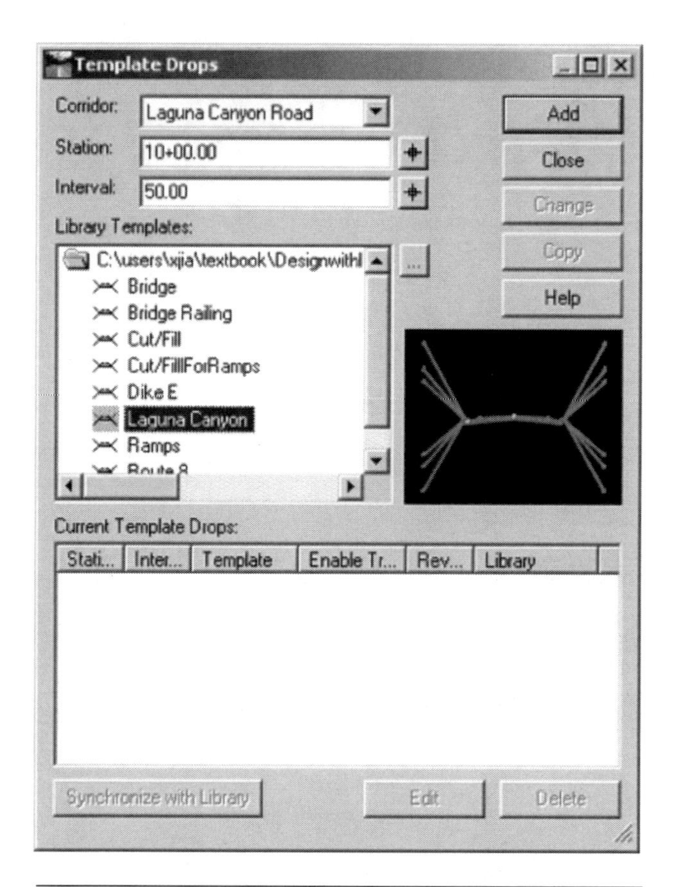

Figure 11-24 Template Drop for Laguna Canyon Road Corridor

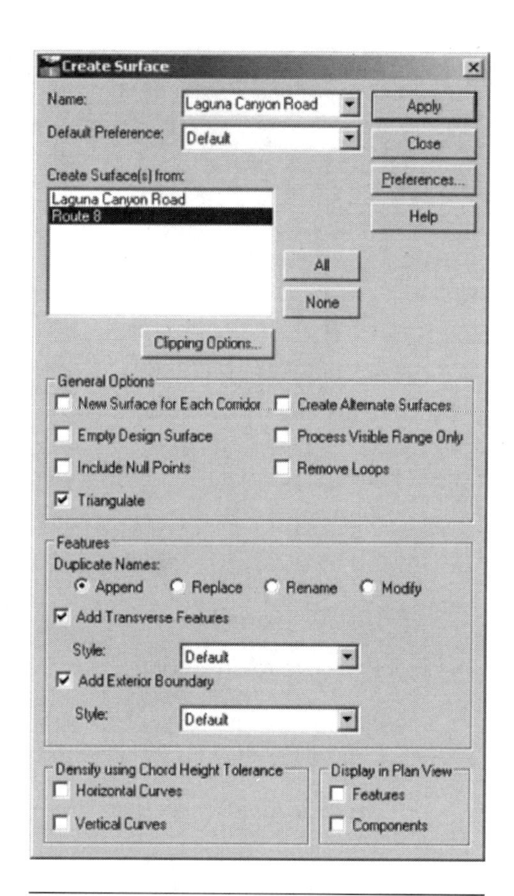

Figure 11-25 Create Surface for Laguna Canyon Road

The **Laguna Canyon** template has been added to the Corridor **Laguna Canyon Road.**

d. From the **Roadway Designer** main menu, select ***File* → *Save*.**

Now the Corridor **Laguna Canyon Road** has been defined, which consists of one template.

In the next subsection we will go over the procedures to develop the design surface for Laguna Canyon Road.

3. Create the design surface for Laguna Canyon Road
   a. Set the surface **OG.dtm** to active.
   b. From the **Roadway Designer** main menu, select ***Corridor* → *Create Surface....***
   c. In the **Create Surface** dialog box, type the information shown in Figure 11-25. Click **Apply**.

      The **Results** dialog box appears.
   d. Review the results and click **Close**.
   e. Select ***InRoads* → *Workspace Bar* → *Surface Tab*.**

      A new surface named **Laguna Canyon Road** appears (see Figure 11-26).
   f. Set this surface to active.

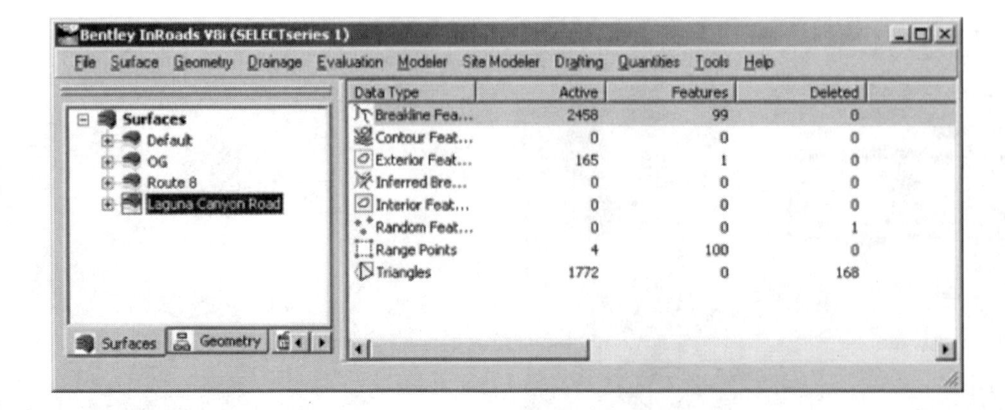

Figure 11-26 Laguna Canyon Road Surface

g. Go to ***InRoads → Surface → View Surface → Triangles....***

The triangles of Laguna Canyon Road in the **Turorial.dgn** file appear (see Figure 11-27).

h. To delete the triangulations, click ***MicroStation → Undo***.

Follow the steps in Chapter 6 to view the Laguna Canyon Road design surface in 3-D mode. Figure 11-28 shows a portion of the 3-D view.

i. Go to ***InRoads → File → Save → Surface.***

A **Save As** dialog box appears.

j. Browse to your project location at **C:/Temp/Tutorial/Chp11/Master Files** folder. Save the file with the name of **Laguna Canyon Road.dtm.**

You have now created the design surface for Laguna Canyon Road.

In the next subsection we will focus on creating the corridors and design surfaces for ramps.

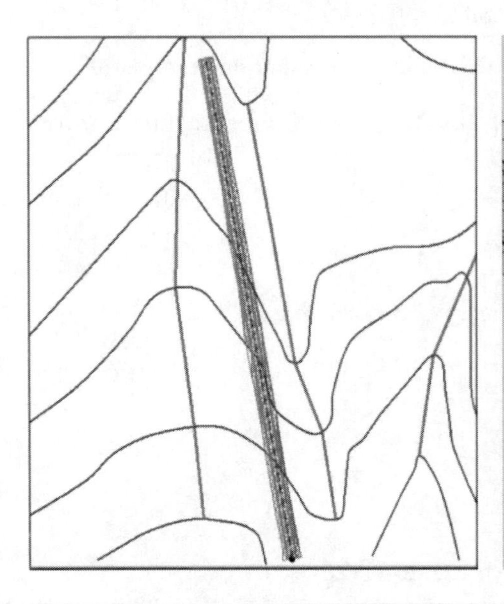

Figure 11-27 Triangles for Laguna Canyon Road

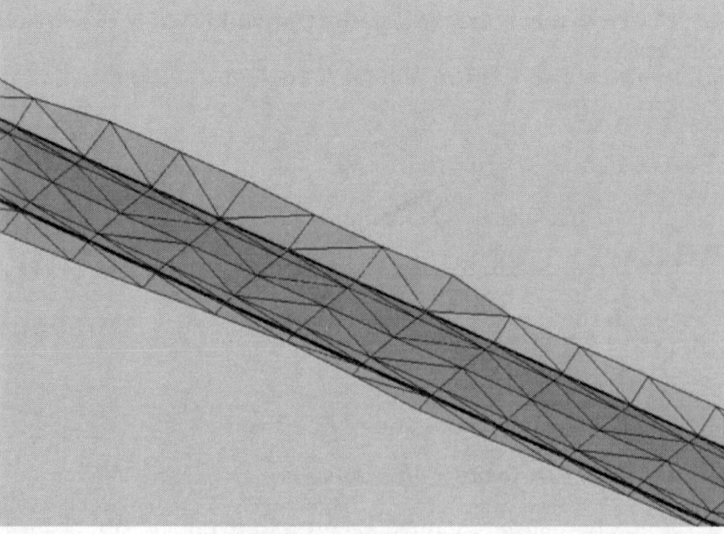

Figure 11-28 A Portion of 3D View of Laguna Canyon Road

11.5 Roadway Corridor and Superelevation for EBOFF Ramp

Similar procedures apply to the EBOFF ramp, which include defining a roadway corridor, adding a template drop, defining superelevation sections, and creating a design surface. The details are described in order as follows.

1. Define a roadway corridor for the EBOFF ramp
   a. From the **C:/Temp/Tutorial/Chp11/Master Files** folder, load **OG.dtm, Tutorial.alg**, and **Tutorial_Chapter11.itl** into InRoads.
   b. Set the surface **OG**, horizontal alignment **EBOFF** and vertical alignment **EBOFF**, and template **Ramps** to active.

      Note: To set template Ramps to active, go to ***InRoads → Modeler → Create Template...***).
   c. Select ***InRoads → Modeler → Roadway Designer.***
   d. From the **Roadway Designer** main menu, select ***Corridor → Corridor Management....***
   e. Type the information shown in Figure 11-29 into the **Manage Corridors** dialog box.

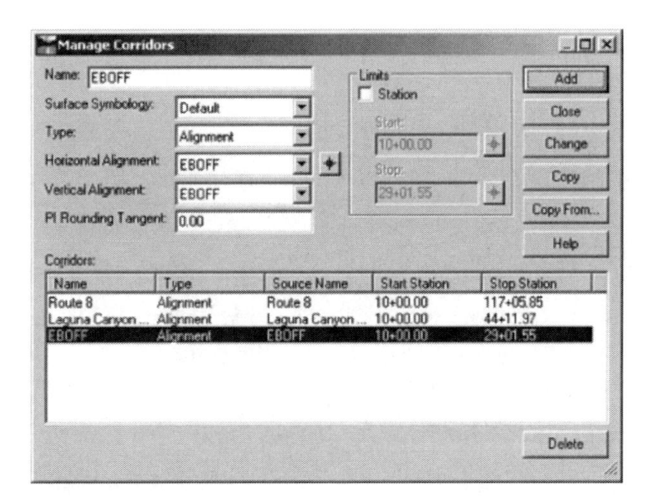

Figure 11-29 Create Corridor for EBOFF Ramp

   f. To create the corridor **EBOFF**, click **Add.** Then click **Close**.

      You have now defined a corridor for the EBOFF ramp. The corridor is empty at this time.

2. Add a template drop to the EBOFF corridor
   a. From the **Roadway Designer** main menu, select ***Corridor → Template Drops....***
   b. Type the information shown in Figure 11-30 into the **Template Drops** dialog box.

      Note: In the **Library Templates** window you need to expand the library folders and select your design template **Ramps**.
   c. Click **Add**.

      The template drop has been added to the corridor **EBOFF**.

d. From the **Roadway Designer** main menu, select ***File* → *Save***.

The corridor **EBOFF** has been defined, which consists of one template.

In the next subsection we will learn how to define superelevation.

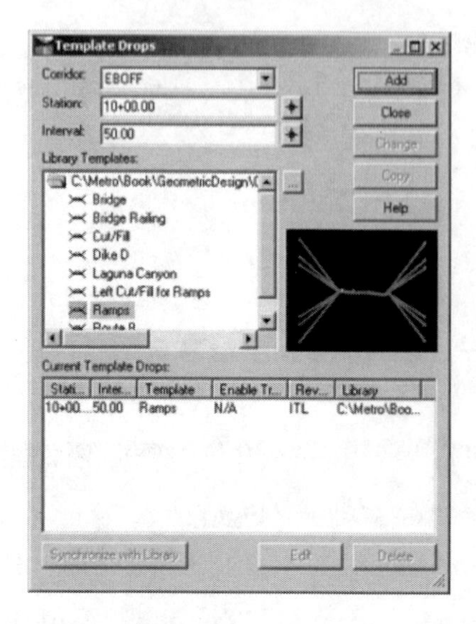

Figure 11-30 Template Drops for EBOFF Ramp

3. Define superelevation in the EBOFF corridor

a. From the **Roadway Designer** main menu, select ***Superelevation* → *Create Superelevation Wizard* → *Table…***.

The **Table Wizard** dialog box appears as illustrated in Figure 11-31.

b. To browse to the project directory (**C:/Temp/Turorial/Chp11/Master Files/** or **C:/Temp/Book/Chp11/Master Files** folder), click […] to the right of the **Table** field. Select **CaltransRamps.sup** for the single-lane ramp.

c. From the **Horizontal Curve Sets** list, select curve 1 and click **Load Values From Table**.

d. From the **Horizontal Curve Sets** list, select curve 2. Click **Load Values From Table**.

e. In the **Table Wizard** dialog box, click **Next>**.

The **Superelevation Section Definitions** dialog box appears as shown in Figure 11-32.

f. In the **Superelevation Section Definitions** dialog box, click **Add…**.

The **Add Superelevation Section** dialog box appears as shown in Figure 11-33.

g. In the **Add Superelevation Section** dialog box, type the information shown in Figure 11-33. Click **OK**.

You have now created the superelevation section for the EBOFF ramp.

Notes: Only one superelevation section is required for the ramp. Your controls for the Ramp template might not be the same as those shown in Figure 11-33; however, you can use your controls to replace the corresponding control names used in the steps below.

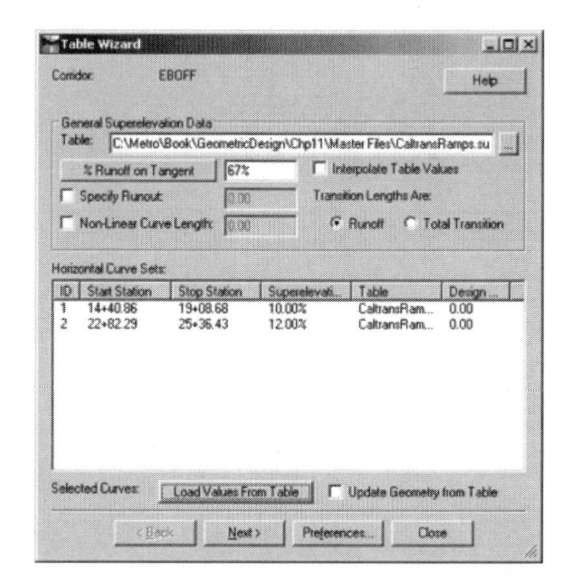

Figure 11-31 Add Caltrans Superelevation Rate Table

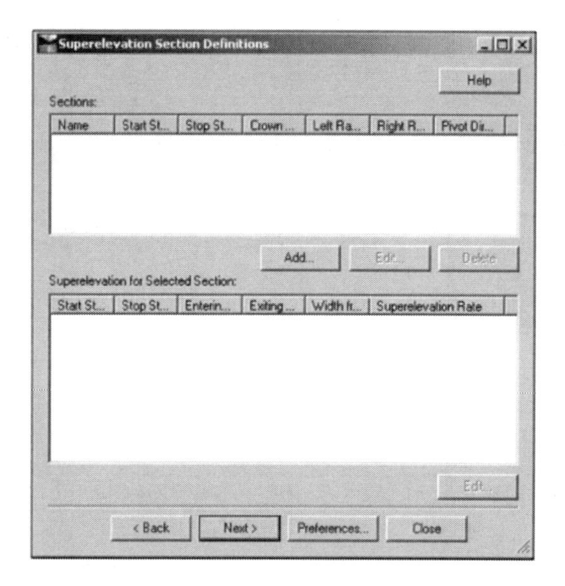

Figure 11-32 Superelevation Section Definitions

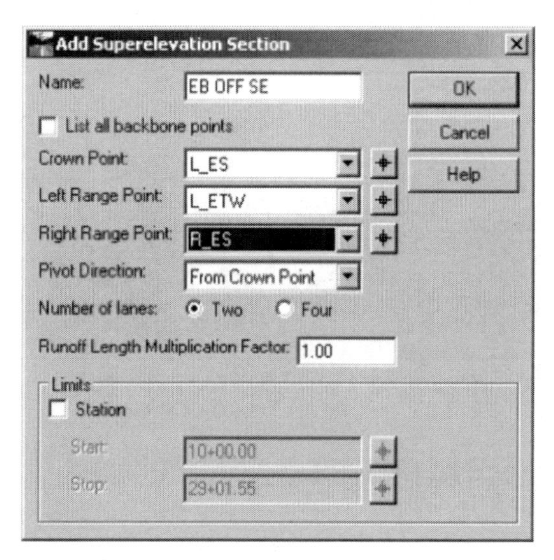

Figure 11-33 Add Superelevation Section Definitions for EBOFF Ramp

h. In the **Superelevation Section Definitions** dialog box shown in Figure 11-34, click **Next>**.

i. In the **Superelevation Controls** dialog box, click **Finish**.

j. In the **Roadway Designer** main menu, select ***File → Save.***

You have now defined the superelevation within the corridor **EBOFF**.

» To see the three views for the superelevation section across the different stations, in the **Roadway Designer** main dialog box, check the **Normal** option to ON in **Display Mode** (see Figure 11-35).

» To see four views for the EBOFF superelevation section across all the stations, click the **Overlay** option of the **Display Mode** (see Figure 11-36).

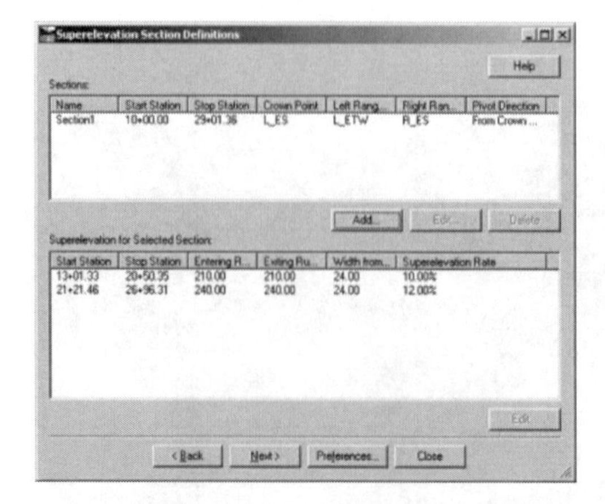

Figure 11-34 Create Superelevation Section

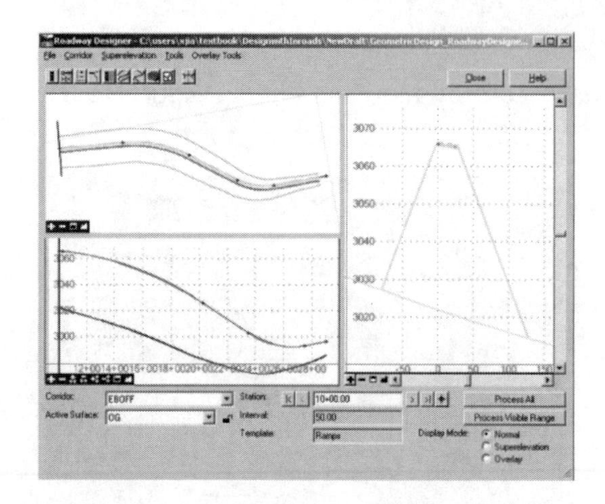

Figure 11-35 Superelevation Sections for EBOFF Ramp (Normal)

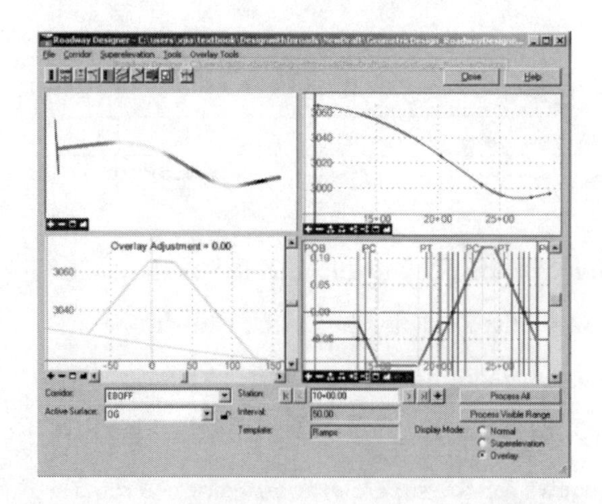

Figure 11-36 Superelevation Sections for EBOFF Ramp (Overlay)

4. Create the EBOFF design surface
   a. Set the surface **OG.dtm** to active.
   b. From the **Roadway Designer** main menu, select ***Corridor*** → ***Create Surface....***
   c. In the **Create Surface** dialog box, type the following information:

Name:	**EBOFF**
Default Preference:	**Default**
Check ON:	**Empty Design Surface**
Check ON:	**Triangulate**
Create Surface(s) from:	**EBOFF**

   d. Click **Apply**.

      The **Results** dialog box appears.
   e. Review the results and click **Close**.
   f. Select the ***InRoads*** → ***Workspace Bar*** → ***Surface Tab***.

      A new surface named **EBOFF.dtm** appears (see Figure 11-37).
   g. Set this surface to active.

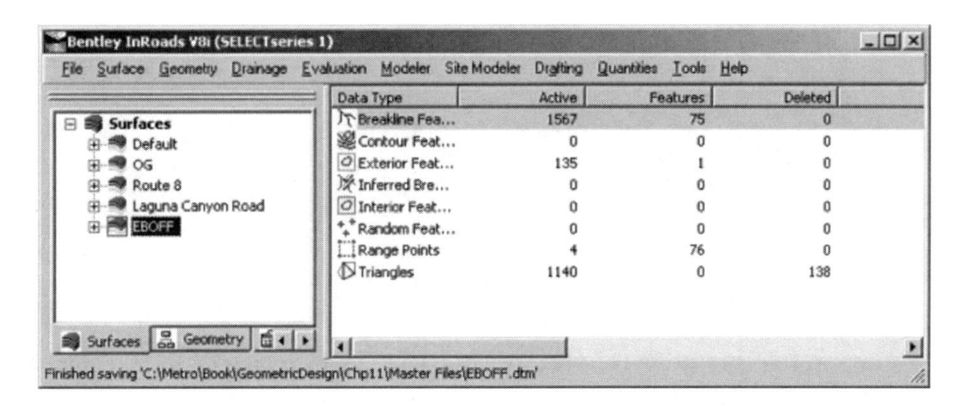

Figure 11-37 EBOFF Design Surface

   h. Right click **EBOFF.dtm** and select **Triangulate**.
   i. To see the triangles of the **EBOFF** ramp in the **Tutorial.dgn** file, go to ***InRoads*** → ***Surface*** → ***View Surface*** → ***Triangles...*** (see Figure 11-38).
   j. To delete the triangulations, click ***MicroStation*** → ***Undo***.

      To view the EBOFF design surface in 3-D mode, follow the steps in Chapter 6. Figure 11-39 shows a portion of the EBOFF design surface in 3-D view.
   k. Go to ***InRoads*** → ***File*** → ***Save*** → ***Surface***.

      A **Save As** dialog box appears.

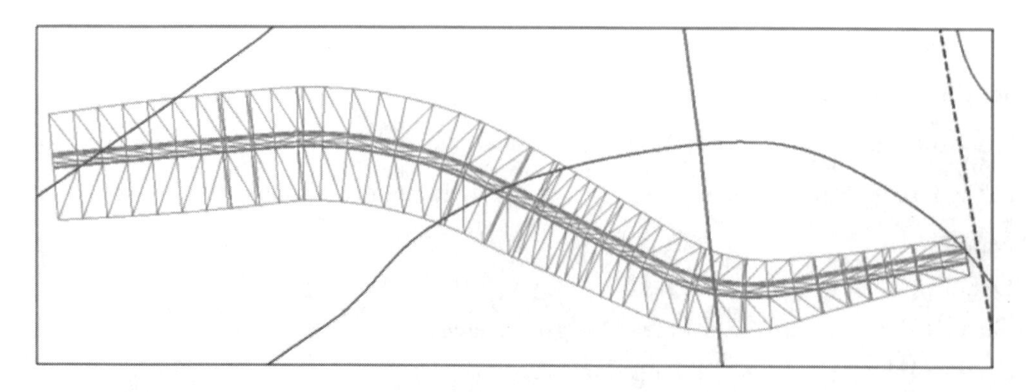

Figure 11-38 Triangles for the EBOFF Ramp Corridor

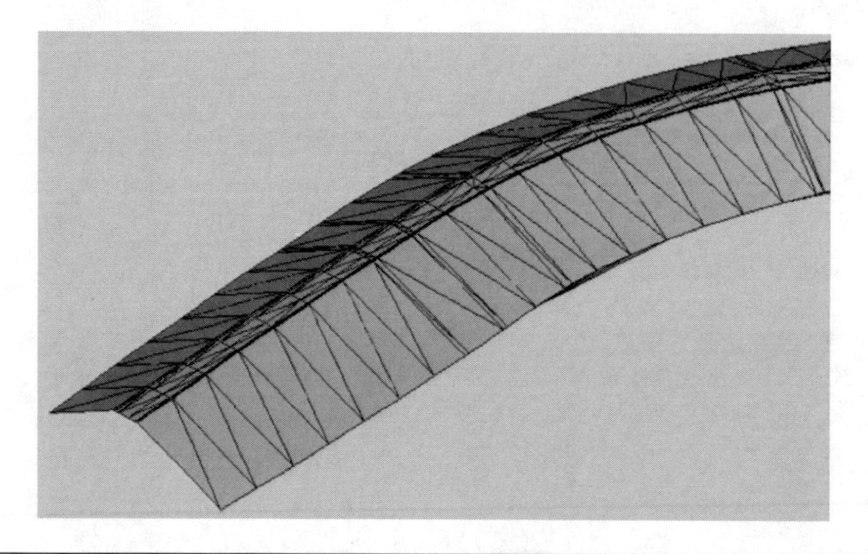

Figure 11-39 A Portion of 3D View of the EBOFF Ramp Corridor

1. Browse to your project location (**C:/Temp/Tutorial/Chp11/Master Files/** folder). Save the file with the name of **EBOFF.dtm**.

   You have now created the design surface for the EBOFF ramp.

In the next subsection we will focus on creating the corridor and design surface for the other three ramps.

11.6 Roadway Corridor and Superelevation for EBON Ramp

Similar procedures apply to the EBON ramp, which include defining a roadway corridor, adding a template drop, defining superelevation sections, and creating a design surface. The details are described in order as follows.

1. Define a roadway corridor for the EBON ramp
   a. From the **C:/Temp/Tutorial/Chp11/Master Files/** folder, load **OG.dtm, Tutorial.alg**, and **Tutorial_Chapter11.itl** into InRoads.

b. Set the surface **OG**, horizontal alignment **EBON** and vertical alignment **EBON** to active.

c. To set the template **Ramps** to active, go to ***InRoads* → *Modeler* → *Create Template....***

d. Select ***InRoads* → *Modeler* → *Roadway Designer*.**

e. From the **Roadway Designer** main menu, select ***Corridor* → *Corridor Management....***

f. Type the information shown in Figure 11-40 into the **Manage Corridors** dialog box.

g. To create the corridor **EBON**, click **Add**. Then click **Close**.

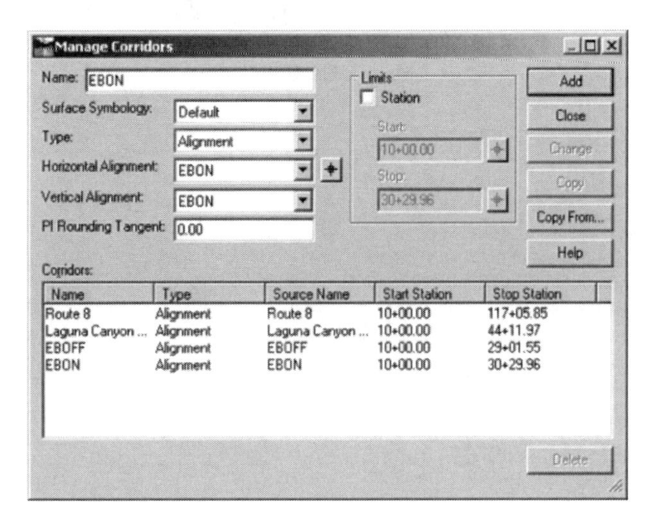

Figure 11-40 Manage Corridors Window

You have now defined a new corridor for the EBON ramp. The corridor is empty at this time.

2. Add template drop to the EBON corridor

   a. From the **Roadway Designer** dialog main menu, select ***Corridor* → *Template Drops....***

   b. Type the information shown in Figure 11-41 into the **Template Drops** dialog box.

   Note: In the **Library Templates** window, expand the library folders and select your design template, **Ramps**.

   c. Click **Add** and **Close**.

   The template drop has been added to the corridor **EBON**.

   d. From the **Roadway Designer** main menu, select ***File* → *Save*.**

   You have now defined the corridor **EBON**, which consists of one template.

In the next subsection we will learn how to define superelevation.

3. Define superelevation in the EBON corridor

   a. From the **Roadway Designer** main menu, select ***Superelevation* → *Create Superelevation Wizard* → *Table....***

   The **Table Wizard** dialog box appears as illustrated in Figure 11-42.

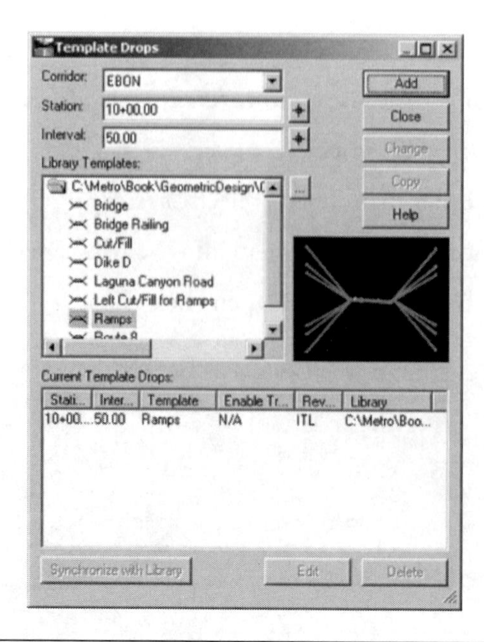

Figure 11-41 Template Drop for EBON Corridor

b. To browse to the project directory (either **C:/Temp/Tutorial/Chp11/Master Files/** or **C:/Temp/Book/Chp11/Master Files** folder), click [...] to the right of the **Table** field. Select **CaltransRamps.sup** for the single-lane ramp.

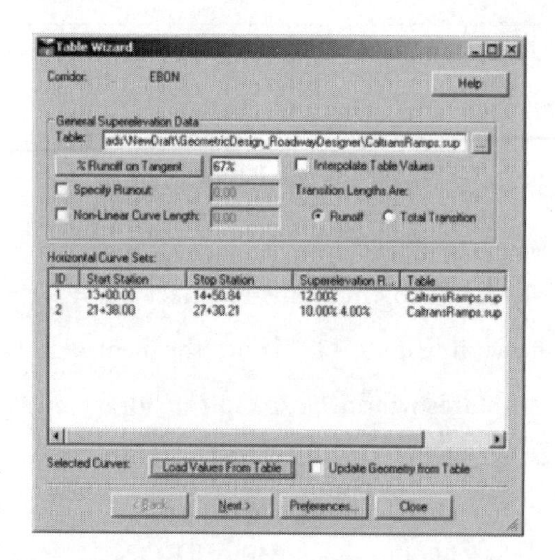

Figure 11-42 Table Wizard for EBON Corridor

c. From the **Horizontal Curve Sets** list, select curve 1. Click **Load Values From Table**.

d. From the **Horizontal Curve Sets** list, select curve 2. Click **Load Values From Table**.

e. In the **Table Wizard** dialog box, click **Next**.

The **Superelevation Section Definitions** dialog box appears.

f. In the **Superelevation Section Definitions** dialog box, click **Add...**.

The **Add Superelevation Section** dialog box appears as shown in Figure 11-43. (Refer to the previous chapter for the details of the control points.)

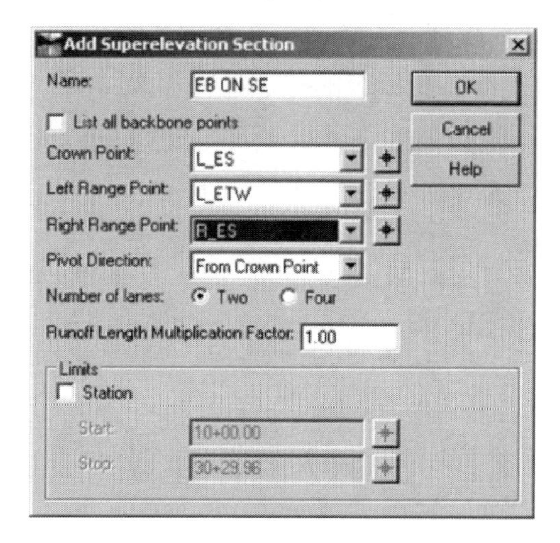

Figure 11-43 Define Control Points for Superelevation Section

g. In the **Add Superelevation Section** dialog box, click **OK**.

You have developed the superelevation section for the **EBON** ramp.

h. In the **Superelevation Section Definitions** dialog box, click **Next>**, then click **Finish**.

i. In the **Roadway Designer** main menu, select ***File → Save***.

You have now defined the superelevation for the corridor **EBON**.

» To see the three views for the superelevation section across the different stations, in the **Roadway Designer** main dialog box, check the **Normal** option to ON in the **Display Mode** (see Figure 11-44).

» To see four views for the EBON superelevation section across all the stations, click on the **Overlay** option of the **Display Mode** (see Figure 11-45).

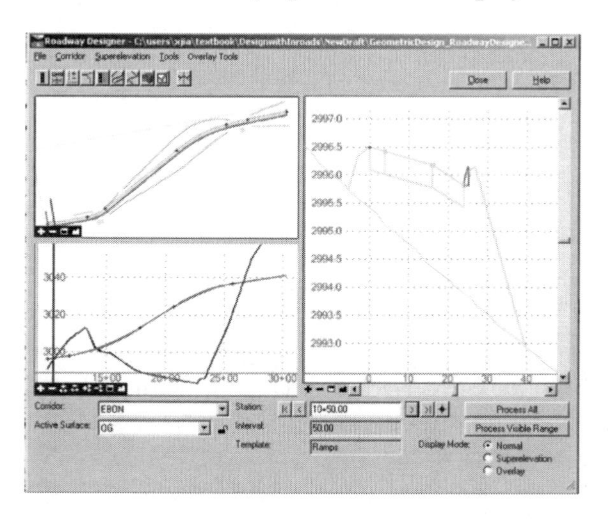

Figure 11-44 Superelevation Sections for EBON Ramp (Normal)

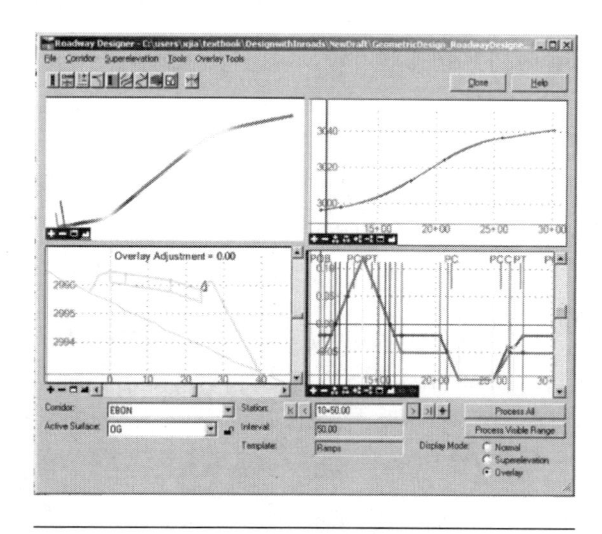

Figure 11-45 Superelevation Sections for EBON Ramp (Overlay)

4. Create the design surface for the EBON ramp

a. Set the surface **OG.dtm** to active.

b. From the **Roadway Designer** main menu, select ***Corridor → Create Surface....***

c. In the **Create Surface** dialog box, type the following information:

Name:	**EBON**
Default Preference:	**Default**
Check ON:	**Empty Design Surface**
Check ON:	**Triangulate**
Create Surface(s) from:	**EBON**

d. Click **Apply**.

The **Results** dialog box appears.

e. Review the results and click **Close**.

f. Select ***InRoads → Workspace Bar → Surface Tab.***

A new surface named **EBON.dtm** appears (see Figure 11-46).

g. Set this surface to active.

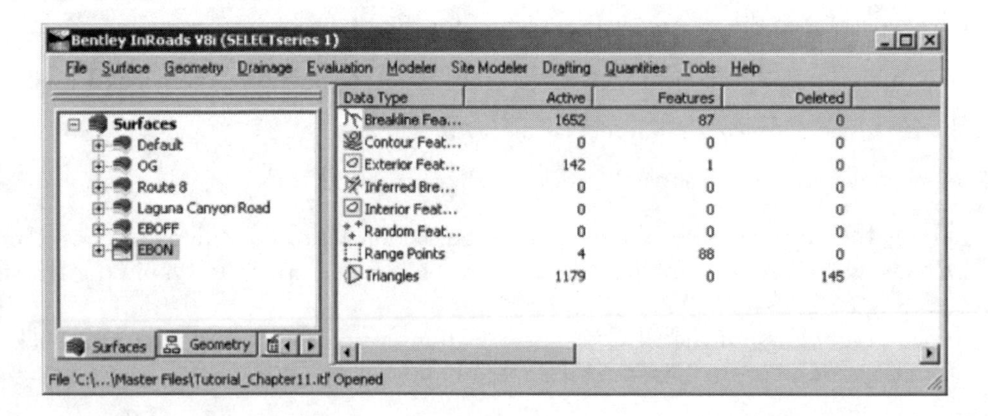

Figure 11-46 EBON Design Surface

h. Right click **EBON.dtm** and select **Triangulate**.

i. Go to ***InRoads → Surface → View Surface → Triangles....***

You will see the triangles of the **EBON** ramp in the **Tutorial.dgn** file (see Figure 11-47).

j. To delete the triangulations, click ***MicroStation → Undo.***

» Follow the steps in Chapter 6 to view the EBON design surface in 3-D mode. Figure 11-48 shows a portion of the 3-D view of the EBON design surface.

k. Go to ***InRoads → File → Save → Surface.***

A **Save As** dialog box appears.

l. Go to your project location (**C:/Temp/Tutorial/Master Files/** folder). Save the file with the name of **EBON.dtm**.

You have now created the design surface for the EBON ramp.

In the next subsection we will focus on creating the corridor and design surface for the other two ramps.

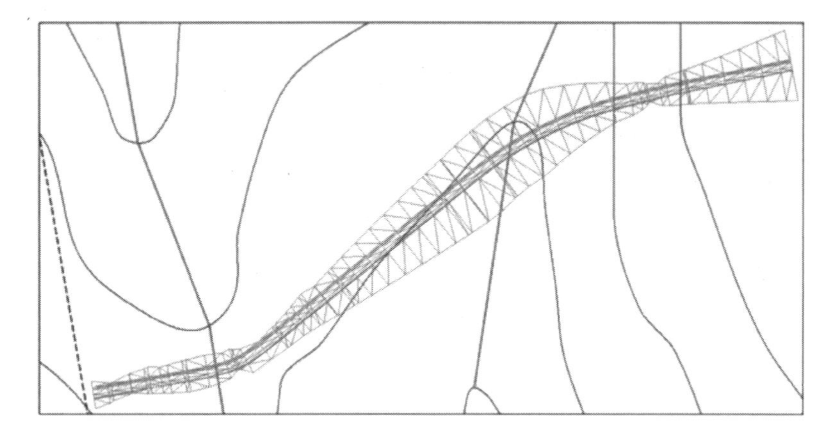

Figure 11-47 Triangles for EBON Ramp Corridor

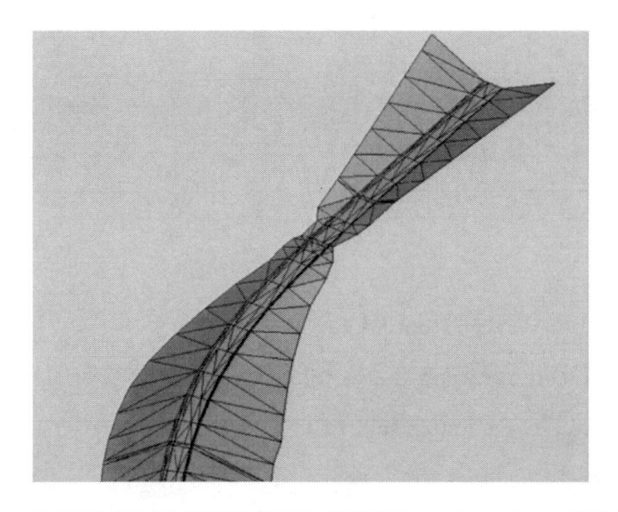

Figure 11-48 A Portion of 3D View of the EBON Ramp Corridor

11.7 Roadway Corridor and Superelevation for WBOFF Ramp

Similar procedures apply to the WBOFF ramp, which include defining a roadway corridor, adding a template drop, defining superelevation sections, and creating a design surface. The details are described in order as follows.

1. Define a roadway corridor for the WBOFF ramp
   a. Load the **OG.dtm, Tutorial.alg,** and **Tutorial_Chapter11.itl** from the **C:/Temp/Tutorial/Chp11/Master Files/** folder into InRoads.
   b. Set the surface **OG**, horizontal alignment **WBOFF** and vertical alignment **WBOFF**, and template **Ramps** to active.

      Note: To set the template Ramps to active, go to ***Modeler → Create Template....***
   c. From the InRoads main menu, select ***Modeler → Roadway Designer***.
   d. From the **Roadway Designer** main menu, select ***Corridor → Corridor Management....***

e. Type the information shown in Figure 11-49 into the **Manage Corridors** dialog box.

f. To create the corridor **WBOFF**, click **Add.** Then click **Close.**

You have now defined a new corridor for the WBOFF ramp. The corridor is empty at this time.

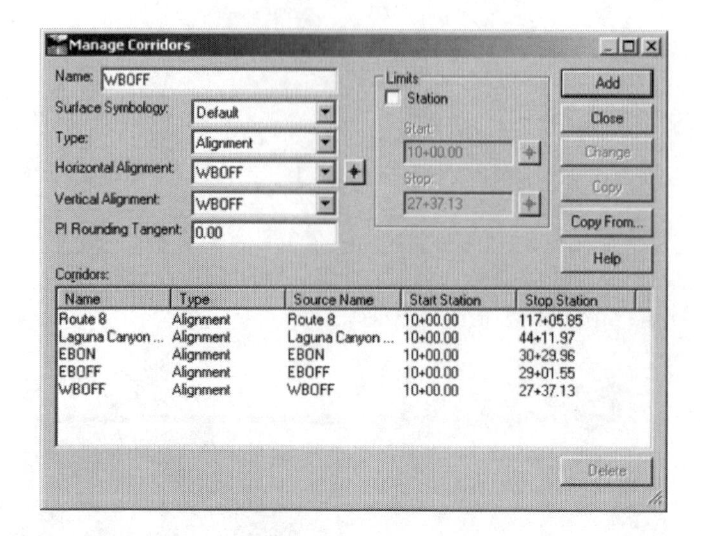

Figure 11-49 Manage Corridor Window for WBOFF Ramp

2. Add a template drop to the WBOFF corridor

a. From the **Roadway Designer** main menu, select ***Corridor → Template Drops....***

b. Within the **Template Drops** dialog, select the **Ramps** template.

Note: In the **Library Templates** window, you need to expand the library folders and select your design template **Ramps.**

c. Click **Add** and then **Close.**

The template drop has been added to the corridor **WBOFF.**

d. From the **Roadway Designer** main menu, select ***File → Save.***

You have now defined the corridor **WBOFF,** which consists of one template.

In the next subsection we will learn how to define superelevation.

3. Define superelevation in the WBOFF corridor

a. From the **Roadway Designer** main menu, select ***Superelevation → Create Superelevation Wizard → Table....***

The **Table Wizard** dialog box appears as illustrated in Figure 11-50.

b. To browse to the project directory (**C:/Temp/Tutorial/Chp11/Master Files/** or **C:/Temp/Book/Chp11/Master Files/** folder), click [...] to the right of the **Table** field. Select the **CaltransRamps.sup** table for the single-lane ramp.

c. From the **Horizontal Curve Sets** list, select curve 1. Click **Load Values From Table.**

d. From the **Horizontal Curve Sets** list, select curve 2. Click **Load Values From Table.**

e. In the **Table Wizard** dialog box, click **Next>.**

The **Superelevation Section Definitions** dialog box appears.

f. From the **Superelevation Section Definitions** dialog box, click **Add....**

The **Add Superelevation Section** dialog box appears as shown in Figure 11-51.

g. In the **Add Superelevation Section** dialog box, type the information shown in Figure 11-51. Click **OK**.

You have developed the superelevation section for the WBOFF ramp. Refer to the previous chapter for the details on control points.

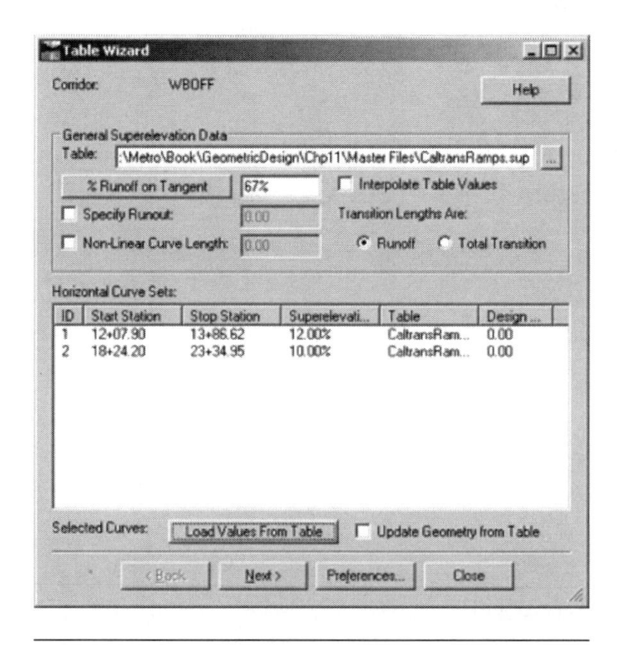

Figure 11-50 Superelevation Rate for the WBOFF Ramp

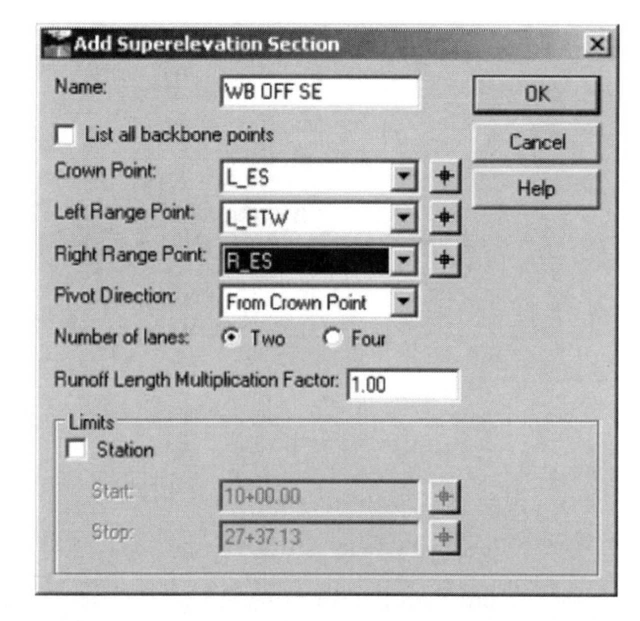

Figure 11-51 Define Information for Superelevation Section

h. Click **Next>**. Then click **Finish**.

i. In the **Roadway Designer** main menu, select ***File → Save***.

You have now defined the superelevation for the corridor **WBOFF**.

» You can see the three views for the superelevation section across the different stations in the **Roadway Designer** main dialog box by clicking on the **Normal** option of **Display Mode** (see Figure 11-52).

» You can see four views for the WBOFF superelevation section across all the stations by clicking on the **Overlay** option of **Display Mode** (see Figure 11-53).

4. Create the design surface for the WBOFF ramp

a. Set the surface **OG.dtm** to active.

b. From the **Roadway Designer** main menu, select ***Corridor → Create Surface....***

c. In the **Create Surface** dialog box, type the following information:

Name:	**WBOFF**
Default Preference:	**Default**
Check ON:	**Empty Design Surface**

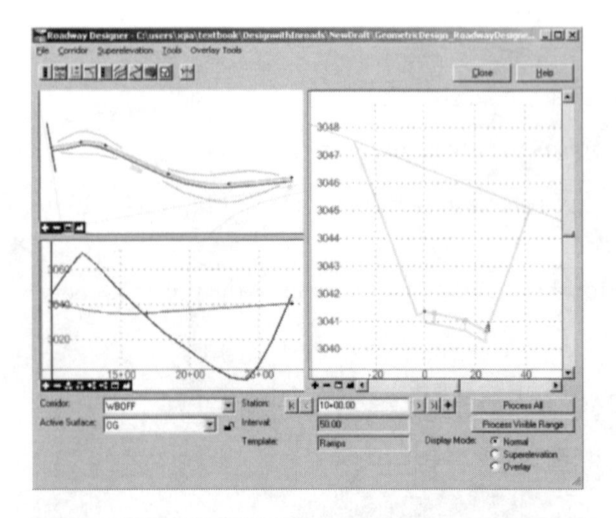

Figure 11-52 Superelevation Sections for WBOFF Ramp (Normal)

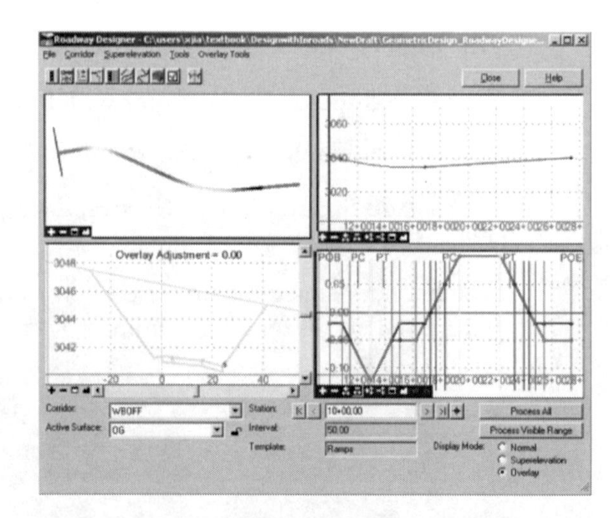

Figure 11-53 Superelevation Sections for WBOFF Ramp (Overlay)

Check ON: **Triangulate**
Create Surface(s) from: **WBOFF**

d. Click **Apply**.

The **Results** dialog box appears.

e. Review the results and click **Close**.

f. Close the **Create Surface** window.

g. From the ***InRoads → Workspace Bar***, select the **Surface Tab**.

You will see a new surface named **WBOFF.dtm** (see Figure 11-54).

h. Set this surface to active.

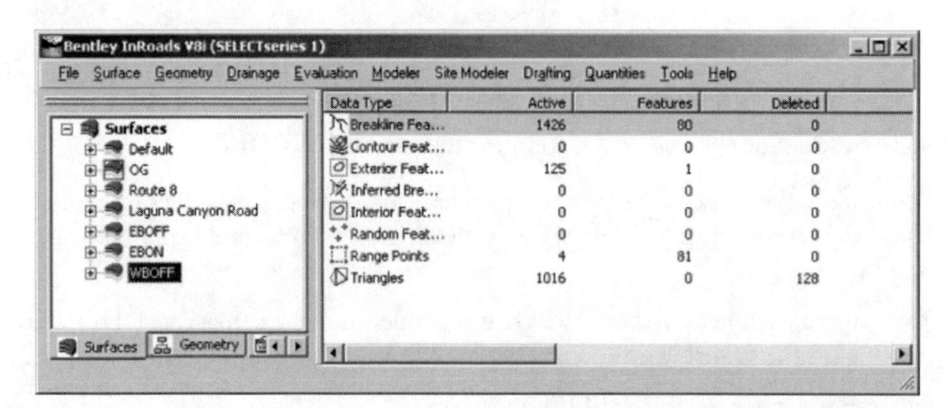

Figure 11-54 WBOFF Design Surface

i. Right click **WBOFF.dtm** and select **Triangulate**.

j. Go to ***InRoads → Surface → View Surface → Triangles...***.

You will see the triangles of the WBOFF ramp in the **Tutorial.dgn** file (see Figure 11-55).

k. To delete the triangulations, click ***MicroStation → Undo***.

Follow the steps in Chapter 6 to view the WBOFF design surface in 3-D mode. Figure 11-56 shows a portion of the 3-D view of WBOFF design surface.

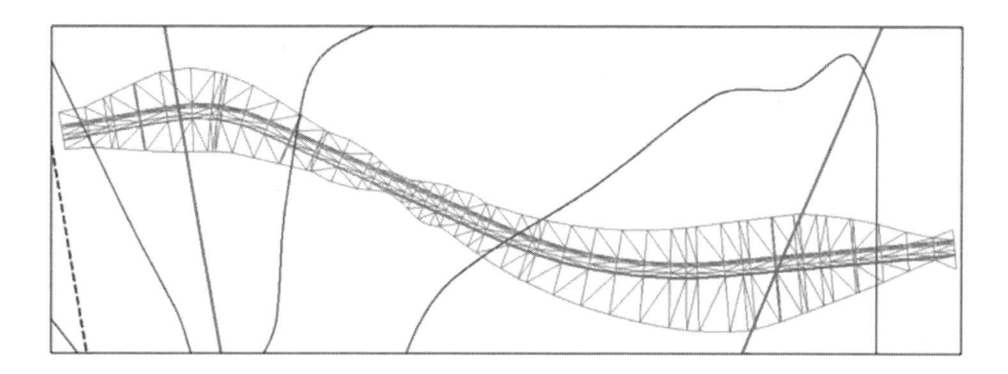

Figure 11-55 Triangles of WBOFF Design Surface

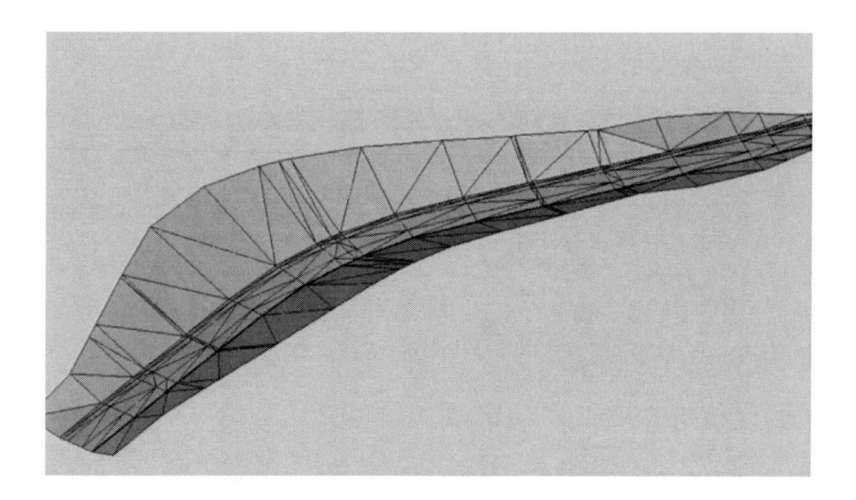

Figure 11-56 A Portion of 3D View of the WBOFF Ramp Corridor

l. From the InRoads main menu, go to ***File → Save → Surface***.

A **Save As** dialog box appears.

m. Browse to your project location (**C:/Temp/Tutorial/Chp11/Master Files/** folder). Save the file with the name of **WBOFF.dtm**.

You have now created the design surface for the WBOFF ramp.

In the next subsection we will focus on creating the corridor and design surface for the last ramp, the WBON ramp.

11.8 Roadway Corridor and Superelevation for WBON Ramp

Similar procedures apply to the WBON ramp, which include defining a roadway corridor, adding a template drop, defining superelevation sections, and creating a design surface. The details are described in order as follows.

5. Define a roadway corridor for the WBON ramp
   a. From the **C:/Temp/Tutroial/Chp11/Master Files/** folder, load **OG.dtm, Tutorial.alg**, and **Tutorial_Chapter11.itl** into InRoads.
   b. Set the surface **OG**, horizontal alignment **WBON** and vertical alignment **WBON**, and template **Ramps** to active.

      Note: To set the template Ramps to active, go to ***Modeler → Create Template....***
   c. From the InRoads main menu, select ***Modeler → Roadway Designer.***
   d. From the **Roadway Designer** main menu, select ***Corridor → Corridor Management....***
   e. Type the information shown in Figure 11-57 into the **Manage Corridors** dialog box.
   f. To create the corridor **WBON**, click **Add.** Then click **Close.**

      You have now defined a new corridor for the WBON ramp. The corridor is empty at this time.

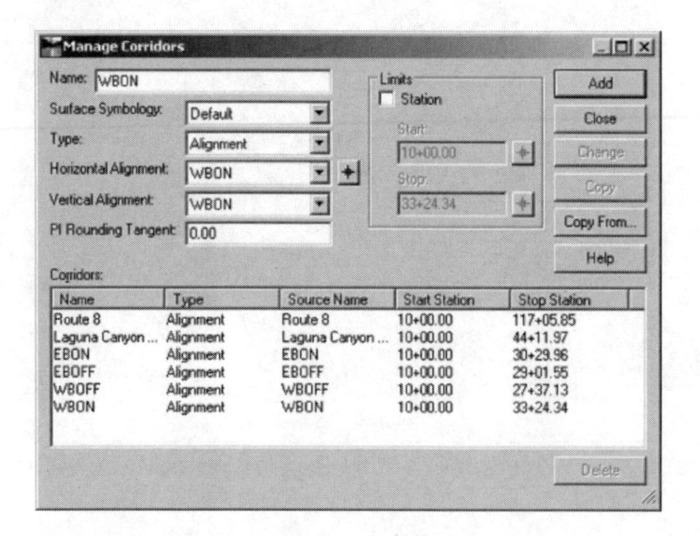

Figure 11-57 Manage Corridor Window for WBON Ramp

6. Add a template drop to the WBON corridor
   a. From the **Roadway Designer** main menu, select ***Corridor → Template Drops....***
   b. From the **Template Drops** dialog box, select the **Ramps** template.

      Note: In the **Library Templates** window you need to expand the library folders and select your design template, **Ramps.**
   c. Click **Add** and then click **Close.**

The template drop has been added to the corridor **WBON**.

d. From the **Roadway Designer** main menu, select ***File*** → ***Save***.

You have now defined the corridor **WBON**, which consists of one template.

In the next subsection we will learn how to define superelevation.

7. Define the superelevation in the WBON corridor

a. From the **Roadway Designer** main menu, select ***Superelevation*** → ***Create Superelevation Wizard*** → ***Table....***

The **Table Wizard** dialog appears as illustrated in Figure 11-58.

b. To browse to the project directory (**C:/Temp/Tutorial/Chp11/Master Files/** or **C:/Temp/Book/Chp11/Master Files/** folder), click [...] to the right of the **Table** field. Select the **CaltransRamps.sup** table for the single-lane ramp.

c. From the **Horizontal Curve Sets** list, select curve 1. Click **Load Values From Table**.

d. From the **Horizontal Curve Sets** list, select curve 2. Click **Load Values From Table**.

e. In the **Table Wizard** dialog box, click **Next>**.

The **Superelevation Section Definitions** dialog box appears.

f. In the **Superelevation Section Definitions** dialog box, click **Add....**

The **Add Superelevation Section** dialog box appears as shown in Figure 11-59.

g. In the **Add Superelevation Section** dialog box, enter the information shown in Figure 11-59 and click **OK**.

You have developed the superelevation section for the **WBON** ramp.

h. Click **Next>**, then click **Finish**.

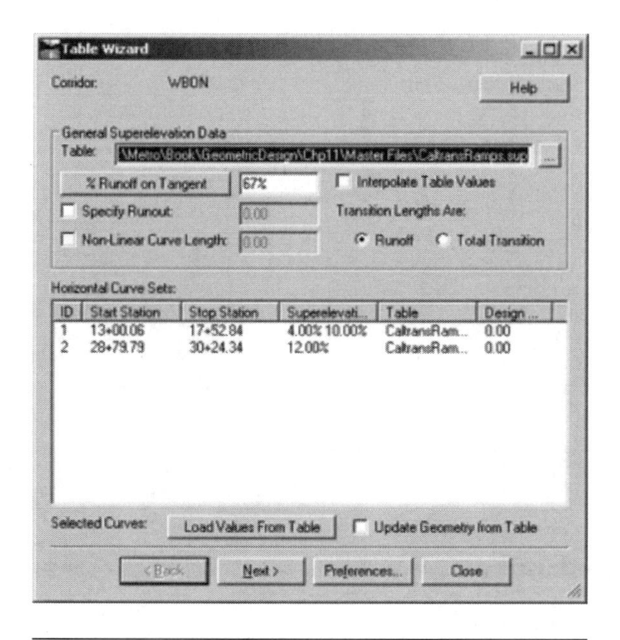

Figure 11-58 Superelevation Rate for the WBON Ramp

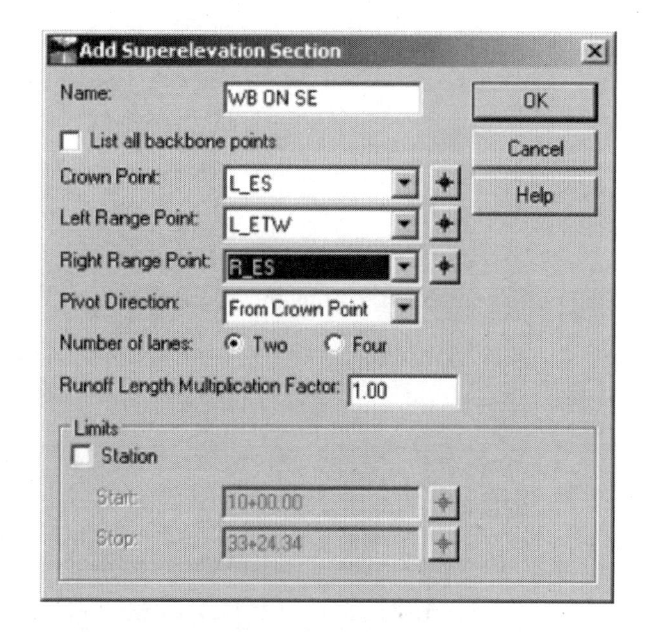

Figure 11-59 Define Information for Superelevation Section

i. In the **Roadway Designer** main menu, select ***File → Save***.

You have now defined the superelevation for the corridor **WBON**.

» To see the three views for the superelevation section across the different stations, in the **Roadway Designer** main dialog box, check the **Normal** option to ON in **Display Mode** (see Figure 11-60).

» To see four views for the **WBON** superelevation section across all the stations, click the **Overlay** option of **Display Mode** (see Figure 11-61).

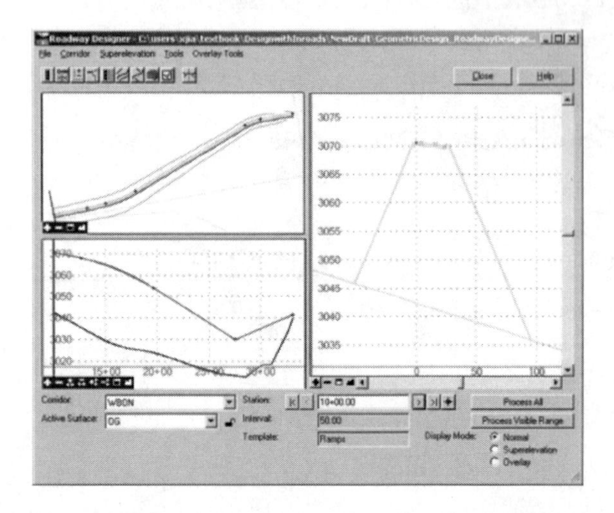

Figure 11-60 Superelevation Sections for WBON Ramp (Normal)

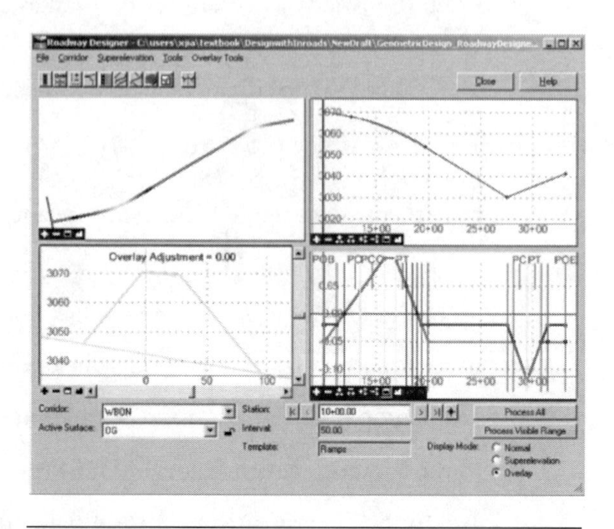

Figure 11-61 Superelevation Sections for WBON Ramp (Overlay)

8. Create the design surface for the **WBON** ramp

a. Set the surface **OG.dtm** to active.

b. From the **Roadway Designer** main menu, select ***Corridor → Create Surface....***

c. In the **Create Surface** dialog box, type the following information:

Name:	**WBON**
Default Preference:	**Default**
Check ON:	**Empty Design Surface**
Check ON:	**Triangulate**
Create Surface(s) from:	**WBON**

d. Click **Apply**.

The **Results** dialog box appears.

e. Review the results and click the **Close** button.

f. **Close** the Create Surface Window.

g. From the ***InRoads → Workspace Bar***, select the **Surface Tab**.

A new surface named **WBON.dtm** appears (see Figure 11-62).

h. Set this surface to active.

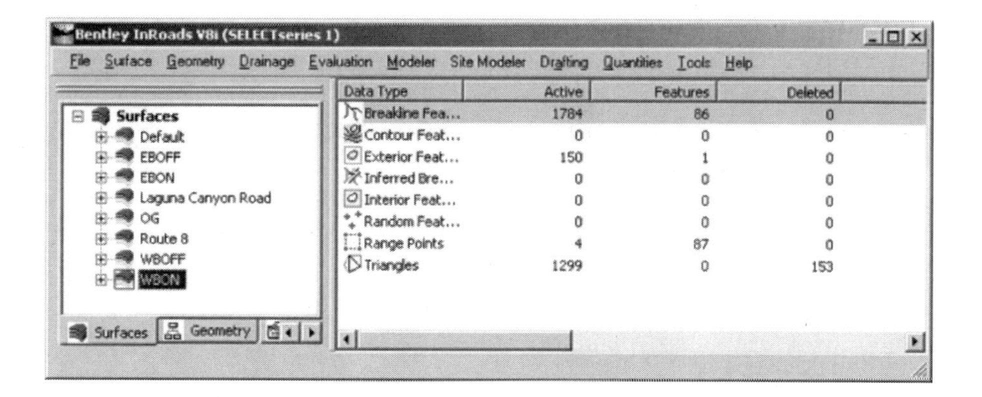

Figure 11-62 WBON Design Surface

i. Right click **WBON.dtm** and select **Triangulate.**

j. Go to ***InRoads* → *Surface* → *View Surface* → *Triangles....***

You will see the triangles of the **WBON** ramp in the **Tutorial.dgn** file (see Figure 11-63).

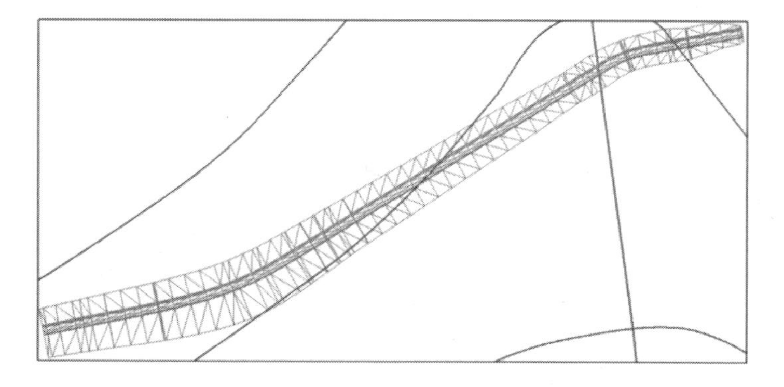

Figure 11-63 Triangles of WBON Design Surface

k. To delete the triangulations, click ***MicroStation* → *Undo*.**

l. Follow the steps in Chapter 6 to view the WBON design surface in 3-D mode.

Figure 11-64 shows a portion of the 3-D view of WBON design surface.

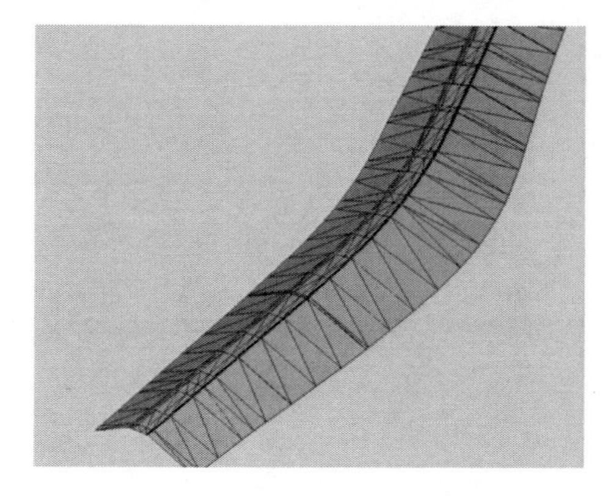

Figure 11-64 A Portion of 3-D View of the WBON Ramp Corridor

m. Go to ***InRoads → File → Save → Surface***.

A **Save As** dialog box appears.

n. Browse to your project location (**C:/Temp/Tutorial/Chp11/Master Files/** folder). Save the file under the name of **WBON.dtm**.

You have now created all the design surfaces required for the tutorial project.

With both the existing surface and design surfaces now available, we can prepare the earthwork estimation and report for the project. The details of the earthwork calculation and other related issues will be described in next chapter.

o. Go to ***MicroStation → File → Save*** to save the **Tutorial_DesignSurface.dgn** file.

The **Tutorial_DesignSurface.dgn** file contains all the displayed DTM surfaces for the project.

Name________________________________ Date______________

11.9 Questions

Q11.1 What is the maximum superelevation rate (e_{max})? Is e_{max} controlled by the type of roadway facility?

Q11.2 What is the superelevation rate (e)? Is the superelevation rate controlled by the type of facility (or design speed) and the radius of the design curve? Which Caltrans *HDM* table should you use to determine an appropriate e for a given curve?

Q11.3 What are superelevation range points? What is superelevation pivot point?

Q11.4 Is the superelevation rate determined by AASHTO methods in the same way as defined by the Caltrans method?

Q11.5 List the steps required to define superelevation for a horizontal alignment.

Q11.6 Why do we need to create vertical control alignments? Can you use the vertical control alignments to create a superelevation diagram?

Q11.7 When we create superelevation sections, we need to specify 67.0 in the Percent Runoff on Tangent field. Why do you need to do that?

Q11.8 Suppose you use a curve of 1,200 ft. to connect two tangents on a ramp. What is the appropriate superelevation rate to use for a roadway project in California?

Q11.9 What is the maximum superelevation rate for an expressway? For a 2-lane conventional highway? For a 4-lane conventional roadway?

Q11.10 What are the prerequisites for creating a roadway design surface?

Q11.11 Is superelevation incorporated into the creation of a roadway design surface?

Q11.12 What is the difference between **Express Modeler** and **Roadway Modeler**? Hint: Refer to InRoads Online Help to answer this question.

Q11.13 How many breaklines are created for the Route 8 design surface?

Q11.14 Can we use Route 8 as the active Horizontal Alignment to create the Westbound Off design surface?

Chapter 12
Earthwork

LEARNING OBJECTIVES

After completing this chapter you should know:

1. How to create cross sections based on given existing ground and design surfaces;
2. How to create earthwork-related reports;
3. How to create and utilize mass diagrams.

SECTIONS

One of the major objectives in a highway design project is to estimate the amount of earthwork necessary for the project. Earthwork calculation involves the determination of earthwork operations (including excavation, embankment, and hauling) of the soil or earth in the project's right-of-way.

In this chapter we will discuss how to use the 3-D design surfaces, the horizontal and vertical alignments, and the existing DTM surface created in previous chapters to calculate earthwork quantities using the **end-area** volume method. First we will create cross sections for Route 8, Laguna Canyon Road, and the four ramps at appropriate intervals along the grade line. Then we will create earthwork-related reports for the project with an assumption that the shrinkage and swell factors are 1.0. In a real design project the designers should get soil testing reports from soil engineers and apply the shrinkage and swell factors accordingly.

With the cut and fill information stored for each cross section in InRoads, we will also create mass diagrams that represent the accumulative earthwork for the project. Finally, we will learn how to utilize the mass diagram to estimate the earthwork and the related costs.

It is important to note that this project does not include the pavement design for Route 8, Laguna Canyon Road, and the four ramps. Therefore, the earthwork related to pavement materials is not included in this chapter.

12.1 Earthwork Estimation for Route 8

This section is dedicated to the earthwork estimation for Route 8, which includes cross section development, earthwork volume report generation, and mass diagram creation.

Before we work on the earthwork estimation, you need to have MicroStation and InRoads installed on your computer. You also need to copy all the files from the **C:/Temp/Book/Chp12/Master Files** folder to the **C:/Temp/Tutorial/Chp12/Master Files/** folder or you can download the files related to this task from the book website (www.csupomona.edu/~xjia/GeometricDesign/Chp12/Master Files/).

The detailed procedures for earthwork estimation for Route 8 are illustrated in the following subsections in order.

1. Create cross sections for Route 8

   a. Start MicroStation and InRoads (if they are not open).

   b. Load **Tutorial_VA.dgn** from the **C:/Temp/Tutorial/Chp12/Master Files/** folder.

   c. Load the **OG.dtm, Tutorial.alg, Tutorial_Chapter11.itl, Tutorial.ird, Route 8.dtm, Laguna Canyon Road.dtm, EBOFF.dtm, EBON.dtm, WBOFF.dtm**, and **WBON.dtm** into InRoads.

   d. Set the surface **OG**, horizontal alignment **Route 8** and vertical alignment **Route 8**, and template **Route 8** to active.

      To set the specific template to active, you need to go to ***InRoads → Modeler → Create Template...***) and the **Route 8** corridor.

   e. Select ***InRoads → Evaluation → Cross Section → Cross Sections....***

      The **Cross Sections** dialog box appears (see Figure 12-1).

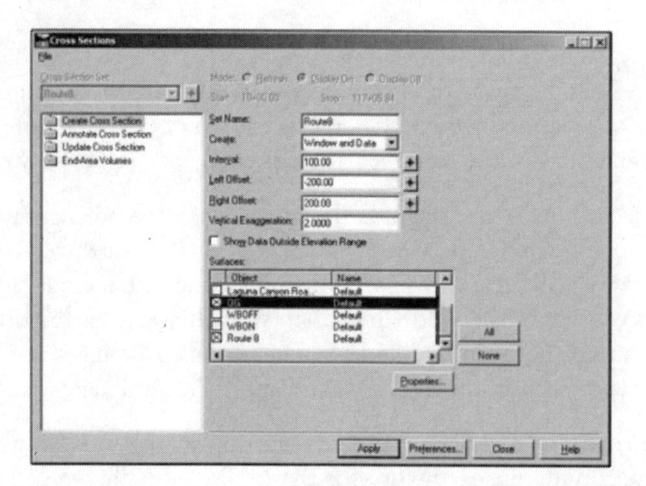

Figure 12-1 Cross Sections Interface

   f. In the **Cross Sections** dialog box, select the ***Create Cross Section → General*** in the left window. Type the following information:

      Set Name: **Route 8**
      Create: **Window and Data**

Interval:	**100.00**	
Left Offset:	**–200.00**	
Right Offset:	**200.00**	
Vertical Exaggeration:	**2.0000**	
Surface:	**Route 8**	**Check ON**
	OG	**Check ON**

Note this important information associated with the settings above:

» **Set Name** is an identifier. InRoads uses it to differentiate one specific group of cross sections from others. Usually it is the same as the alignment name.

» **Create** specifies which information will be created by InRoads. Here the **Window and Data** option indicates that both surface data and the cross section window will be created.

» **Interval** determines the distance among adjacent cross sections. Here 100' means that a cross section will be cut every 100' along the corridor. You can change this value to meet your specific needs.

» **Left offset** determines the offset value to the left of the centerline. **Right offset** determines the offset value to the right of the centerline. InRoads assigns a negative value to the left of the centerline and a positive value to the right.

» **Vertical exaggeration** is a scaling factor for the vertical orientation. Here a vertical exaggeration of 2 means the cross section is scaled up vertically by 2. This setting is usually done for visual convenience, considering the relatively smaller magnitude of the vertical elevation in comparison with the horizontal distance.

» **Surfaces** section determines the surfaces displayed on the cross sections. Here the OG and Route 8 are selected to ensure only the traverse elements of the original ground and the Route 8 design surface are shown in the Route 8 cross sections.

g. In the **Cross Sections** dialog box, in the left window, select ***Layout → General***. Check **Stacked** to ON.

h. Select the ***Create Cross Section → Include → Components***.

i. Once the settings are established, click **Apply**.

   Similar to the profiles created previously, InRoads prompts you to place a data point in the MicroStation view to specify the location of the left corner of the group of cross sections.

   » Move the cursor to a location in the view with the **Tutorial_VA.dgn** file and press the **Data** button on your mouse.

   A group of cross sections appears (see Figure 12-2).

   » You can change the cross section symbology of the OG and/or Route 8 surface from **default** to **design** or other symbologies and display a cross section line with the line color and weight of your choice.

   Take the cross sections at Station 24+00 and Station 65+00 as an example (see Figure 12-3, Figure 12-4). You can see that the cross section at Station 24+00 is located on the curve and the superelevation is considered in the cross section design. The cross section at Station 65+00 is located on the bridge.

j. From the **Cross Sections** dialog box, click **Close**.

   The cross sections for Route 8 are created.

In the next subsection we will focus on the development of an earthwork-related report for each cross section.

Figure 12-2 Stacked Cross Sections

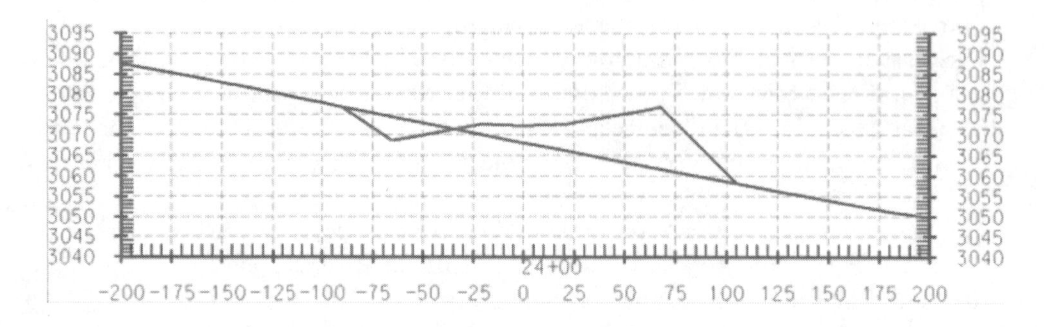

Figure 12-3 Cross Section at Station 24+00 of Route 8

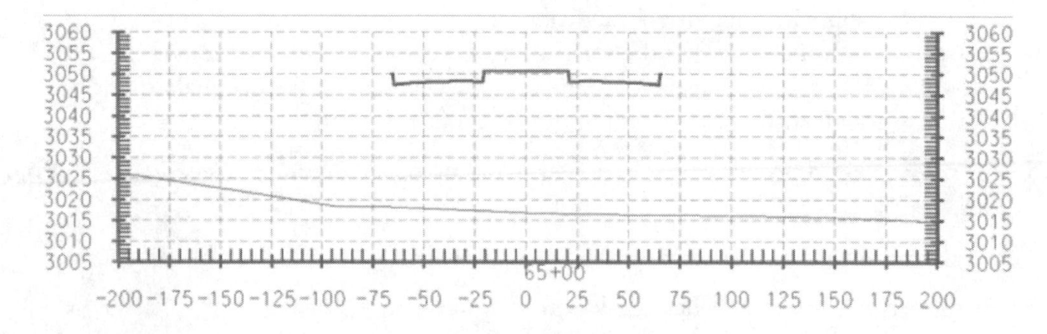

Figure 12-4 Cross Section at Station 65+00 of Route 8

2. Create the earthwork volume report for Route 8

One of the powerful functions of InRoads lies in its ability to create various earthwork-related reports, including average cross slope area, basic volume, end area volume, triangle volumes, and so on. As mentioned before, the end-area volume method will be used in the project to estimate the earthwork quantities. This subsection focuses on the steps to create the end area volume report for Route 8, which later can be used to create the mass haul diagram.

a. Select ***InRoads → Evaluation → Cross Section → Cross Sections…***.

   The **Cross Sections** dialog box appears (see Figure 12-5).

b. Select **End-Area Volumes** as shown in the left panel.

c. In the **Cross Sections** window, establish the settings as follows:

   » From the drop-down list of the **Cross Section Set,** select **Route 8**.
   » Under the **End-Area Volumes** folder in the left window, select the **General** leaf. Enter the following information:

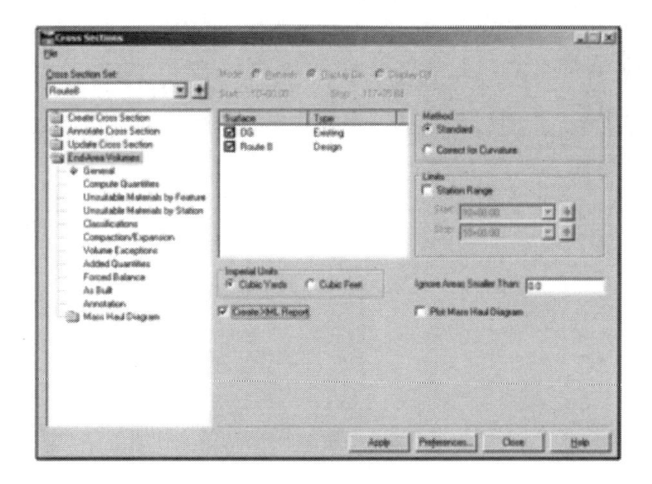

Figure 12-5 End-Area Volumes Interface

Surface:	**Route 8**	**Check ON**
	OG	**Check ON**
Imperial Units:	**Cubic Yards**	
Create XML Report:	**Check ON**	
Method:	**Standard**	

d. Once the settings are established, click **Apply**.

The **Bentley Civil Report Browser Window** (see Figure 12-6) appears, illustrating the end-area volume report for various cross sections of Route 8.

e. Select ***File → Save as....***

The **Save As** dialog box appears (see Figure 12-7).

f. Establish the settings as follows:

» In the **Save In** field, browse to the project directory (**C:/Temp/Tutorial/Chp12/Master Files/** folder).
» In the **File Name** filed, type **Route 8**.
» From the **Save as type** drop-down list, select **XML File** (***.xml**).

g. Once the setting is established, click **Save.**

The **Save As** window disappears.

h. In the **Cross Sections** window, click **Close.**

Now the end area volume report for Route 8 has been created. It will be used to develop the mass haul diagram for the **Route 8** corridor.

3. Create the mass haul diagram for Route 8

The detailed procedures for creating the mass haul diagram for the Route 8 corridor are shown as follows:

a. Select ***InRoads → Evaluation → Cross Section → Cross Sections....***

The **Cross Sections** dialog box appears. Make sure **Create XML Report** is checked OFF and **Plot Mass Haul Diagram** is checked ON.

b. Select ***End-Area Volumes → Mass-Haul Diagram → General.***

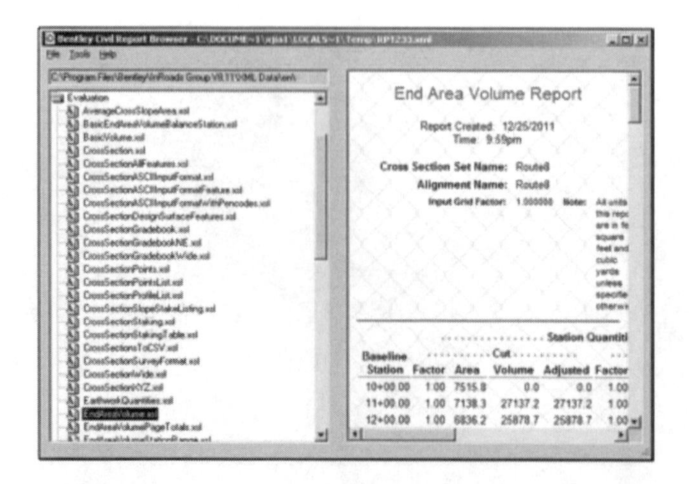

Figure 12-6 Bentley Civil Report Browser

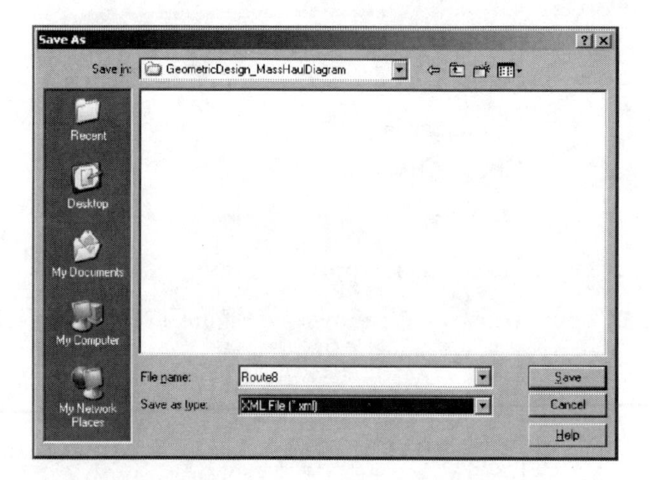

Figure 12-7 Route 8 XML Report

The **Mass Haul Diagram** window appears (see Figure 12-8).

c. Establish the following settings:

Direction:	**Left to Right**	
Exaggeration:	**Horizontal**	**1.0000**
	Vertical	**0.0010**

The **Vertical Exaggeration** is a scaling factor for the vertical orientation. Adjust the value by trial and error until a preferred mass haul diagram is created.

d. Double click the **Data Line**. Select the **Color** and **Weight** of your choice. Click **OK** to close the **Line Symbology** window.

e. Go to ***Mass Haul Diagram → Horizontal Center Axis***.

Figure 12-9 appears.

f. Establish the following settings (the spacing of major ticks represents 100 ft. or 1 station:

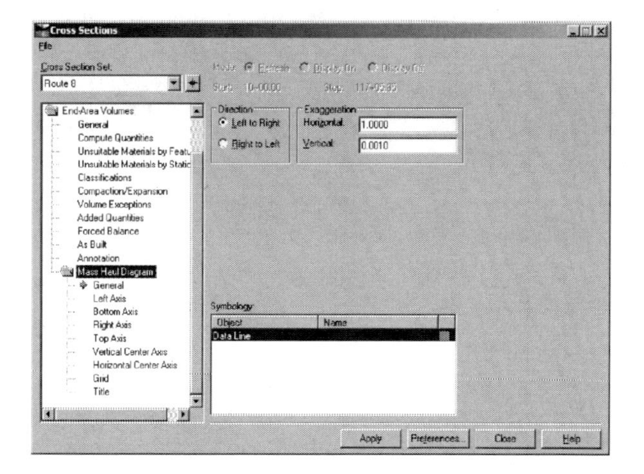

Figure 12-8 Mass Haul Diagram Interface

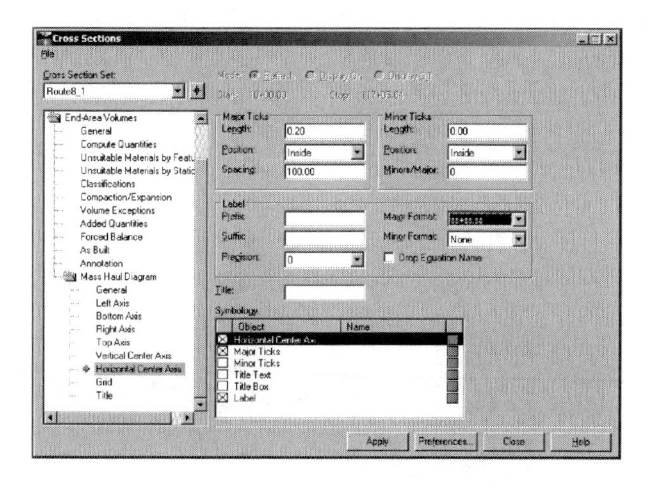

Figure 12-9 General Information for Mass Haul Diagram

Major Ticks:	Length:	**0.2**	
	Position:	**Inside**	
	Spacing:	**100.00**	
Major Format:	**SS+SS.SS**		
Horizontal Center Axis:	**Check ON**		
	Double click the **Color** icon. Select the color and weight of your choice for the horizontal center axis.		
Major Ticks:	**Check ON**		
Label:	**Check ON**		
	Double click the **Color** icon and establish the settings shown in Figure 12-10.		
	Set offsets:	Horizontal:	**–0.3**
		Vertical:	**–0.3**

g. Go to ***Mass Haul Diagram → Left Axis.***

Figure 12-11 appears.

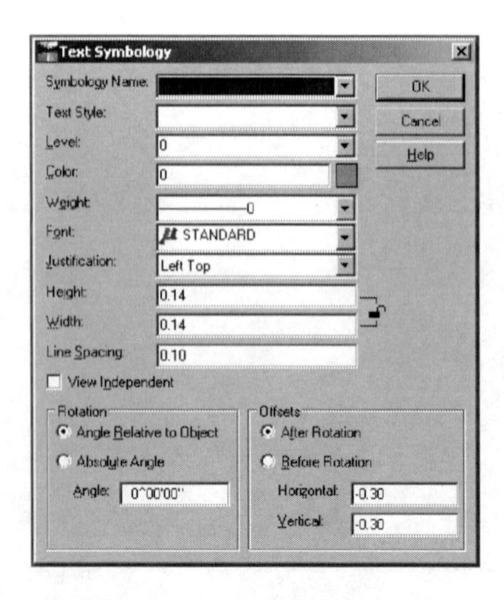

Figure 12-10 Set Offsets for Route 8 Mass Haul Diagram

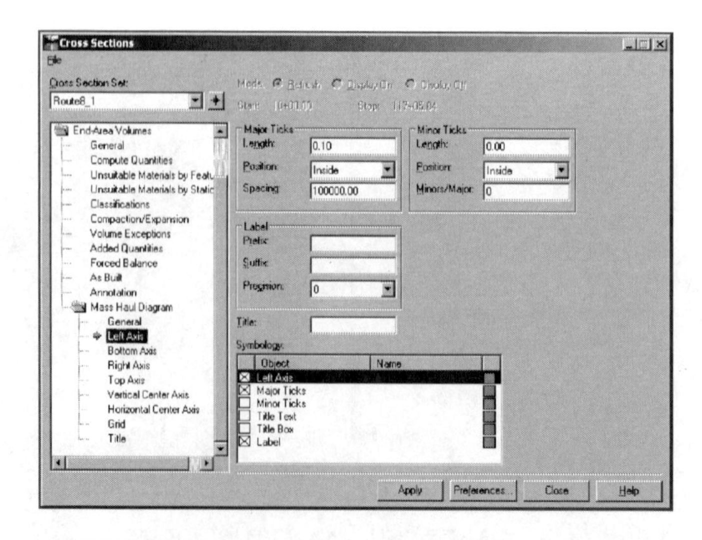

Figure 12-11 Left Axis Settings

h. Establish the following settings (you can adjust the spacing of major ticks based on the magnitude of your earthwork volume amount):

Major Ticks:	Length:	**0.1**	
	Position:	**Inside**	
	Spacing:	**100000.00**	
Left Axis:	**Check ON**		
	Double click the **Color** icon. Select the color and weight of your choice for the horizontal center axis.		
Major Ticks:	**Check ON**		
Label:	**Check ON**		
	Double click the **Color** icon and establish these settings.		
	Set offsets:	Horizontal:	**–1.0**
		Vertical:	**0.0**
	Others:	**Default**	

i. Go to ***Mass Haul Diagram*** → ***Right Axis***. In the **Symbology** section, check **Right Axis** to ON.

j. Go to ***Mass Haul Diagram*** → ***Top Axis***. In the **Symbology** section, check **Top Axis** to ON.

k. Go to ***Mass Haul Diagram*** → ***Bottom Axis.*** In the **Symbology** section, check **Bottom Axis** to ON.

l. Once the settings are established, click **Apply**.

m. Move the cursor to a location in the **MicroStation** view and press the **Data** button on your mouse.

A mass haul diagram appears on the screen (see Figure 12-12).

n. In the **Cross Sections** window, click **Close**.

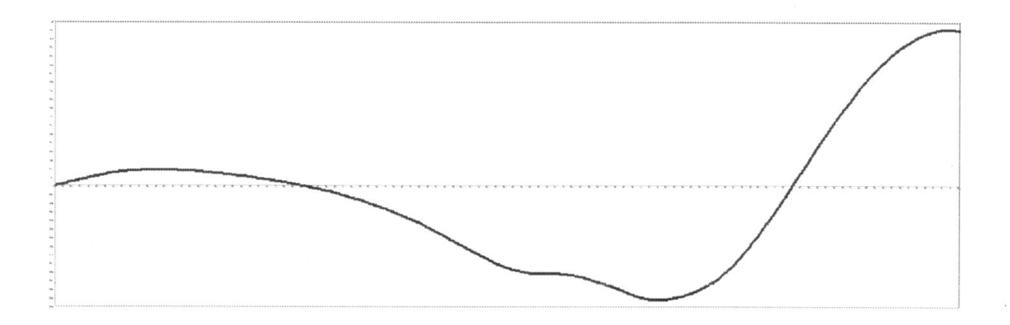

Figure 12-12 Mass Diagram for Route 8

12.2 Earthwork Estimation for Laguna Canyon Road

This section focuses on the earthwork estimation for Laguna Canyon Road. The related procedures are similar to those for Route 8.

1. Create cross sections for Laguna Canyon Road
   a. Load the **OG.dtm, Tutorial.alg, Tutorial_Chapter11.itl, Tutorial.ird, Route 8.dtm, Laguna Canyon Road.dtm, EBOFF.dtm, EBON.dtm, WBOFF.dtm,** and **WBON.dtm** into InRoads.
   b. Set the surface **OG**, horizontal alignment **Laguna Canyon Road** and vertical alignment **Laguna Canyon Road**, and template **Laguna Canyon Road** to active.

      To set the specific template to active, go to ***Modeler*** → ***Create Template...*** and the **Laguna Canyon Road** corridor.
   c. From the InRoads main menu, select ***Evaluation*** → ***Cross Section*** → ***Create Cross Section....***

      The **Create Cross Section** dialog box appears (see Figure 12-13).
   d. In the **Create Cross Section** dialog box, under the **Create Cross Section** folder in the left window, select **General** leaf. Type the following information:

Set Name:	**Laguna Canyon Road**
Create:	**Window** and Data
Interval:	**100.00**
Left Offset:	**–200.00**
Right Offset:	**200.00**
Vertical Exaggeration:	**2.0000**
Surface:	**Laguna Canyon Road; OG**

   e. In the **Cross Sections** dialog box, in the left window, select ***Layout*** → ***General.*** Check **Stacked** to ON.
   f. Select the ***Create Cross Section*** → ***Include*** → ***Components.***

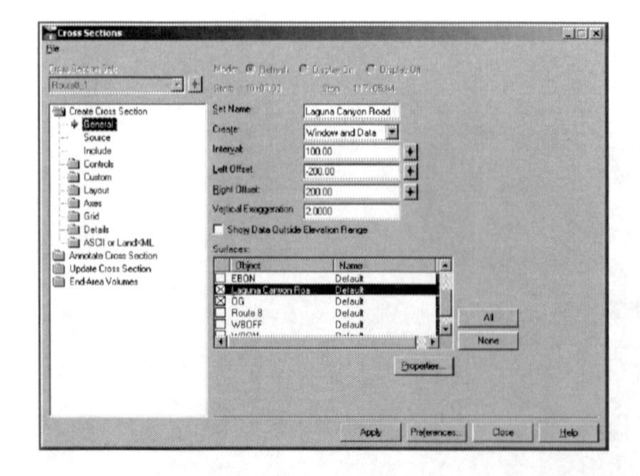

Figure 12-13 General Settings for Laguna Canyon Road

g. Once the settings are established, click **Apply**.

   InRoads will prompt you to place a data point in the MicroStation view. This specifies the location of the left corner of the group of cross sections.

h. Move the cursor to a location in the view with the design file and press Data button on your mouse.

   A group of cross sections appears (see Figure 12-14).

   An example of a cross section at Station 26+00 is illustrated in Figure 12-15.

i. In the **Create Cross Section** dialog box, click **Close**.

   Now the cross sections for Laguna Canyon Road are created.

In the next subsection we will focus on the development of earthwork-related report for each cross section.

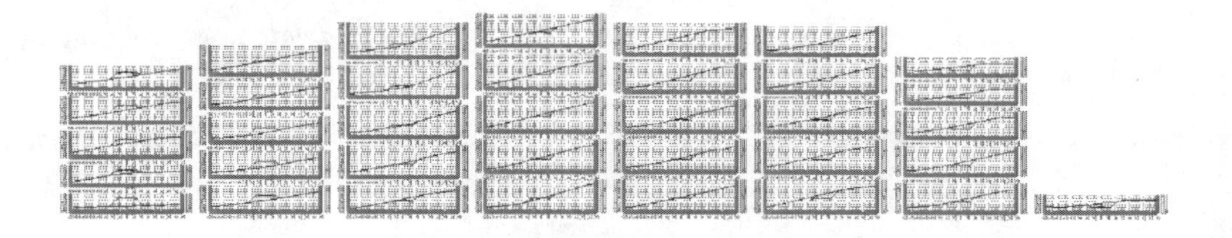

Figure 12-14 Cross Sections for Laguna Canyon Road

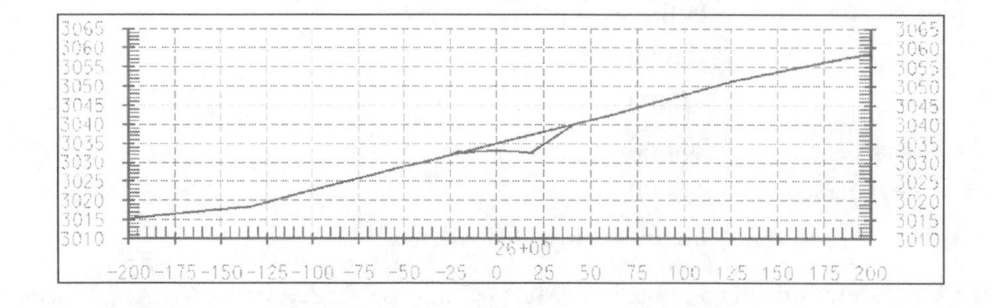

Figure 12-15 Cross Section at Station 26+00 of Laguna Canyon Road

2. Create earthwork volume report for Laguna Canyon Road

This subsection focuses on the steps to create the end area volume report for Laguna Canyon Road, which you can later use to create the mass haul diagram.

a. Select ***InRoads → Evaluation → Cross Section → Cross Sections....***

   The **Cross Sections** dialog box appears.

b. Select **End-Area Volumes** and establish the settings as follows:

   » Select **Laguna Canyon Road** from the drop-down list of the **Cross Section** Set.
   » Select the **General** leaf under the **End-Area Volumes** folder in the left window, and then input the following information:

Surface:	**Laguna Canyon Road**	**Check on**
	OG	**Check on**
Imperial Units:	**Cubic Yards**	
Create XML Report:	**Check on**	
Method:	**Standard**	

c. Once the settings are established, click **Apply.**

   The **Bentley Civil Report Browser** window appears (see Figure 12-16) where the end-area volume report for various cross sections of Laguna Canyon Road is illustrated.

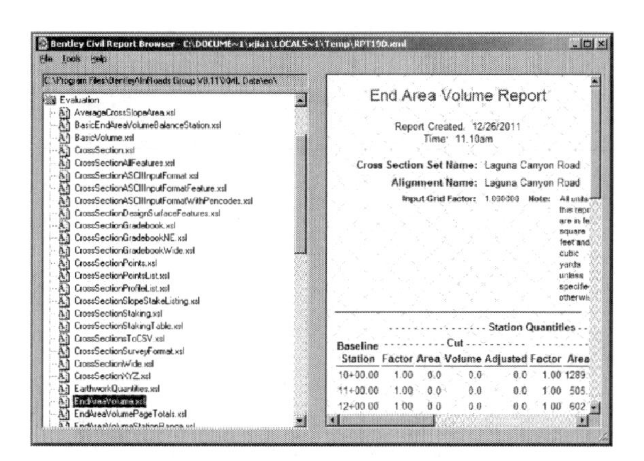

Figure 12-16 Bentley Civil Report Browser

d. Select ***File → Save As....***

   The **Save As** dialog box appears.

e. Establish the settings as follows:

   » In the **Save in** field, browse to the project directory (**C:/Temp/Tutorial/Chp12/Master Files/** folder).
   » In the **File Name** field, type **Laguna Canyon Road.**
   » From the **Save as** type drop-down list, select **XML File** (***.xml**).

f. Once the setting is established, click **Save.**

g. In the **End Area Volumes** window, click **Close.**

Now the end area volume report for Laguna Canyon Road has been created. It will be used to develop the mass haul diagram for the **Laguna Canyon Road** corridor.

3. Create a mass haul diagram for Laguna Canyon Road

The detailed procedures of creating the mass haul diagram for the Laguna Canyon Road corridor are shown as below:

a. Select ***InRoads → Evaluation → Cross Section → Cross Sections....***

   The **Cross Sections** dialog box appears. Make sure **Create XML Report** is checked OFF and **Plot Mass Haul Diagram** is checked ON.

b. Select ***End-Area Volumes → Mass-Haul Diagram → General.***

   The **Mass Haul Diagram** window appears.

c. Establish the following settings:

Direction:	**Left to Right**	
Exaggeration:	**Horizontal**	**1.0000**
	Vertical	**0.1000**

   Again, the **Vertical Exaggeration** serves as a scaling factor for the vertical orientation. You can adjust the value by trial and error until a preferred mass haul diagram is created.

d. Double click the **Data Line**. Select the **Color** and **Weight** of your choice. Select **OK.**

e. Go to ***Mass Haul Diagram → Horizontal Center Axis.***

f. Establish the following settings (the spacing of major ticks represents 100 ft., or 1 station):

Major Ticks:	Length:	**0.2**	
	Position:	**Inside**	
	Spacing:	**100.00**	
Major Format:	**SS+SS.SS**		
Horizontal Center Axis:	**Check ON**		
	Double click the **Color** icon. Select the color and weight of your choice for horizontal center axis.		
Major Ticks:	**Check ON**		
Label:	**Check ON**		
	Double click the color icon and set up the settings as follows:		
	Set offsets:	Horizontal:	**–0.3**
		Vertical:	**–0.3**
	Others:	**Default**	

g. Go to ***Mass Haul Diagram → Left Axis.*** Establish the following settings (you can adjust the spacing of major ticks based on the magnitude of your earthwork volume amount):

Major Ticks:	Length:	**0.1**
	Position:	**Inside**
	Spacing:	**1000.00**
Left Axis:	**Check ON**	
	Double click the **Color** icon.	

	Select the color and weight of your choice for horizontal center axis		
Major Ticks:	**Check ON**		
Label:	**Check ON**		
	Double click the color icon and set up the settings as follows:		
	Set offsets:	Horizontal:	**–1.0**
		Vertical:	**0.0**
	Others:	**Default**	

h. Go to ***Mass Haul Diagram → Right Axis*** and in the **Symbology** section, click **Right Axis.**

i. Go to ***Mass Haul Diagram → Top Axis*** and in the **Symbology** section, click **Top Axis.**

j. Go to ***Mass Haul Diagram → Bottom Axis*** and in the **Symbology** section, click **Bottom Axis.**

k. Once the settings are established, click **Apply**.

l. Move the cursor to a location in the MicroStation view and press the **Data** button on your mouse.

   A mass haul diagram appears (see Figure 12-17).

m. In the **Cross Sections** window, click **Close.**

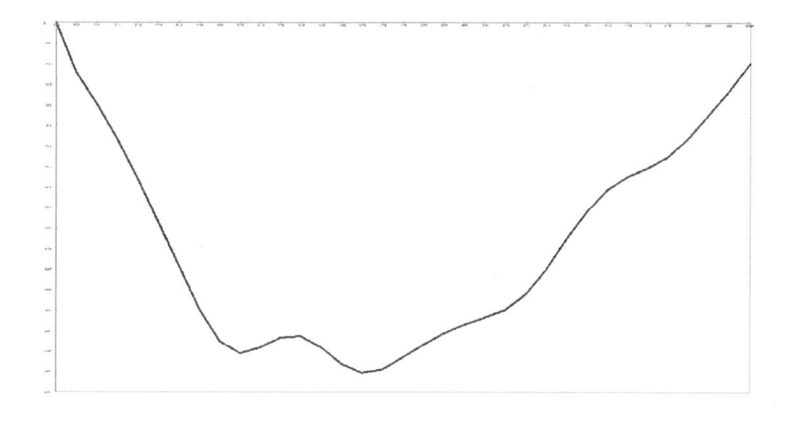

Figure 12-17 Mass Diagram for Laguna Canyon Road

12.3 Earthwork Estimation for the EBOFF Ramp

This section focuses on the earthwork estimation for the EBOFF ramp. The related procedures are as follows:

1. Create cross sections for the EBOFF ramp

   a. Load the **OG.dtm, Tutorial.alg, Tutorial_Chapter11.itl, Tutorial.ird, Route 8.dtm, Laguna Canyon Road.dtm, EBOFF.dtm, EBON.dtm, WBOFF.dtm,** and **WBON.dtm** into InRoads.

   b. Set the surface **OG**, horizontal alignment **EBOFF** and vertical alignment **EBOFF**, and template **Ramps** to active.

Note: To set the specific template to active, go to ***Modeler* → *Create Template...***) and the **EBOFF** corridor.

c. Select ***InRoads* → *Evaluation* → *Cross Section* → *Cross Sections....***

The **Cross Sections** dialog box appears.

d. In the **Create Cross Section** dialog box, under the **Create Cross Section** folder in the left window, select the **General** leaf. Type the following information:

Set Name:	**EBOFF**
Create:	**Window and Data**
Interval:	**100.00**
Left Offset:	**-200.00**
Right Offset:	**200.00**
Vertical Exaggeration:	**2.0000**
Surface:	**EBOFF; OG**

e. In the **Cross Sections** dialog box, in the left window, select the ***Layout* → *General.*** Click **Stacked.**

f. Select the ***Create Cross Section* → *Include* → *Components.***

g. Once the settings are established, click **Apply**.

InRoads will prompt you to place a data point in the MicroStation view to specify the location of the left corner of the group of cross sections.

h. Move the cursor to a location in the view with the design file and press the **Data** button on your mouse.

A group of cross sections appears (see Figure 12-18).

An example of a cross section at Station 26+00 is illustrated in Figure 12-19.

i. In the Create **Cross Section** dialog box, click **Close**.

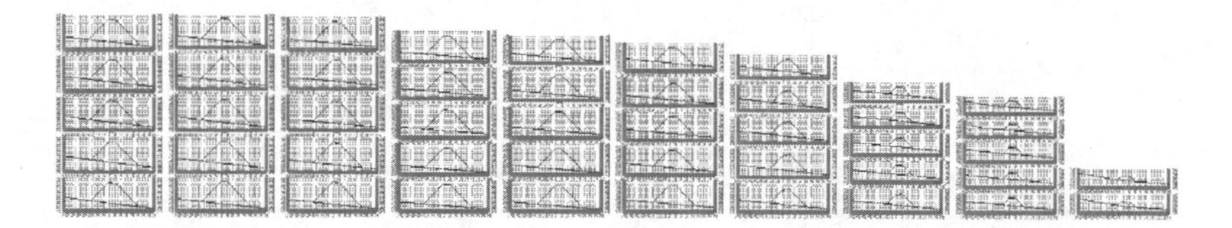

Figure 12-18 Cross Sections for the EBOFF Ramp

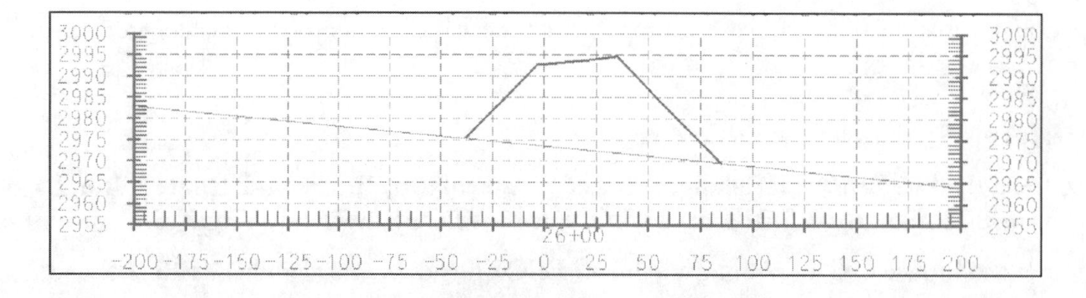

Figure 12-19 Cross Section at Station 26+00 of the EBOFF Ramp

Now the cross sections for the EBOFF Ramp have been created.

In the next subsection we will focus on the development of an earthwork-related report for each cross section.

2. Create an earthwork volume report for the EBOFF ramp

This subsection focuses on the steps to create the end area volume report for the EBOFF ramp. Later it can be used to create the mass haul diagram.

a. Select ***InRoads → Evaluation → Cross Section → Cross Sections....***

   The **Cross Sections** dialog box appears.

b. Select **End-Area Volumes** and establish the settings as follows:

   » Select **EBOFF** from the drop-down list of the **Cross Section Set**.
   » Under the **End-Area Volumes** folder in the left window, select the **General** leaf. Type the following information:

Surface:	**EBOFF**	**Check on**
	OG	**Check on**
Imperial Units:	**Cubic Yards**	
Create XML Report:	**Check on**	
Method:	**Standard**	

   » Once the settings are established, click **Apply**.

   The **Bentley Civil Report Browser** window appears (see Figure 12-20). The end-area volume report for various cross sections of the EBOFF ramp is illustrated.

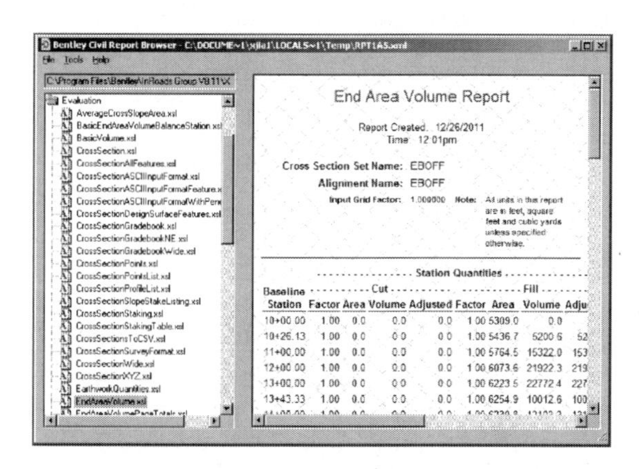

Figure 12-20 Bentley Civil Report Browser for the EBOFF Ramp

c. Select ***File → Save As....***

   The **Save As** dialog box appears.

d. Establish the settings as follows:

   » In the **Save in** field, browse to the project directory (**C:/Temp/Tutorial/Chp12/Master Files/** folder).
   » In the **File Name** field, type **EBOFF**.
   » From the **Save as** type drop-down list, select **XML File (*.xml)**.

e. Once the setting is established, click **Save**.

f. In the **End Area Volumes** window, click **Close**.

You have created the end area volume report for the EBOFF ramp. It will be used to develop the mass haul diagram for the **EBOFF** corridor.

3. Create the mass haul diagram for the EBOFF ramp

The detailed procedures for creating the mass haul diagram for the EBOFF ramp corridor are shown as below:

a. Select ***InRoads → Evaluation → Cross Section → Cross Sections…***.

The **Cross Sections** dialog box appears.

b. Make sure **Create XML Report** is checked OFF and **Plot Mass Haul Diagram** is checked ON.

c. Select ***End-Area Volumes → Mass-Haul Diagram → General.***

The **Mass Haul Diagram** window appears.

d. Establish the following settings:

Direction:	**Left to Right**	
Exaggeration:	**Horizontal**	**1.0000**
	Vertical:	**0.0100**

The **Vertical Exaggeration** is a scaling factor for the vertical orientation. Adjust the value through trial and error until a preferred mass haul diagram is created.

e. Double click the **Data Line** and select the **Color** and **Weight** of your choice. Select **OK**.

f. Go to ***Mass Haul Diagram → Horizontal Center Axis***. Establish the following settings:

Major Ticks	Length:	**0.2**	
	Position:	**Inside**	
	Spacing:	**100.00**	
Major Format:	**SS+SS.SS**		
Horizontal Center Axis:	**Check ON**		
	Double click the **Color** icon. Select the **Color** and **Weight** of your choice for the horizontal center axis.		
Major Ticks:	**Check ON**		
Label:	**Check ON**		
	Double click the color icon and establish the settings as follows:		
	Set offsets:	Horizontal:	**–0.3**
		Vertical:	**–0.3**
	Others:	**Default**	

g. Go to ***Mass Haul Diagram → Left Axis***. Establish the following settings:

Major Ticks:	Length:	**0.1**
	Position:	**Inside**
	Spacing:	**50000.00**

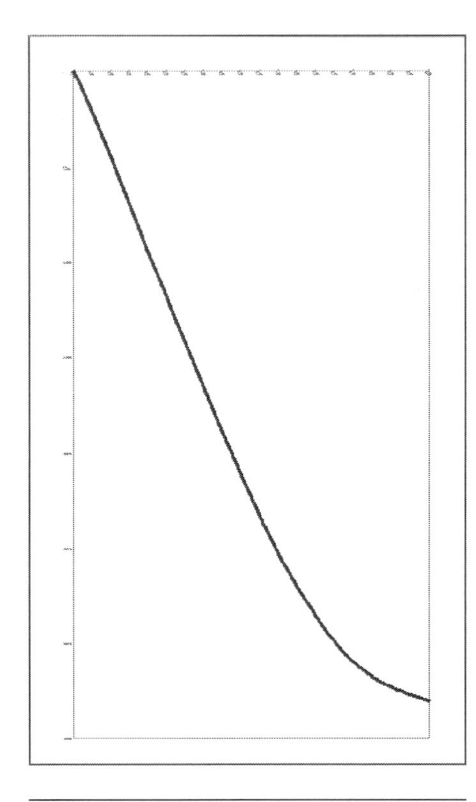

Figure 12-21 Mass Diagram for the EBOFF Ramp

Left Axis:	**Check ON**
	Double click the **Color** icon. Select the **color** and **weight** of your choice for the horizontal center axis.
Major Ticks:	**Check ON**
Label:	**Check ON**
	Double click the **Color** icon and establish the settings as follows:

Set offsets:	Horizontal:	**–1.0**
	Vertical:	**0.0**
Others:	**Default**	

h. Go to ***Mass Haul Diagram → Right Axis.*** In the **Symbology** section, click **Right Axis.**

i. Go to ***Mass Haul Diagram → Top Axis***. In the **Symbology** section, click **Top Axis.**

j. Go to ***Mass Haul Diagram → Bottom Axis***. In the **Symbology** section, click **Bottom Axis.**

k. Once the settings are established, click **Apply**.

l. Move the cursor to a location in the MicroStation view and press the **Data** button on your mouse.

A mass haul diagram appears (see Figure 12-21).

m. In the **Cross Sections** window, click **Close**.

12.4 Earthwork Estimation for the EBON Ramp

This section focuses on the earthwork estimation for the EBON ramp. The related procedures are as follows:

1. Create cross sections for the EBON ramp

   a. Load the **OG.dtm, Tutorial.alg, Tutorial_Chapter11.itl, Tutorial.ird, Route 8.dtm, Laguna Canyon Road.dtm, EBOFF.dtm, EBON.dtm, WBOFF.dtm,** and **WBON.dtm** into InRoads.

   b. Set the surface **OG**, horizontal alignment **EBON** and vertical alignment **EBON**, and template **Ramps** to active.

   To set active the specific template, go to ***Modeler → Create Template...*** and the **EBON** corridor.

   c. Select ***InRoads → Evaluation → Cross Section → Cross Sections....***

   The **Cross Sections** dialog box appears.

   d. In the **Create Cross Section** dialog box, under the **Create Cross Section** folder in the left window, select the **General** leaf. Type the following information:

Set Name:	**EBON**
Create:	**Window and Data**
Interval:	**100.00**

Left Offset:	**–200.00**
Right Offset:	**200.00**
Vertical Exaggeration:	**2.0000**
Surface:	**EBON; OG**

e. In the **Cross Sections** dialog box in the left window, select ***Layout → General.*** Click **Stacked**.

f. Select the ***Create Cross Section → Include → Components***.

g. Once the settings are established, click **Apply**.

   InRoads will prompt you to place a data point in the MicroStation view to specify the location of the left corner of the group of cross sections.

h. Move the cursor to a location in the view with the design file and press the **Data** button on your mouse.

   A group of cross sections appear (see Figure 12-22). An example of a cross section at Station 23+00 is illustrated in Figure 12-23.

i. In the **Create Cross Section** dialog box, click **Close**.

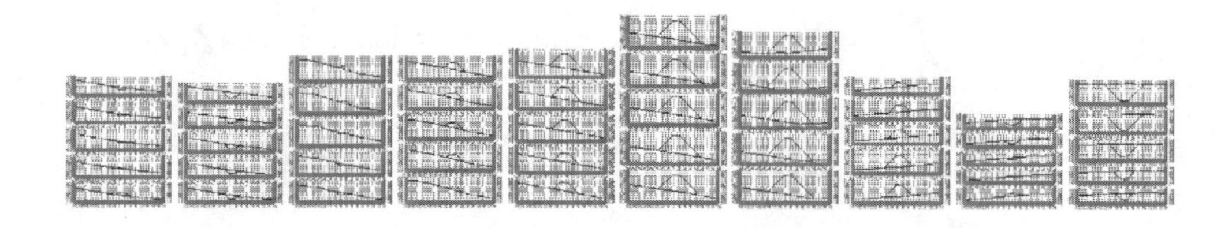

Figure 12-22 Cross Sections for the EBON Ramp

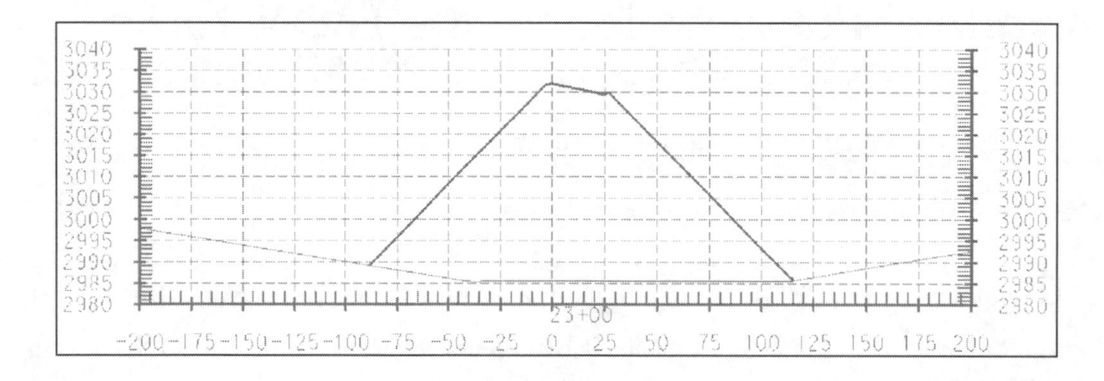

Figure 12-23 Cross Section at Station 23+00 of the EBON Ramp

You have now created the cross sections for the EBON ramp.

In the next subsection we will focus on the development of earthwork-related report for each cross section.

2. Create an earthwork volume report for the EBON ramp

This subsection focuses on the steps to create the end area volume report for the EBON ramp, which later can be used to create the mass haul diagram.

a. Select ***InRoads → Evaluation → Cross Section → Cross Sections....***

The **Cross Sections** dialog box appears.

b. Select **End-Area Volumes** and establish the settings as follows:

» From the drop-down list of the **Cross Section Set,** select **EBON**.
» Under the **End-Area Volumes** folder in the left window, select the **General** leaf. Type the following information:

Surface:	**EBON**	**Check ON**
	OG	**Check ON**
Imperial Units:	**Cubic Yards**	
Create XML Report:	**Check on**	
Method:	**Standard**	

c. Once the settings are established, click **Apply**.

The **Bentley Civil Report Browser** window appears (see Figure 12-24) where the end-area volume report for various cross sections of the EBON ramp is illustrated.

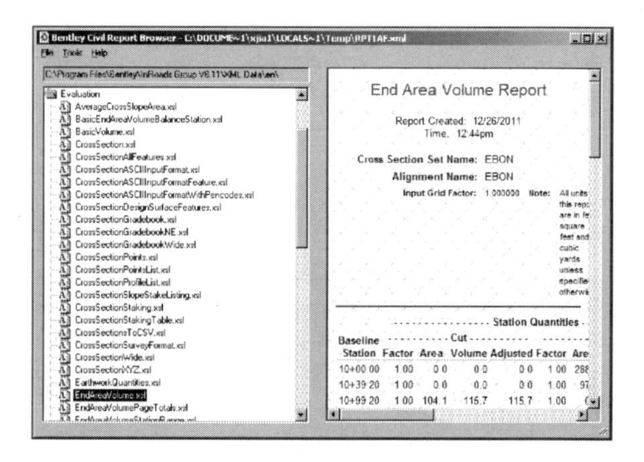

Figure 12-24 Bentley Civil Report Browser for the EBON Ramp

d. Select ***File → Save As....***

The **Save As** dialog box appears.

e. Establish the settings as follows:

» In the **Save in** field, browse to the project directory (**C:/Temp/Tutorial/Chp12/Master Files/** folder).
» In the **File Name** filed, type **EBON**.
» From the **Save as** type drop-down list, select **XML File (*.xml)**.

f. Once the setting is established, click **Save**.

g. In the **End Area Volumes** window, click **Close**.

You have now created the end area volume report for the EBON ramp. It will be used to develop the mass haul diagram for the **EBON** corridor.

3. Create mass haul diagram for the EBON ramp

The detailed procedures for creating the mass haul diagram for the EBON ramp corridor are shown below:

a. Select ***InRoads → Evaluation → Cross Section → Cross Sections....***

The **Cross Sections** dialog box appears. Make sure **Create XML Report** is checked OFF and **Plot Mass Haul Diagram** is checked ON.

b. Select ***End-Area Volumes → Mass-Haul Diagram → General.***

The **Mass Haul Diagram** window appears.

c. Establish the following settings:

Direction:	**Left to Right**	
Exaggeration:	**Horizontal**	**1.0000**
	Vertical	**0.0100**

The **Vertical Exaggeration** is a scaling factor for the vertical orientation. You can adjust the value by trial and error until a preferred mass haul diagram is created.

d. Double click **Data Line**. Select the Color and Weight of your choice. Select **OK**.

e. Go to ***Mass Haul Diagram → Horizontal Center Axis***.

f. Establish the following settings (the spacing of major ticks represents 100 ft. or 1 station):

Major Ticks:	Length:	**0.2**	
	Position:	**Inside**	
	Spacing:	**100.00**	
Major Format:	**SS+SS.SS**		
Horizontal Center Axis:	**Check ON**		
	Double click the **Color** icon and set up the **color** and **weight** of your choice for horizontal center axis.		
Major Ticks:	**Check ON**		
Label:	**Check ON**		
	Double click the color icon and set up the settings as follows:		
	Set offsets:	Horizontal:	**–0.3**
		Vertical:	**–0.3**
	Others:	**Default**	

g. Go to ***Mass Haul Diagram → Left Axis.*** Establish the following settings (you can adjust the spacing of major ticks based on the magnitude of your earthwork volume amount):

Major Ticks:	Length:	**0.1**
	Position:	**Inside**
	Spacing:	**10000.00**
Left Axis:	**Check ON**	
	Double click the **Color** icon. Select the color and weight of your choice for the horizontal center axis.	
Major Ticks:	**Check ON**	
Label:	**Check ON**	
	Double click the color icon and	

set up the settings as follows:

Set offsets:	Horizontal:	**–1.0**
	Vertical:	**0.0**
Others:	**Default**	

h. Go to ***Mass Haul Diagram → Right Axis.*** In the **Symbology** section, click **Right Axis.**

i. Go to ***Mass Haul Diagram → Top Axis.*** In the **Symbology** section, click **Top Axis.**

j. Go to ***Mass Haul Diagram → Bottom Axis.*** In the **Symbology** section, click **Bottom Axis.**

k. Once the settings are established, click **Apply**.

l. Move the cursor to a location in the MicroStation view and press the **Data** button on your mouse.

   A mass haul diagram appears on the screen (see Figure 12-25).

m. In the **Cross Sections** window, click **Close.**

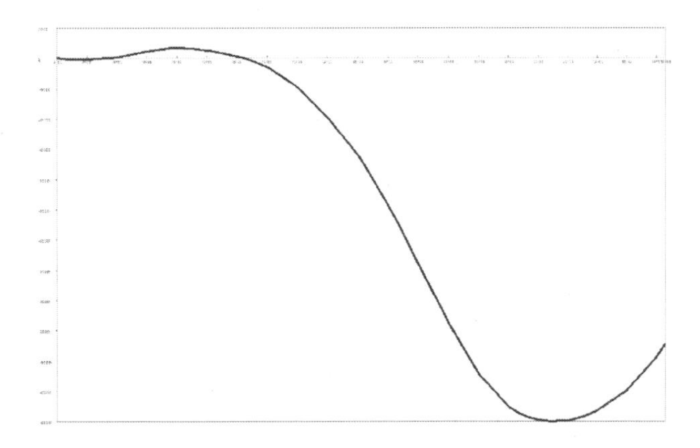

Figure 12-25 Mass Diagram for the EBON Ramp

12.5 Earthwork Estimation for the WBOFF Ramp

This section focuses on the earthwork estimation for the WBOFF road. The related procedures are similar to those for Route 8.

1. Create cross sections for the WBOFF ramp

   a. Load the **OG.dtm, Tutorial.alg, Tutorial_Chapter11.itl, Tutorial.ird, Route 8.dtm, Laguna Canyon Road.dtm, EBOFF.dtm, EBON.dtm, WBOFF.dtm,** and **WBON.dtm** into InRoads.

   b. Set the surface **OG**, horizontal alignment **WBOFF** and vertical alignment **WBOFF**, and template **Ramps** to active.

      To set the specific template to active, go to ***Modeler → Create Template...***) and the **WBOFF** corridor.

   c. Select ***InRoads → Evaluation → Cross Section → Cross Sections....***

      The **Cross Sections** dialog box appears.

d. In the **Create Cross Section** dialog box, under the **Create Cross Section** folder in the left window, select the **General** leaf. Type the following information:

Set Name:	**WBOFF**
Create:	**Window and Data**
Interval:	**100.00**
Left Offset:	**–200.00**
Right Offset:	**200.00**
Vertical Exaggeration:	**2.0000**
Surface:	**WBOFF; OG**

e. In the **Cross Sections** dialog box, in the left window, select the ***Layout → General***. Then click **Stacked**.

f. Select the ***Create Cross Section → Include → Components.***

g. Once the settings are established, click **Apply**.

InRoads will prompt you to place a data point in the MicroStation view to specify the location of the left corner of the group of cross sections.

h. Move the cursor to a location in the view with the design file and press the **Data** button on your mouse.

A group of cross sections appear (see Figure 12-26).

An example of a cross section at Station 18+00 is illustrated in Figure 12-27.

i. In the **Create Cross Section** dialog box, click **Close**.

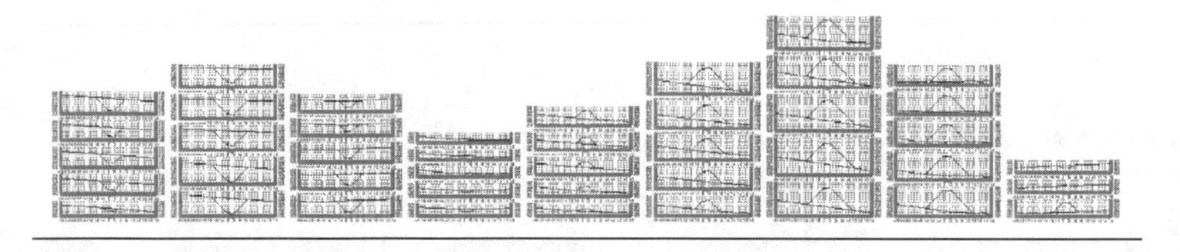

Figure 12-26 Cross Sections for the WBOFF Ramp

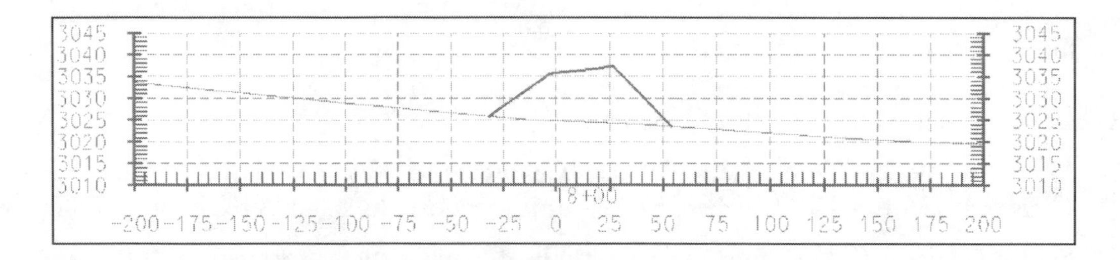

Figure 12-27 Cross Section at Station 18+00 of the WBOFF Ramp

The cross sections for the WBOFF ramp have been created.

In the next subsection we will focus on the development of earthwork-related report for each cross section.

2. Create an earthwork volume report for the WBOFF ramp

This subsection focuses on the steps to create the end area volume report for the WBOFF ramp, which later can be used to create the mass haul diagram.

a. Select ***InRoads → Evaluation → Cross Section → Cross Sections….***

   The **Cross Sections** dialog box appears.

b. Select **End-Area Volumes** and establish the settings as follows:

   » From the drop-down list of the **Cross Section Set**, select **WBOFF**.
   » Under the **End-Area Volumes** folder in the left window, select the **General** leaf and type the following information:

Surface:	**WBOFF**	**Check ON**
	OG	**Check ON**
Imperial Units:	**Cubic Yards**	
Create XML Report:	**Check on**	
Method:	**Standard**	

c. Once the settings are established, click **Apply**.

   The **Bentley Civil Report Browser** window appears (see Figure 12-28) and the end-area volume report for various cross sections of the **WBOFF** ramp is illustrated.

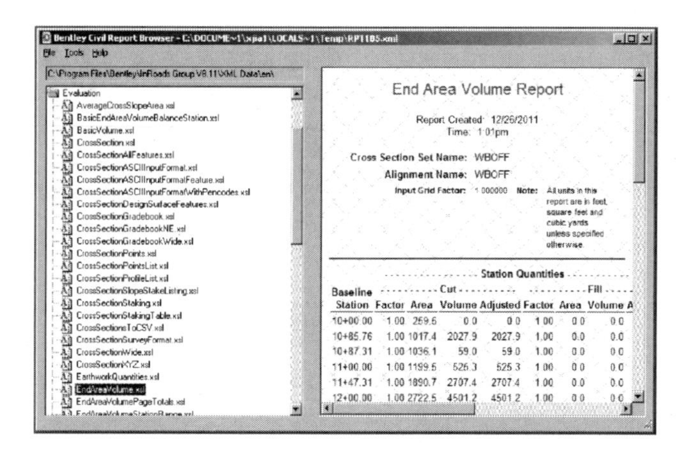

Figure 12-28 Bentley Civil Report Browser for the WBOFF Ramp

d. Select ***File → Save As….***

   The **Save As** dialog box appears.

e. Establish the settings as follows:

   » In the **Save in** field, browse to the project directory (**C:/Temp/Tutorial/Chp12/Master Files/** folder).
   » In the **File Name** field, type **WBOFF**.
   » From the **Save as** type drop-down list, select **XML File** (***.xml**).

f. Once the setting is established, click **Save.**

g. In the **End Area Volumes** window, click **Close**.

The end area volume report for the WBOFF ramp has been created. It will be used to develop the mass haul diagram for the **WBOFF** corridor.

3. Create the mass haul diagram for the WBOFF ramp

The detailed procedures of creating the mass haul diagram for the WBOFF ramp corridor are shown here:

a. Select ***InRoads → Evaluation → Cross Section → Cross Sections….***

   The **Cross Sections** dialog box appears. Make sure **Create XML Report** is OFF and **Plot Mass Haul Diagram** is ON.

b. Select ***End-Area Volumes → Mass-Haul Diagram → General.***

   The **Mass Haul Diagram** window appears.

c. Establish the following settings:

Direction:	**Left to Right**	
Exaggeration:	**Horizontal**	**1.0000**
	Vertical	**0.0100**

   The **Vertical Exaggeration** is a scaling factor for the vertical orientation. Adjust the value by trial and error until a preferred mass haul diagram is created.

d. Double click the **Data Line.** Select the **Color** and **Weight** of your choice. Click **OK.**

e. Go to ***Mass Haul Diagram → Horizontal Center Axis.***

f. Establish the following settings (the spacing of major ticks represents 100 ft. or 1 station):

Major Ticks:	Length:	**0.2**	
	Position:	**Inside**	
	Spacing:	**100.00**	
Major Format:	**SS+SS.SS**		
Horizontal Center Axis:	**Check ON**		
	Double click the **Color** icon. Select the **Color** and **Weight** of your choice for the horizontal center axis.		
Major Ticks:	**Check ON**		
Label:	**Check ON**		
	Double click the color icon and establish the settings as follows:		
	Set offsets:	Horizontal:	**–0.3**
		Vertical:	**–0.3**
	Others:	**Default**	

g. Go to ***Mass Haul Diagram → Left Axis.*** Establish the following settings (you can adjust the spacing of major ticks based on the magnitude of your earthwork volume amount):

Major Ticks:	Length:	**0.1**
	Position:	**Inside**
	Spacing:	**10000.00**
Left Axis:	**Check ON**	
	Double click the **Color** icon.	

	Select the color and weight of your choice for the horizontal center axis.		
Major Ticks:	**Check ON**		
Label:	**Check ON**		
	Double click the **Color** icon and set up the settings as follows:		
	Set offsets:	Horizontal:	**–1.0**
		Vertical:	**0.0**
	Others:	**Default**	

h. Go to ***Mass Haul Diagram → Right Axis.*** In the **Symbology** section, click **Right Axis.**

i. Go to ***Mass Haul Diagram → Top Axis***. in the **Symbology** section, click **Top Axis.**

j. Go to ***Mass Haul Diagram → Bottom Axis.*** In the **Symbology** section, click **Bottom Axis.**

k. Once the settings are established, click **Apply**.

l. Move the cursor to a location in the MicroStation view and press the **Data** button on your mouse.

   A mass haul diagram appears (see Figure 12-29).

m. In the **Cross Sections** window, click **Close**.

12.6 Earthwork Estimation for the WBON Ramp

This section focuses on the earthwork estimation for the WBON Ramp. The related procedures are as follows.

1. Create cross sections for the WBON ramp

   a. Load the **OG.dtm, Tutorial.alg, Tutorial_Chapter11.itl, Tutorial.ird, Route 8.dtm, Laguna Canyon Road.dtm, EBOFF.dtm, EBON.dtm, WBOFF.dtm,** and **WBON.dtm** into InRoads.

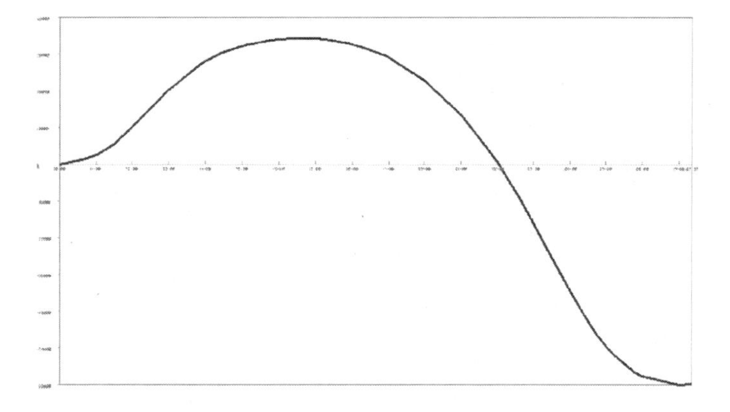

Figure 12-29 Mass Diagram for the WBOFF Ramp

b. Set the surface **OG**, horizontal alignment **WBON** and vertical alignment **WBON**, and template **Ramps** to active.

   To set the specific template to active, go to ***Modeler → Create Template...*** and the **WBON** corridor.

c. Select ***InRoads → Evaluation -> Cross Section → Cross Sections....***

   The **Cross Sections** dialog box appears.

d. In the **Create Cross Section** dialog box, under the **Create Cross Section** folder in the left window, select the **General** leaf. Then type the following information:

Set Name:	**WBON**
Create:	**Window and Data**
Interval:	**100.00**
Left Offset:	**–200.00**
Right Offset:	**200.00**
Vertical Exaggeration:	**2.0000**
Surface:	**WBON; OG**

e. In the **Cross Sections** dialog box, in the left window, select ***Layout → General.*** Then check **Stacked** to ON.

f. Select the ***Create Cross Section → Include → Components.***

g. Once the settings are established, click **Apply**.

   InRoads will prompt you to place a data point in the MicroStation view, which specifies the location of the left corner of the group of cross sections.

h. Move the cursor to a location in the view with the design file and press the **Data** button on your mouse.

   A group of cross sections appear (see Figure 12-30).

   An example of cross section at Station 16+00 is illustrated in Figure 12-31.

i. In the **Create Cross Section** dialog box, click **Close**.

   The cross sections for the WBON Ramp have been created.

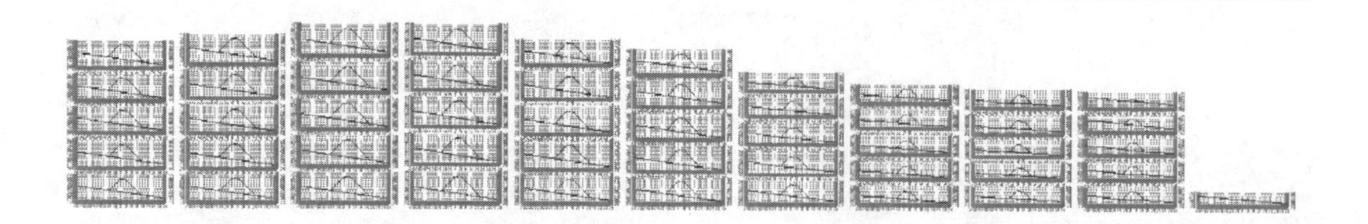

Figure 12-30 Cross Sections for the WBON Ramp

In the next subsection we will focus on the development of earthwork-related report for each cross section.

2. Create an earthwork volume report for the WBON ramp

This subsection focuses on the steps to create the end area volume report for the WBON ramp, which can be used to create the mass haul diagram.

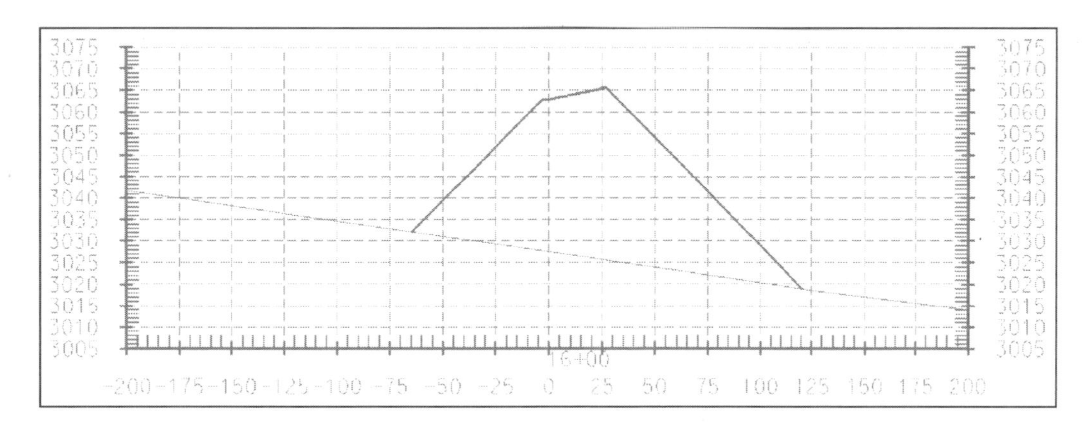

Figure 12-31 Cross Section at Station 16+00 of the WBON Ramp

a. Select ***InRoads → Evaluation → Cross Section → Cross Sections....***

The **Cross Sections** dialog box appears.

b. Select **End-Area Volumes** and establish these settings:

» From the drop-down list of the **Cross Section Set**, select **WBON**.

» Under the **End-Area Volumes** folder in the left window, select the **General** leaf. Type the following information:

Surface	**WBON**	**Check ON**
	OG	**Check ON**
Imperial Units:	**Cubic Yards**	
Create XML Report:	**Check on**	
Method:	**Standard**	

c. Once the settings are established, click **Apply**.

The **Bentley Civil Report Browser** window appears (see Figure 12-32) where the end-area volume report for various cross sections of the **WBON** ramp is illustrated.

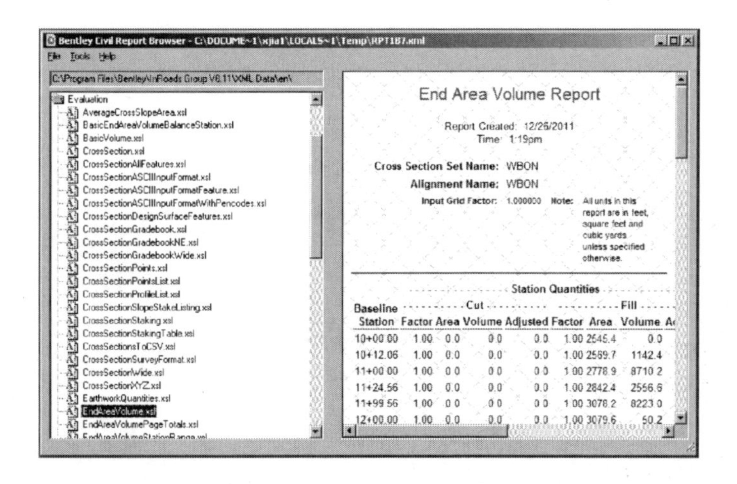

Figure 12-32 Bentley Civil Report Browser for the WBON Ramp

d. Select ***File → Save As....***

e. The **Save As** dialog box appears. Establish the settings as follows:

   » In the **Save in** field, browse to the project directory (**C:/Temp/Tutorial/Chp12/Master Files/** folder).
   » In the **File Name** field, type **WBON**.
   » From the **Save as** type drop-down list, select **XML File** (***.xml**).

f. Once the setting is established, click **Save**.

g. In the **End Area Volumes** window, click **Close**.

   The end area volume report for the WBON Ramp has been created. It will be used to develop the mass haul diagram for the **WBON** corridor.

3. Create the mass haul diagram for the WBON ramp

Follow these detailed procedures to create the mass haul diagram for the WBON ramp corridor below:

a. Select ***InRoads → Evaluation → Cross Section → Cross Sections....***

   The **Cross Sections** dialog box appears.

b. Make sure **Create XML Report** is checked off and **Plot Mass Haul Diagram** is checked ON.

c. Select ***End-Area Volumes → Mass-Haul Diagram → General.***

   The **Mass Haul Diagram** window appears.

d. Establish the following settings:

Direction:	**Left to Right**	
Exaggeration:	**Horizontal**	**1.0000**
	Vertical	**0.0100**

   The **Vertical Exaggeration** is a scaling factor for the vertical orientation. Adjust the value by trial and error until a preferred mass haul diagram is created.

e. Double click **Data Line**. Select the **Color** and **Weight** of your choice. Select **OK**.

f. Go to ***Mass Haul Diagram → Horizontal Center Axis.***

g. Establish the following settings (the spacing of major ticks represents 100 ft. or 1 station):

Major Ticks:	Length:	**0.2**	
	Position:	**Inside**	
	Spacing:	**100.00**	
Major Format:	**SS+SS.SS**		
Horizontal Center Axis:	**Check ON**		
	Double click the **Color** icon. Select the **Color** and **Weight** of your choice for horizontal center axis.		
Major Ticks:	**Check ON**		
Label:	**Check ON**		
	Double click the Color icon and establish the settings as follows:		
	Set offsets:	Horizontal:	**–0.3**

		Vertical:	**–0.3**
	Others:	**Default**	

h. Go to ***Mass Haul Diagram → Left Axis***. Establish the following settings (you can adjust the spacing of major ticks based on the magnitude of your earthwork volume amount):

Major Ticks:	Length:	**0.1**	
	Position:	**Inside**	
	Spacing:	**10000.00**	
Left Axis:	**Check ON**		
	Double click the **Color** icon. Select the **Color** and **Weight** of your choice for the horizontal center axis.		
Major Ticks:	**Check ON**		
Label:	**Check ON**		
	Double click the Color icon and establish the settings as follows:		
	Set offsets:	Horizontal:	**–1.0**
		Vertical:	**0.0**
	Others:	**Default**	

i. Go to ***Mass Haul Diagram → Right Axis***. In the **Symbology** section, check **Right Axis** to ON.

j. Go to ***Mass Haul Diagram → Top Axis***. In the **Symbology** section, check **Top Axis** to ON.

k. Go to ***Mass Haul Diagram → Bottom Axis***. In the **Symbology** section, check **Bottom Axis** to ON.

l. Once the settings are established, click **Apply**.

m. Move the cursor to a location in the MicroStation view and press the **Data** button on your mouse.

   A mass haul diagram appears (see Figure 12-33).

n. In the **Cross Sections** window, click **Close**.

o. Go to ***MicroStation → File → Save As***. In the **C:/Temp/Tutorial/Chp12/Master Files** folder, save **Tutorial_VA.dgn** to **Tutorial_CrossSection.dgn**.

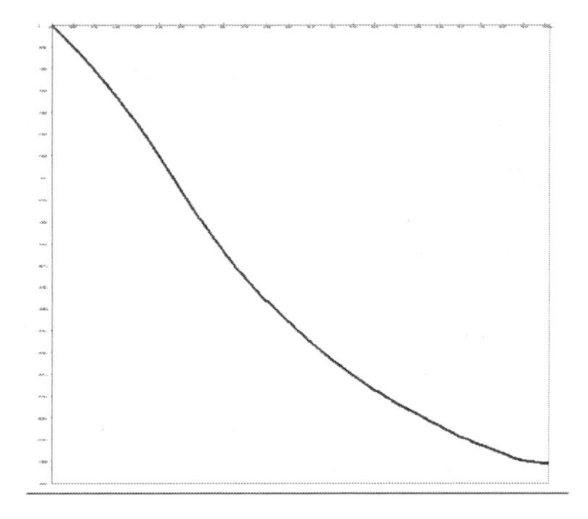

Figure 12-33 Mass Diagram for the WBON Ramp

12.7 Utilization of Mass Haul Diagram to Estimate Earthwork

From the previous sections you know that InRoads is powerful in creating the mass haul diagrams for a highway project. In this section we will learn how to use the previously developed mass haul diagrams to conduct the earthwork estimation, including amount of borrow, amount of waste, overhaul distance, etc. To help you better understand the relative procedures, we will review the basics of a mass haul diagram before we cover the details of the mass diagrams for each roadway corridor.

12.7.1 Background information of mass haul diagram

A mass haul diagram plots the cumulative volumes of cut-and-fill along a highway alignment. The diagram describes the earth to be distributed or hauled. You create a mass diagram by setting a chart with the "Station" (x-axis) and the "Cumulative Volumes" or the mass ordinate (y-axis). To help your understanding, a mass haul diagram is illustrated in Figure 12-34. The primary concepts associated with a mass haul diagram are as follows:

Cut/Fill Section. Sections where the slope of a curve in the mass diagram is positive are cut (excavation) sections. Sections where the slope of a curve is negative are *fill* (embankment) sections. In the mass haul diagram in Figure 12-34, the cut section is from Station 0+00 to Station 7+40 (the peak point in the graph), whereas the section from Station 7+40 to Station 14+00 is the fill section. Notice there is a set of small circles along the mass haul curve. They are required by the method of moments to determine the centers of mass as shown in the figure. You will find them useful if you use the same method to determine the location of centers of mass. Here you can ignore them since the location of centers of mass has been calculated and is shown in the figure.

Balance line and balance points. A *balance line* is a horizontal line that intersects the mass diagram at two points, which are usually called balance points. The volumes of cut and fill (excavation and embankment) between the two balance points are equal. For example, in Figure 12-34, the solid horizontal line that goes through the origin point (0, 0) is a balance line. The two intersecting points with the mass haul diagram are balance points. The volumes of cut and fill (around 8,200 cubic yards) are equal to each other.

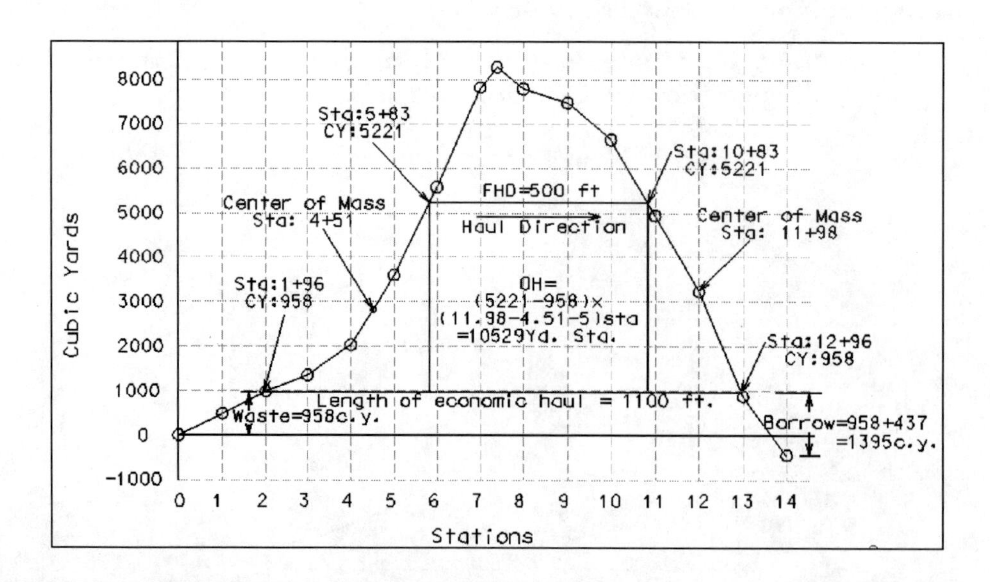

Figure 12-34 Example Mass Diagram

Free haul and free haul distance (**FHD**). A construction contract, after the contractor receives the contract from the transportation agency for a highway project, normally involves earthwork (excavation and/or embankment) and hauling work. The contractor is paid for the earthwork. The contractor does not receive full pay from the hauling work. Transporting "dirt" within a certain distance or between stations is done without payment. This free transporting distance is referred to as the free haul distance.

Normally transportation agencies set up free hauling distances. In practice, the free hauling distance usually ranges from 500 ft. to 3,000 ft. In Figure 12-34, the free haul distance is 500 ft.

Overhaul (**OH**) **and overhaul distance.** *Overhaul* is defined as any dirt transported beyond a pre-defined free haul distance. Correspondingly, the length of a haul more than the free haul distance is called the *overhaul distance.* Overhaul quantities are usually expressed in cubic-yard station.

From the mass diagram, the overhaul distance can be calculated as the distance from the center of mass (C.M.) of the excavation to the C.M. of the embankment. In practice, there are alternative methods to determine the C.M., including the graphical method and the method of moments. In Figure 12-34 the overhaul volume between the free haul distance and the economical haul distance (which is described in the next item) is 5,221 − 958 = 4,263 c.y. The C.M. of the excavation is at Station 4+51, whereas the C.M. of the embankment is at Station 11+98. Therefore, the overhaul distance is Station (11+98) – Station (4+51) − 500 ft. = 247 ft. (2.47 stations). Note that the free haul distance of 500 ft. needs to be subtracted since the overhaul distance is defined as the distance beyond the free haul distance. The overhaul quantity is 4,263 c.y. × 2.47 stations = 10,529 cubic-yard station.

Economic haul distance (**EHD**). Sometimes it is uneconomical to haul material beyond a certain long distance. Instead of hauling the material, it is more economical to waste the material excavated from the roadway section and borrow material from a borrow pit within the free hauling distance. This distance is called *economical hauling distance*, or EHD. The EHD is calculated as follows:

Assume:

h	Distance of haul in stations beyond the free hauling distance
e	Unit cost of earth excavation
o	Unit cost of the overhaul (or the haul beyond free haul)

Then: $e + ho = 2e$

he/o stations

$EHD = h + FHD$

Assume a transportation agency sets up its free hauling distance to be 2,000 ft., or 20 stations. The cost for excavation is \$1.80/yd^3, while the unit price of overhaul is \$0.20/yd^3−station. The economical hauling distance is 20 stations plus 1.8/0.2 = 29 stations. The total distance is 49 stations.

In the example shown in Figure 12-34, the EHD is assumed to be 1,100 ft.

Hauling direction. *Hauling direction* within a roadway project is from cut to fill section. In Figure 12-34, the hauling direction of the free haul distance is from left to right.

Waste and borrow. Waste indicates there is a surplus of excavation material for a highway project. Usually the waste situation exists under two conditions:

1. The excavation volume is greater than the embankment volume, and additional excavation volume follows in the EHD.
2. Excavation material is beyond the EHD.

Conversely, *borrow* indicates there is a net shortage of soil for a highway project. Usually the borrow situation occurs under two conditions:

1. The excavation volume is less than the embankment volume, and additional embankment volume follows in the EHD.
2. Embankment material is beyond the EHD.

In the example shown in Figure 12-34, the waste amount comes to 958 c.y. as the excavation material (the positive slope) is beyond the EHD. The amount of borrow consists of two parts. The first part of 958 c.y. exists since the embankment material (the negative slope) is beyond the EHD, while the second part of borrow (437 c.y.) is because the overall fill volume is 437 c.y. greater than the overall cut volume, indicating a net shortage of earthwork.

12.7.2 Assumptions associated with mass haul diagrams

Before we go into the details of earthwork estimation by using mass haul diagrams, we must make some assumptions about the diagrams that can be used to illustrate the relative procedures.

We assume the following information:

Unit Cost of Excavation:	$4.00/CY
Unit Cost of Borrow:	$6.00/CY
Unit Cost of Overhaul:	$0.20/station-yd
Free Haul Distance:	40 stations

Based on the above information, the economical hauling distance is 40 stations plus 4/0.2 = 60 stations.

Note that the unit costs are presented for illustrative purpose only. In actuality they could significantly differ from the prices adopted by transportation agencies.

12.7.3 Earthwork estimation for Route 8 using a mass haul diagram

This subsection focuses on the earthwork estimation for Route 8 using a mass haul diagram. Note that there is a bridge over Laguna Canyon Road in the Route 8 corridor. No earthwork is required for the bridge section. However, while we created the mass haul diagram via InRoads, the earth for the bridge part is not taken into consideration. Hence, before we conduct the estimation of earthwork quantities, the mass haul diagram for Route 8 should be modified. The detailed procedures for the mass haul diagram modification and earthwork evaluation are presented as follows:

a. Identify the locations of the bridge beginning and the end in the Route 8 mass haul diagram.

   The bridge beginning and end stations along Route 8 are 64+30.67 and 65+54.67, respectively (see Chapter 9). As shown in the Figure 12-35, the curve section between BB (beginning bridge) and EB (end bridge) are on a down slope, which indicates the excavation material. As discussed before, there is no earthwork needed for this part and earthwork volume should remain the same from BB to EB.

b. Replace the curve section between BB and EB with a horizontal line. Move up the curve section to the right of the bridge (see Figure 12-36). The curve section to the right of the EB line is moved up to connect with one end of the horizontal line.

c. Draw a balance line equal in length to the free haul distance of 4,000 ft. (see Figure 12-37).

d. Draw a balance line equal in length to the economic haul distance of 6,000 ft. (see Figure 12-38). The curve sections between the EHD and FHD represent the overhauled materials.

e. Draw other balance line(s) on the curve sections beyond the economic haul line. In Figure 12-39, there is one balance line to the left of the economic haul line whose length is less than the FHD. This hauling between the ends of the balance line should be done for free. The cut and fill volumes are balanced out on the balance line.

f. Identify the information of the critical points (e.g., troughs, domes, balance points, end points, etc.) in the diagram and determine the center of mass points for overhaul. In Figure 12-40, the station and cumulative volume for various critical points are exhibited. The C.M.s of the overhaul are calculated using the method of moments.

g. Conduct the earthwork estimation and the cost calculation based on the information of critical points. Looking at Figure 12-41, it is easy to obtain the earthwork quantities.

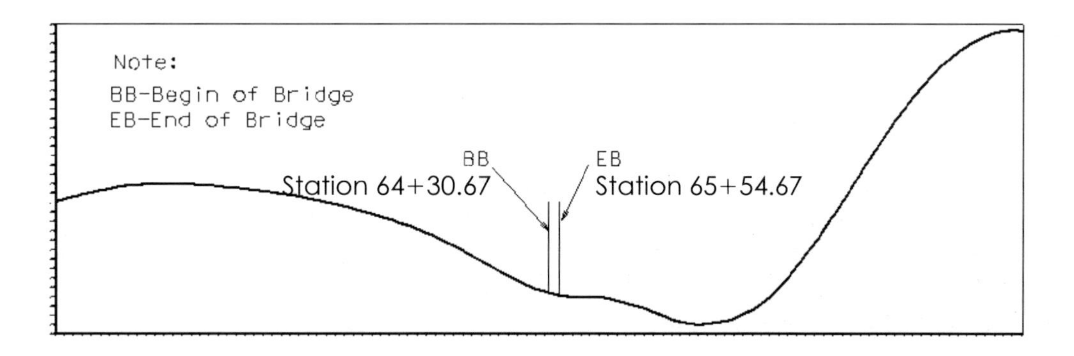

Figure 12-35 Location of Bridge

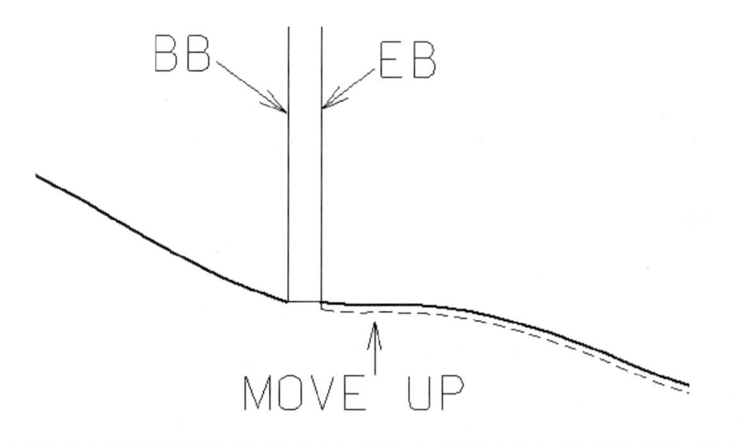

Figure 12-36 Curve to the Right of the Bridge

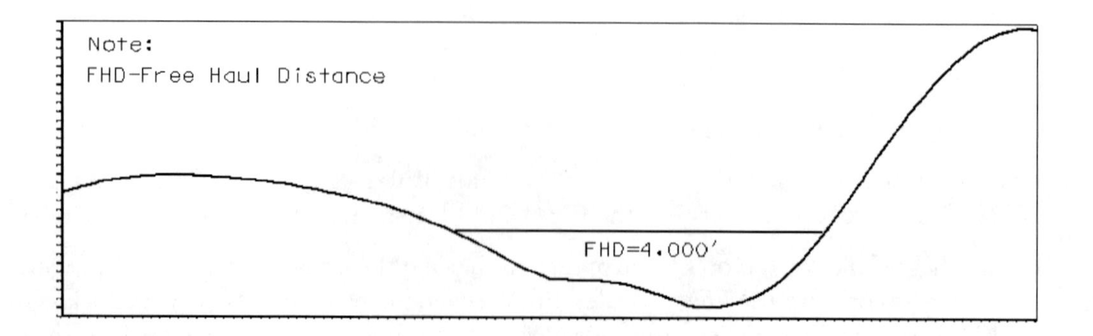

Figure 12-37 Free Haul Distance

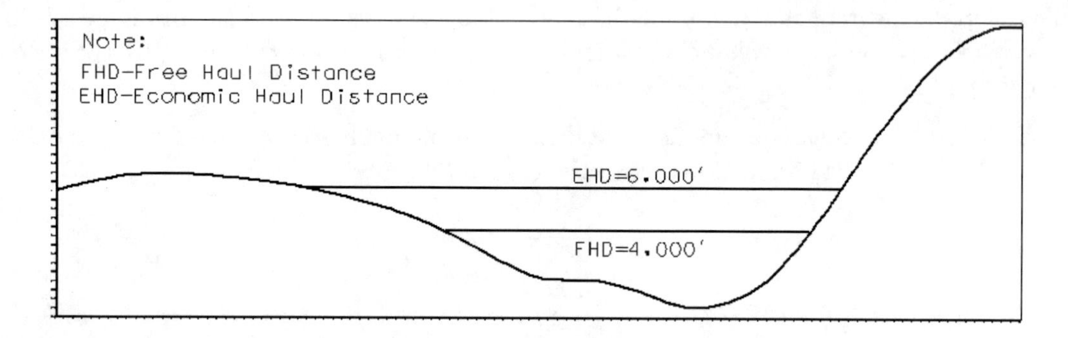

Figure 12-38 Free Haul and Economical Haul Distances

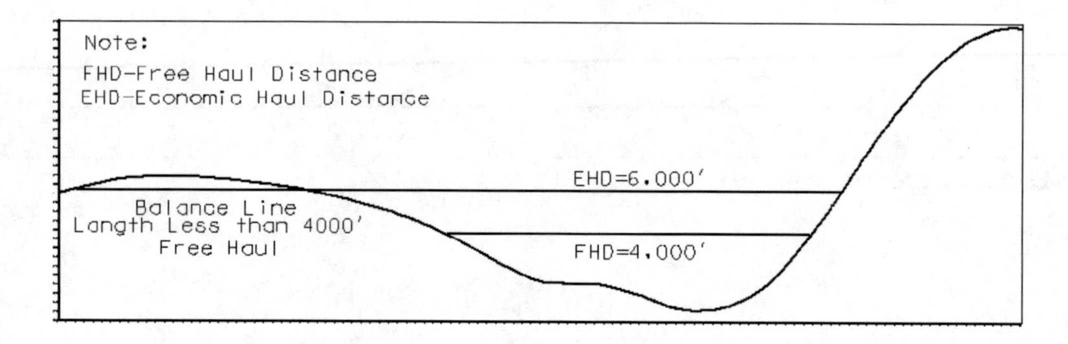

Figure 12-39 Free Haul and Economical Haul Distances

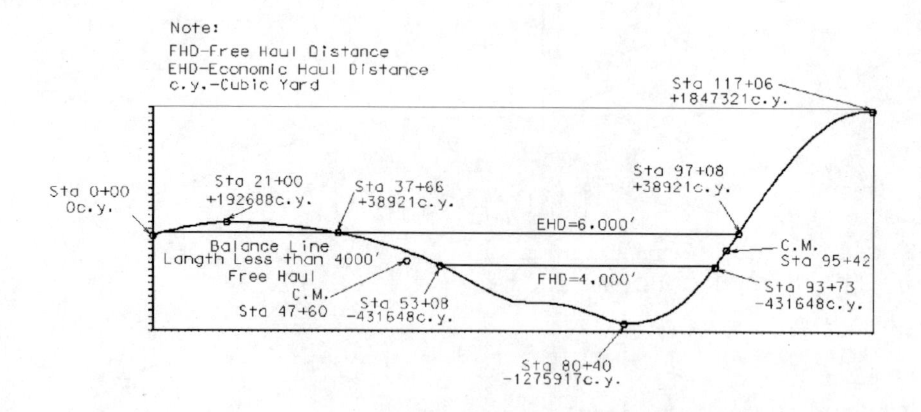

Figure 12-40 Mass Haul Diagram and Relative Information

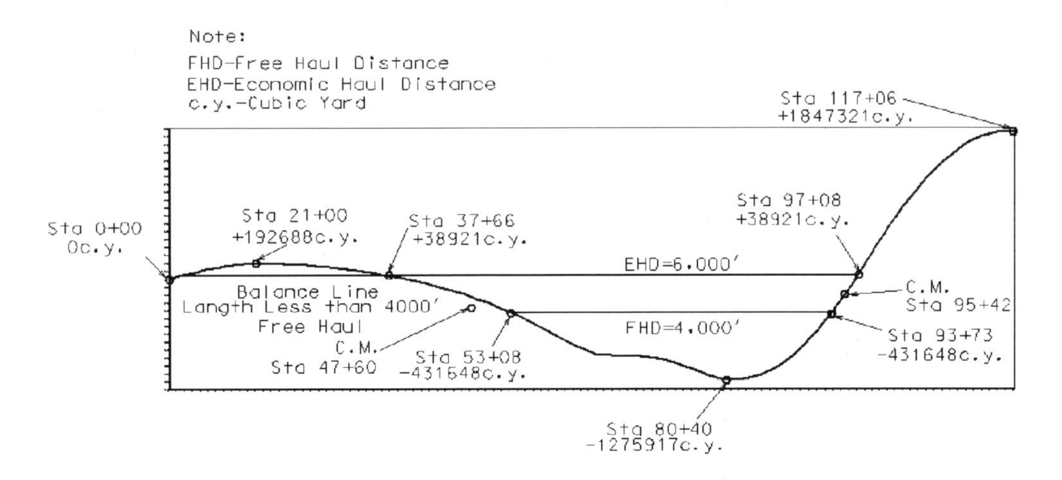

Figure 12-41 Mass Haul Diagram and Relative Information

Overhaul	
Overhaul volume:	38921 − (−431648) = 470569 (c.y.)
Overhaul distance:	Station (95+42) − Station (47+60)−40 stations = 7.82 stations
Overall overhaul quantity:	470569 (c.y.) × 7.82 stations = 3.68 × 10^6 sta-yd
Excavation	3123238 + 192688 = 3315926 (c.y.)
Borrow and Waste	Borrow: None Waste: 1808400 c.y.

Table 12-1 Earthwork Cost

Items	Unit Costs	Quantities	Overall Costs
Excavation	$4.0/c.y.	3.32 × 10^6 c.y.	$13.3 × 10^6
Borrow	$6.0/c.y.	0	0
Overhaul	$0.2/station-yd.	3.68 × 10^6 station-yd.	$0.7 × 10^6
		Total	**$14.0 × 10^6**

12.7.4 Earthwork estimation for Laguna Canyon Road using a mass haul diagram

The procedures for earthwork estimation for Laguna Canyon Road by using a mass haul diagram are similar to those for Route 8. However, there are also some major differences (see Figure 12-42). There is no bridge in the Laguna Canyon Road section; therefore, no need to modify the mass haul diagram created for Laguna Canyon Road.

The balance line with the FHD cannot fit in the mass haul diagram. This will affect the quantities of overhaul and borrow/waste.

Based on Figure 12-42, the relative earthwork quantities and costs are shown below (note that some small cut and fill volumes are ignored for ease of calculation).

Overhaul:	None
Excavation:	27,719 (c.y.)
Borrow and Waste:	Waste: 4,136 c.y.
	Borrow: None

Table 12-2 Earthwork Cost

Items	Unit Costs	Quantities	Overall Costs
Excavation	$4.0/c.y.	27,719 c.y.	$110,876
Borrow	$6.0/c.y.	0	0
Overhaul	$0.2/sta-yd	0	0
		Total	**$110,876**

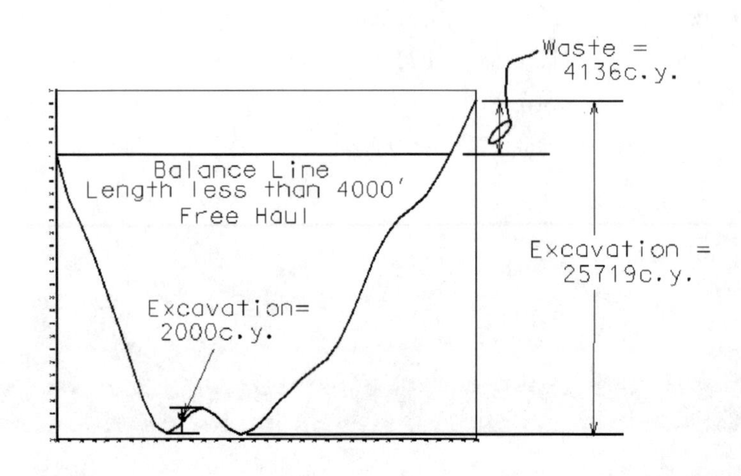

Figure 12-42 Mass Haul Diagram and Relative Information

12.7.5 Earthwork estimation for the EBOFF ramp using a mass haul diagram

The procedures of the earthwork estimation for the EBOFF ramp using a mass haul diagram are similar to those for Route 8. The whole curve section is on a down slope, which indicates there are no excavation materials.

Based on Figure 12-43, the relative earthwork quantities and costs are shown below.

Overhaul:	None
Excavation:	None
Borrow and Waste:	Waste: none
	Borrow: 306,700 c.y.

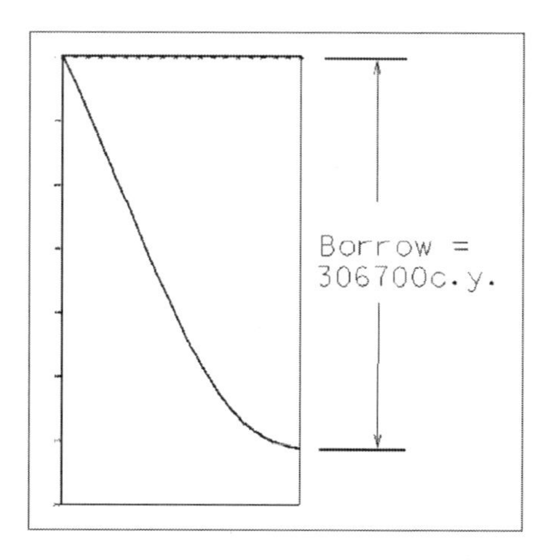

Figure 12-43 Mass Haul Diagram and Relative Information

Table 12-3 Earthwork Cost

Items	Unit Costs	Quantities	Overall Costs
Excavation	\$4.0/c.y.	0	0
Borrow	\$6.0/c.y.	306,700 c.y.	$\$1.8 \times 10^6$
Overhaul	\$0.2/sta-yd	0	0
		Total	$\mathbf{\$1.8 \times 10^6}$

12.7.6 Earthwork estimation for the EBON ramp using a mass haul diagram

The procedures for the earthwork estimation for the EBON ramp using a mass haul diagram are similar to those for Route 8. The balance line with the FHD or EHD cannot fit in the mass haul diagram. This will affect the quantities of overhaul and borrow/waste.

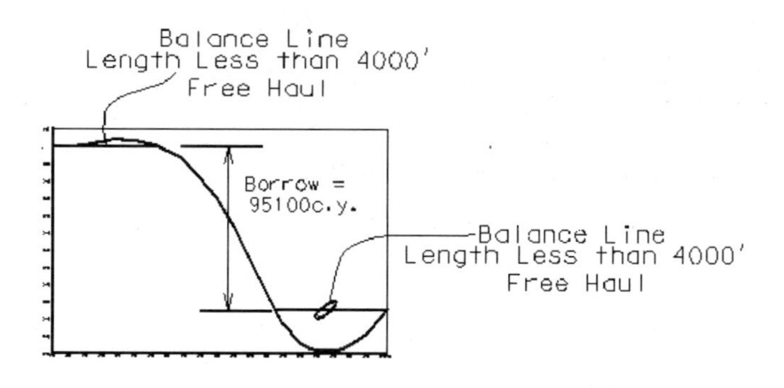

Figure 12-44 Mass Haul Diagram and Relative Information

Based on Figure 12-44, the relative earthwork quantities and costs are shown below. Note that some small cut and fill volumes are ignored for ease of calculation:

Overhaul:	None
Excavation:	None
Borrow and Waste:	Waste: None
	Borrow: 95,100 c.y.

Table 12-4 Earthwork Cost

Items	Unit Costs	Quantities	Overall Costs
Excavation	$4.0/c.y.	0	0
Borrow	$6.0/c.y.	95,100 c.y.	$570,600
Overhaul	$0.2/sta-yd	0	0
		Total	**$570,600**

12.7.7 Earthwork estimation for the WBOFF ramp using a mass haul diagram

The procedures for the earthwork estimation for the WBOFF ramp using a mass haul diagram are similar to those for Route 8. The balance line with the FHD or EHD cannot fit in the mass haul diagram. This will affect the quantities of overhaul and borrow/waste.

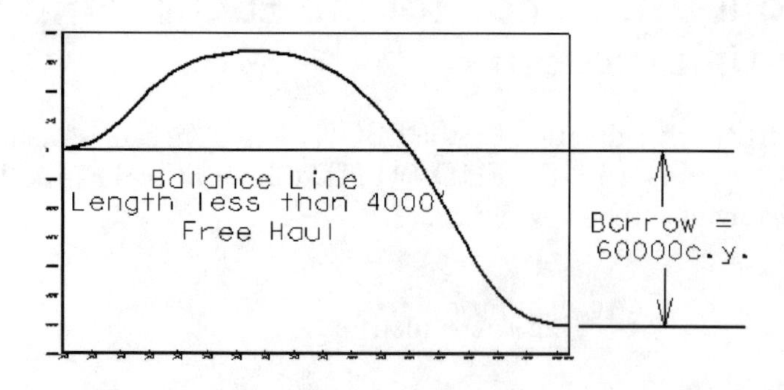

Figure 12-45 Mass Haul Diagram and Relative Information

Based on Figure 12-45, the relative earthwork quantities and costs are shown below.

Overhaul:	None
Excavation:	None
Borrow and Waste:	Waste: None
	Borrow: 60,000 c.y.

Table 12-5 Earthwork Cost

Items	Unit Costs	Quantities	Overall Costs
Excavation	$4.0/c.y.	0	0
Borrow	$6.0/c.y.	60,000 c.y.	$360,000
Overhaul	$0.2/sta-yd	0	0
		Total	**$360,000**

12.7.8 Earthwork estimation for the WBON ramp using a mass haul diagram

The procedures for the earthwork estimation for the WBON ramp using a mass haul diagram are similar to those for Route 8. The whole curve section is on a down slope. This indicates there are no excavation materials.

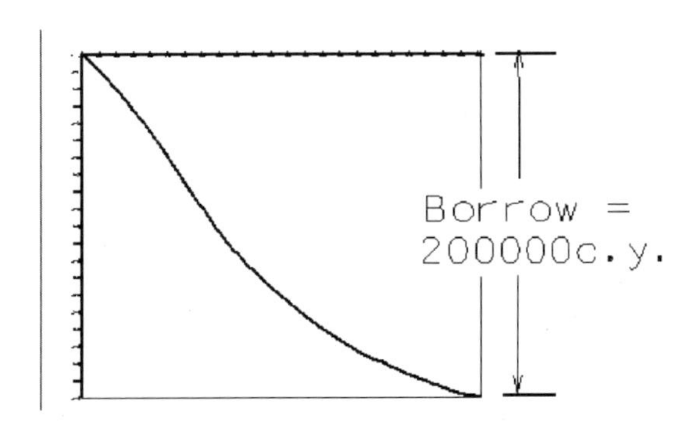

Figure 12-46 Mass Haul Diagram and Relative Information

Based on Figure 12-46, the relative earthwork quantities and costs are shown below (note that some small cut and fill volumes are ignored for ease of calculation).

Overhaul:	None
Excavation:	None
Borrow and Waste:	Waste: None
	Borrow: 200,000 c.y.

Table 12-6 Earthwork Cost

Items	Unit Costs	Quantities	Overall Costs
Excavation	\$4.0/c.y.	0	0
Borrow	\$6.0/c.y.	200,000 c.y.	$\$1.2 \times 10^6$
Overhaul	\$0.2/sta-yd	0	0
		Total	**$\$1.2 \times 10^6$**

Name____ Date_______________

12.7 Questions

Q12.1 What is the End-Area method?

Q12.2 What is the interval used in the End-Area method for this tutorial project? Can we use the interval of 25 ft. for the earthwork calculation?

Q12.3 Suppose you have two consecutive stations whose cross sections have the following information:

Table 12-7

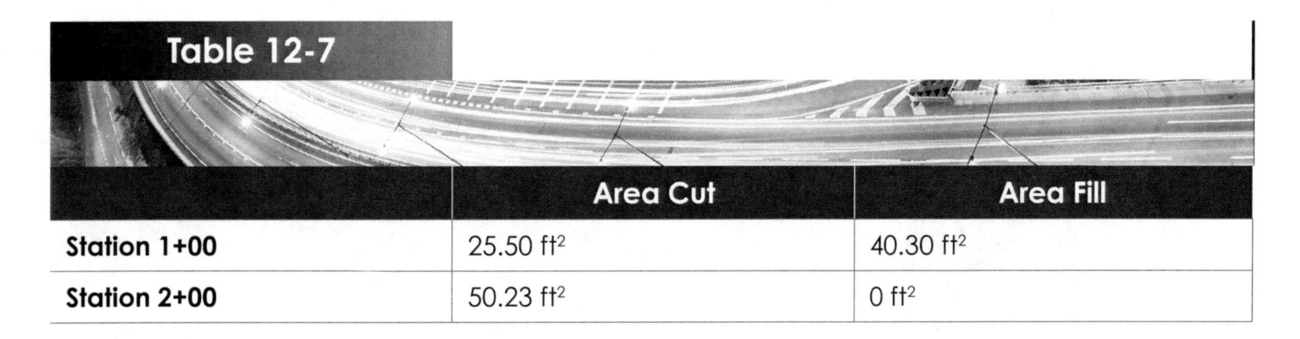

	Area Cut	Area Fill
Station 1+00	25.50 ft^2	40.30 ft^2
Station 2+00	50.23 ft^2	0 ft^2

What is the cut and fill volume between the two stations? What is the net volume between the two stations?

Q12.4 What is a mass diagram? How do you calculate the mass ordinate of a diagram?

Q12.5 In this tutorial project, we calculate the earthwork for Route 8 and four ramps independently. In real life, we could merge these four design surfaces (Route 8 and the four ramp design surfaces) and compare them to the original surface to calculate the earthwork. Describe the surface merging method. (Hint: Refer to the InRoads Help for details).

Chapter 13
PS & E Preparation

LEARNING OBJECTIVES

After completing this chapter you should know:

1. Procedures used to create title sheets, cross section sheets, and key map sheets using MicroStation;
2. Procedures related to the generation of plan and profile sheets using InRoads;
3. General procedures applicable for developing specifications and contract documents for transportation projects.

SECTIONS

The transportation design process, as described in Chapters 2 and 3, produces plans, specifications, and cost estimates (PS & E). *Plans* are drawings and technical documents prepared to convey physical information so that designers, reviewers, and the public can understand both existing and proposed conditions of a project. Using plans, contractors transform design features on paper drawings to the physical ground.

Specifications, which prescribe scope, materials, workmanship, acceptance criteria, methods of measurement, and payment for a project, provide directions, provisions, and requirements for a project construction. Many transportation agencies have developed and used standard specifications as a baseline for projects. The purpose of using standard specifications, along with supplemental and/or special provisions for a project, is to provide a lowest cost solution to transportation needs.

Costs needed for the construction of a project are estimated and summarized. They act as the financial references to the procurement of a project.

This chapter, considered as the last step of the project development process, assembles all the design elements (including the horizontal alignment, the vertical alignment, the cross section design, and the corridor design) discussed in previous chapters for the tutorial project.

13.1 Preparation of Plans

The tutorial project is located in California. We need to use Caltrans' *Plans Preparation Manual* (*PPM*) and *CADD User's Manual* to prepare the plans for the tutorial project.

The latest versions of these two manuals are located on the following Caltrans websites:

CADD User's Manual:
www.dot.ca.gov/hq/oppd/cadd/usta/caddman/default.htm

Plans Preparation Manual:
www.dot.ca.gov/hq/oppd/cadd/usta/ppman/default.htm

Caltrans requires the following components in a complete set of construction plans:

Title Sheet
Typical Cross Section
Project Orientation
Layouts, Profiles, and Superelevation Diagrams
Construction Details
Drainage Details
Pavement Delineation (Stripping) and Sign Plans
Signal, Lighting, and Electrical System Plans
Construction Staging Plans
Temporary Traffic Control Plans
Planting and Irrigation

This chapter concentrates only on the preparation of the title sheet, typical cross section sheets, project orientation, layouts, and profiles for the tutorial project. The plans for construction details, drainage details, pavement delineation, construction staging, planting, and irrigation for the tutorial project are beyond the scope of this book.

13.1.1 Title Sheet Preparation

A *title sheet*, which provides a neat, clear, and concise presentation of a project, consists of title block, index of plans, vicinity map, north arrow, and small-scale map. The *title block* lists major work items, name of the project highway, county, and a signature block for the engineer's signature. The *index of plans* identifies all plan sheets. The *vicinity map* is placed at the center of a title sheet to show the length as well as the beginning and end of a project. The *north arrow* indicates the project orientation. Lastly, the arrow pointing to a small-scale map indicates the project's general location.

A list of sample title sheet drawings is available on this Caltrans website: www.dot.ca.gov/hq/oppd/cadd/usta/ppman/examples/examples_titles.htm.

The title sheet of the tutorial project includes the project name and description, index of plans (or a list of project plan sheets), project construction limits, location map, and north arrow.

Steps involved in developing the title sheet in MicroStation are as follow:

Step 1. Create a new file

a. Start MicroStation (if it is not open).

b. Identify the Tutorial cell library **Tutorialcelib.cel** (or Caltrans' cell library file **ctcellib.cel**[1]) and the seed file **V8eSeed.dgn** from **C:/Temp/Tutorial/Chp13/Master Files** folder or **C:/Temp/Book/Chp13/Master Files** folder.

   You can also download these two files from the book website (www.csupomona.edu/~xjia/GeometricDesign/Chp13/Master Files).

c. Create a new file, **01-TitleSheet.dgn**, under the working project directory **C:\Temp\Tutorial\Chp13\Plan Sheets**[2] (see Figure 13-1).

d. Browse to the **C:\Temp\Tutorial\Chp13\Master Files** directory and select **V8eSeed.dgn** as the seed file. Click **Save** and **Open**.

   Note: Your directory may be different from the location shown in Figure 13-1.

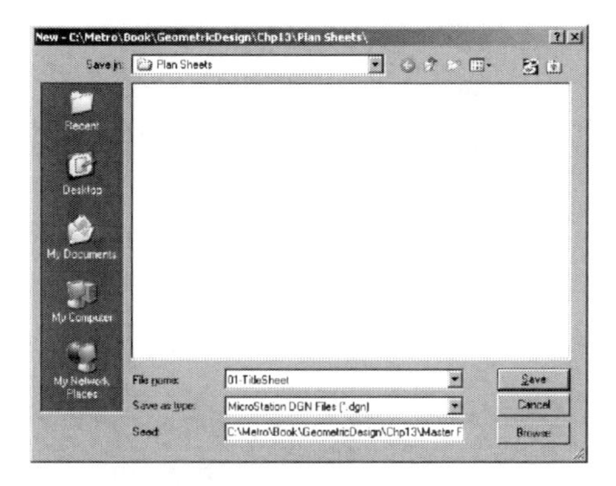

Figure 13-1 Saving New File 01_TitleSheet.dgn

Step 2. Set reference to the design file

a. Attach the **tutorial_2D_HA.dgn** design file in the directory **C:\Temp\Tutorial\Chp13\Master Files** to the newly created **01-TitleSheet.dgn** by doing the following:

   » Select ***MicroStation → File → Reference → Tools → Attach....***
   » Browse to the **C:\Temp\Tutorial\Chp13\Master Files** directory and select **tutorial_2D_HA.dgn.** Click **Open** (see Figure 13-2).

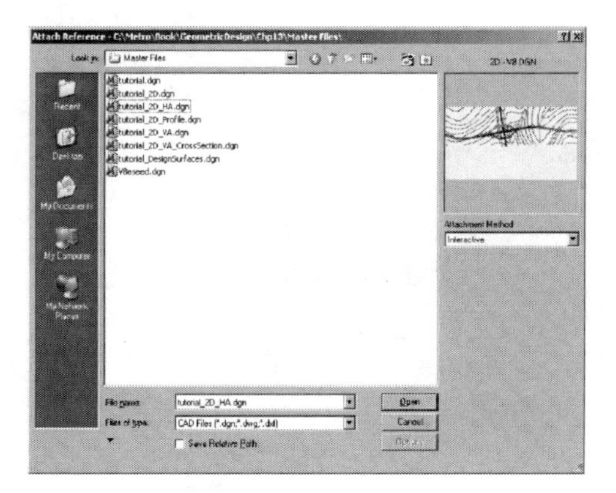

Figure 13-2 Attach Tutorial_2D_HA.dgn

[1] *When you use Caltrans library, the title sheet may look different.*
[2] *C:\Temp\Tutorial\ ... and C:/Temp/Tutorial/... are interchangeable.*

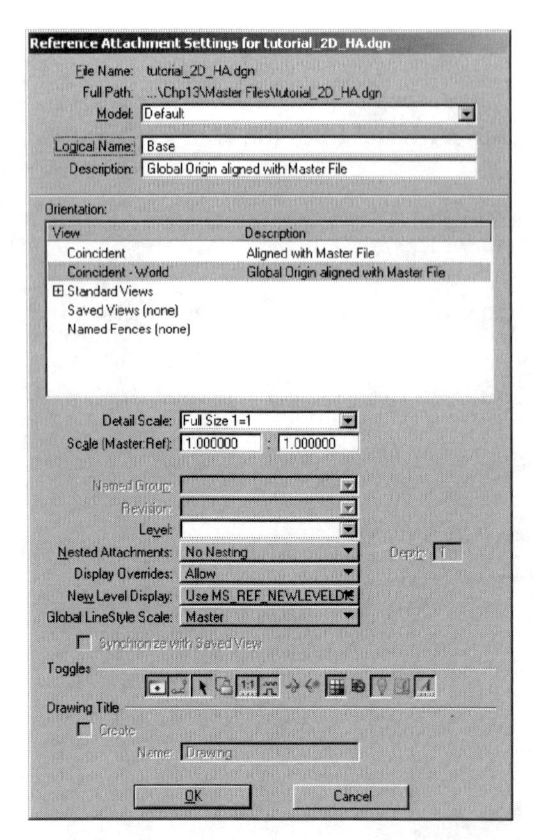

Figure 13-3 Reference Attachment Dialog Box

The **Reference Attachment Settings** dialog box appears.

» Enter **Base** for logical name, and select **Coincident-World** for orientation. Click **OK** and **Close** (see Figure 13-3).

b. To view the reference file on screen, click on the **FitView** icon.

Step 3. Load and place plan sheet cells

a. Select ***MicroStation*** → ***Element***, select **Cells**, and attach the Caltrans cell library.

b. Select ***File*** → ***Attach File…*** and locate the **Tutorialcel lib.cel** or **ctcellib.cel** file from the **C:\Temp\Tutorial\Chp13\Master Files** folder. Click **Open**.

A list of cells with name and description is shown.

c. Select **TITLE** for the title sheet border, click on **Placement**, and close the dialog box.

d. Go to ***Tools*** → ***Cells*** → ***Place Active Cell***. Change **X Scale** and **Y Scale** to **10.0**.

The Caltrans cells are created in the base scale of 1:50. Therefore, 10:1 here means the title sheet would have a scale of 1:500. With the appropriate scales, you can place the title border around the reference file by clicking on the drawing window (see Figure 13-4). The title border encloses the project site (see Figure 13-5).

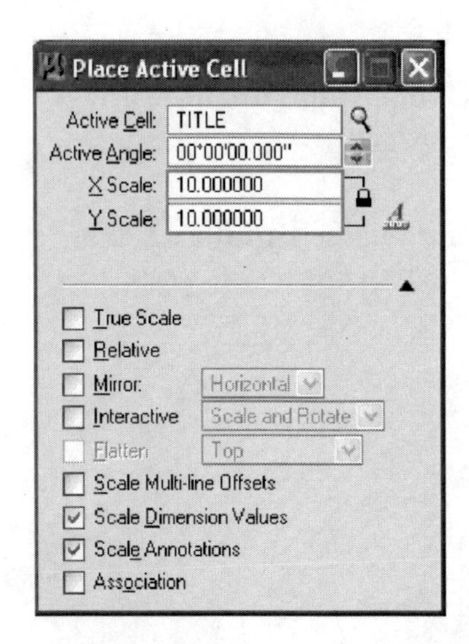

Figure 13-4 Place Active Cell Dialog Box

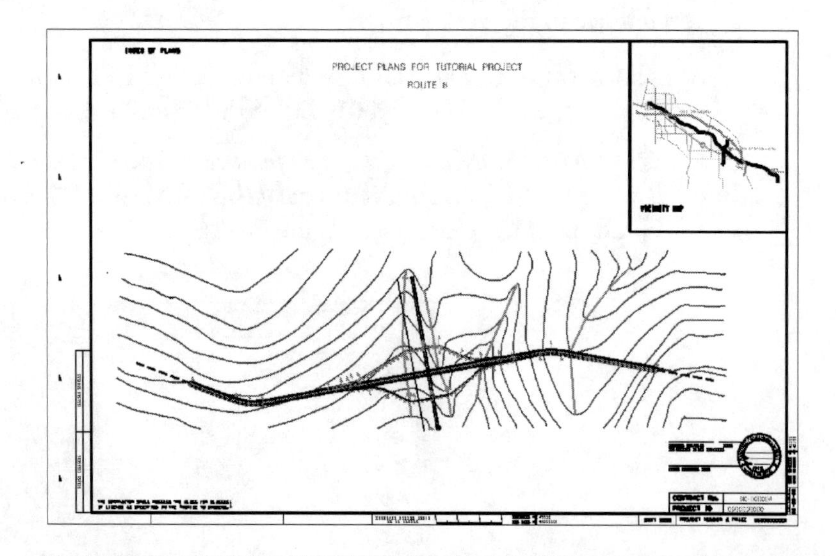

Figure 13-5 Title Border for the Tutorial Project

e. Go to ***Tools*** → ***Cells*** → ***Place Active Cell***. Select **CFTITL** from **Active Cell** by clicking (see Figure 13-6). Snap to the title block as shown in Figure 13-7.

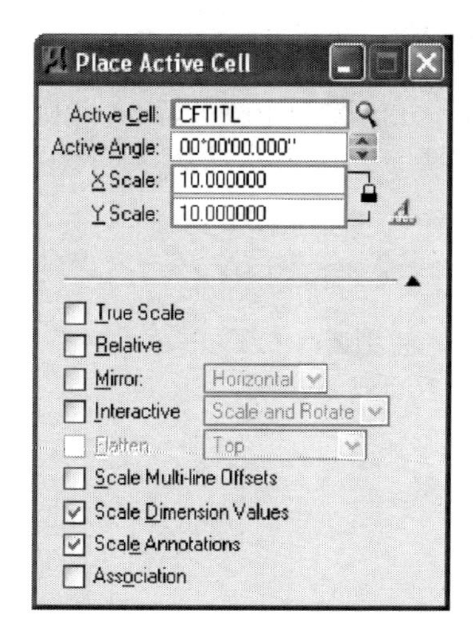

Figure 13-6 Place Active Cell Dialog Box for CFTITL

As shown in Figure 13-7, the contour line of the reference base file is located outside the left upper corner of the title clip frame. Therefore, the reference file (or **Tutorial_2D_HA.dgn**) shall be clipped.

f. Go to ***File* → *Reference* → *Tools* → *Clip Boundary***, select **Method: Element**, and click on the red clip frame in the drawing area (see Figure 13-8).

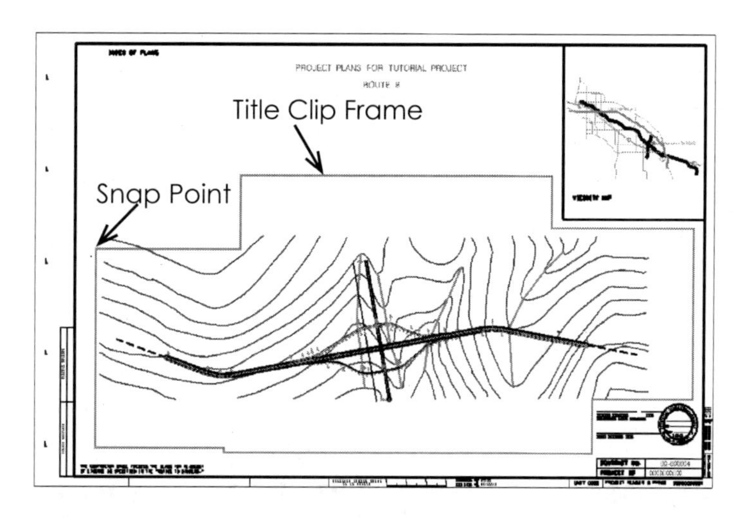

Figure 13-7 Drawing Window View

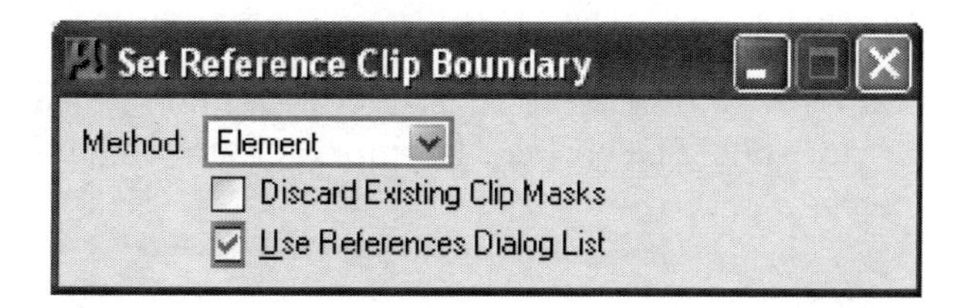

Figure 13-8 Reference Clip Boundary Setting

Step 4. Complete the title sheet

a. Use ***MicroStation* → *Text Tools*** to edit the project description text, complete an index of plans, and label the construction limits.

b. Set the text size large enough (may be 150 for Height and Width).

c. Add **ROUTE 8** at the end of the **CONSTRUCTION ON** (if you use a Caltrans Title sheet).

d. Add the following texts below the **Index of Plans:**

- » **Title and Location Map**
- » **Typical Cross Sections**
- » **Key Map and Line Index Sheet**
- » **Layouts**
- » **Profiles**

e. Add **Beginning of Construction** and **End of Construction** as shown in Figure 13-9.

f. Repeat the previous steps and place the north arrow cell, **NARR**, and location arrow cell, **LOCARR**, on the plan accordingly. Figure 13-9 shows the title sheet for the tutorial project.

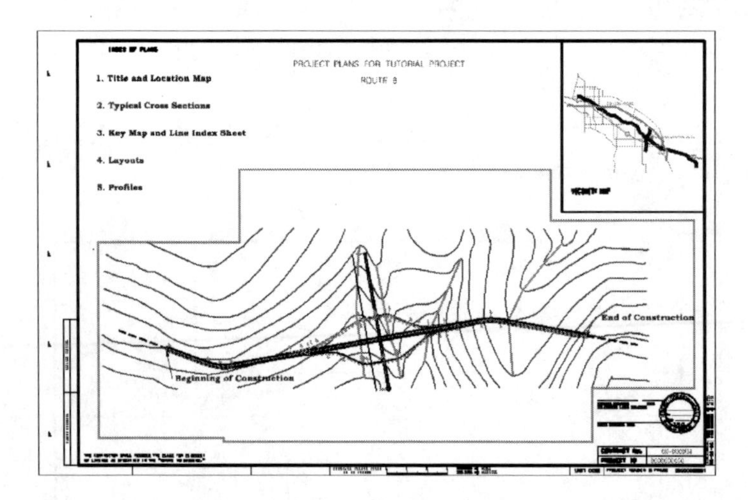

Figure 13-9 Title Sheet for the Tutorial Project

13.1.2 Typical Cross Section Sheet Preparation

A typical cross section consists of items such as existing conditions, proposed roadway lanes, shoulders, and medians; surfacing materials and thickness; roadbed slopes; and curbs, gutters and sidewalks for a roadway. These items are placed on a cross section sheet according to Caltrans requirements.

Prior to developing the typical cross section sheet for Route 8, Laguna Canyon Road, and four ramps, the typical cross section master file should be created. The procedure for creating the typical cross section master file in InRoads is as follows:

1. Start MicroStation and InRoads (if they are not open).
2. Create a new file **Tutorial_2D-Xsec.dgn** under the **C:\Temp\Tutorial\Chp13\Plan Sheets** directory. Browse to the **C:\Temp\Tutorial\Chp13\Master Files** directory and select **V8eSeed.dgn** as the seed file. Click **Save** and **Open**.

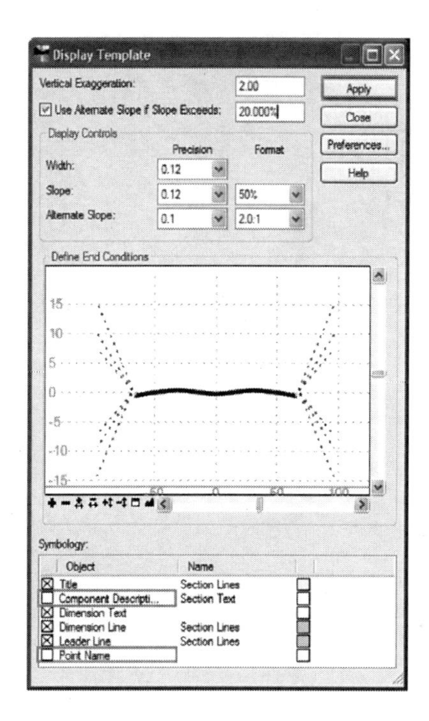

Figure 13-10 Template Display Settings

3. Load the template file.
   a. Select ***InRoads → Files → Open.***
   b. Browse to the directory **C:\Temp\Tutorial\Chp13\Master Files**. Double-click on the **Tutorial_Chapter11.itl** file, and click **Close**.
4. Verify the global scale factor.
   a. Select ***InRoads → Tools → Application Add-in → Check on Global Scale Factor → OK.***
   b. Select ***Tools → Global Scale Factor*** and set the scale factors to **10, 1, 1** for text, cell, and style, respectively. Then click **Apply** and **Close**.
5. Open the template library
   a. Open the template library of the tutorial project. Go to ***InRoads → Modeler → Create Template.***
6. Create the templates
   a. Expand the **Tutorial** library folder, right click on **Route 8**, and select **Display....**
   b. Use the default display template settings.
   c. Uncheck **Component** description and point name, and click **Apply** (see Figure 13-10).
7. Click any place on the drawing window.

   A typical cross section for Route 8 is shown in Figure 13-11.
8. Repeat 5 to 7 for all other typical sections of the project.

   Typical cross sections for ramps and Laguna Canyon Road are shown in Figures 13-12 and 13-13, respectively.

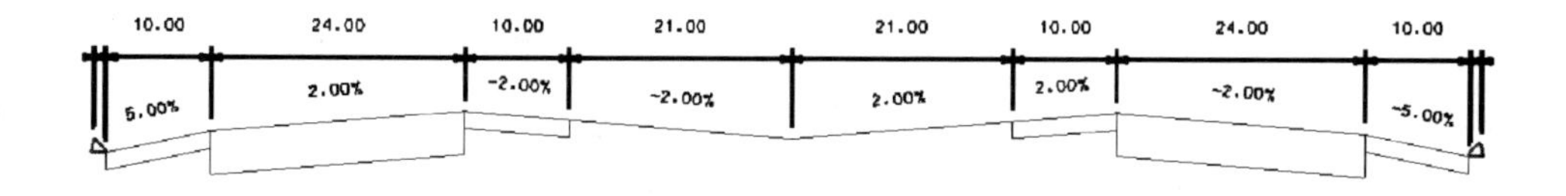

Figure 13-11 Typical Cross Section for Route 8

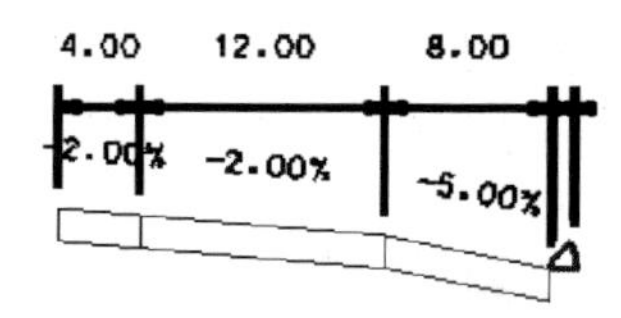

Figure 13-12 Typical Cross Section for Ramps

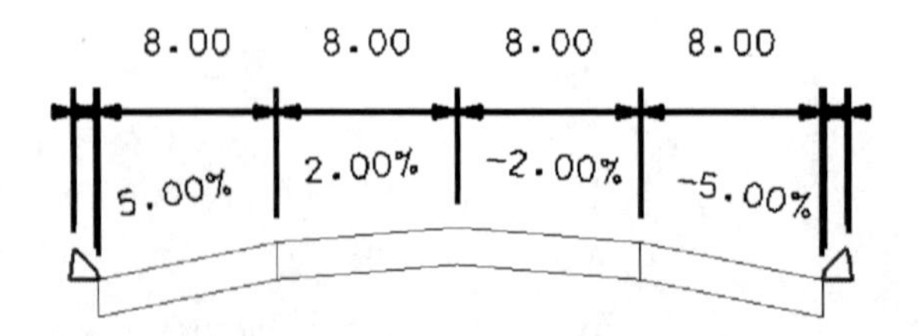

Figure 13-13 Typical Cross Section for Laguna Canyon Road

With the cross section file **tutorial_2D-Xsec.dgn** created, the typical cross section plan sheet can be developed by the similar procedure used for the title sheet.

Step 1. Create a new file

1. Open **MicroStation** if it is not open.
2. Create a new cross section sheet file **02_XSection.dgn** under the **C:\Temp\Tutorial\Chp13\Plan Sheets** directory.
3. Browse to the **C:\Temp\Tutorial\Chp13\Master Files** directory. Select **V8eSeed.dgn** as the seed file. Click **Save** and **Open**.

Step 2. Set the reference to the cross section file.

1. Reference the previously created **tutorial_2D-Xsec.dgn** file.
2. Set the logical name to be **Base**. Select the **Coincident–World ...**, and click **OK**.

Step 3. Load and place plan sheet cells.

1. Use the full plan sheet cell **FULPLN** to insert a plan border. Use the settings shown in Figure 13-14 to place the plan border around the typical cross sections.
2. Select the full clip frame cell **CFFULL** as the active cell. Place a cell inside the plan border and use it for the reference boundary clip.

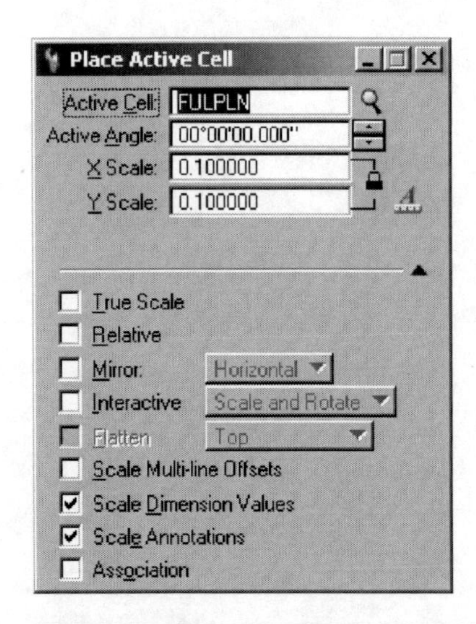

Figure 13-14 Settings for Placing the Plan Border Cell

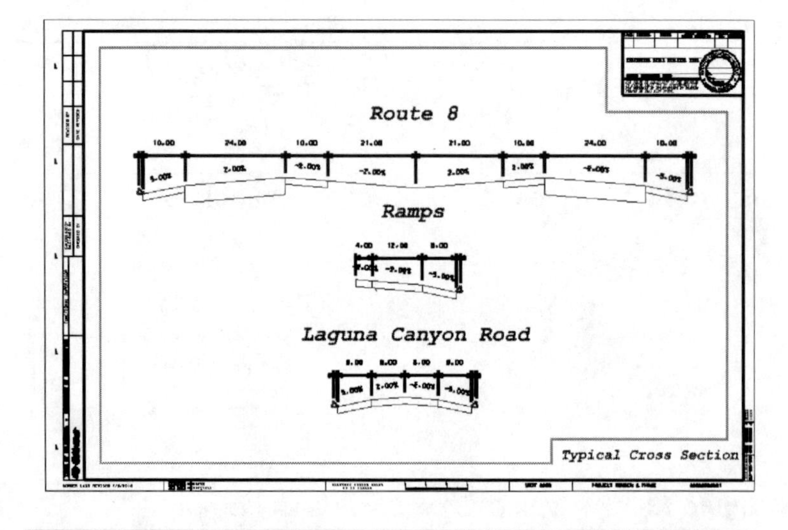

Figure 13-15 Typical Cross Section Sheet

Step 4. Complete the cross section sheet.

Since we do not have this type of information, we skip them in the cross section sheet.

1. Place a set of texts in the cross section (see Figure 13-15).

Note the tutorial cross section sheet does not include notes, design designation table, station limits, and materials as required by Caltrans.

13.1.3 Project Orientation Sheet Preparation

A *project orientation* sheet (or *key map*) is an index sheet that provides a spatial reference to layout sheets. A layout sheet describes a portion of alignments in detail. Prior to developing layout sheets, a key map master file should be created. The procedure for creating the key map master file in MicroStation is as follows:

1. Create a new file.
   a. Start **MicroStation** if it is not open.
   b. Under the directory **C:\Temp\Tutorial\Chp13\Plan Sheets**, create a file **KepMap.dgn**.
   c. Browse to the **C:\Temp\Tutorial\Chp13\Master Files** directory and select **V8eSeed.dgn** as the seed file. Click **Save** and **Open**.

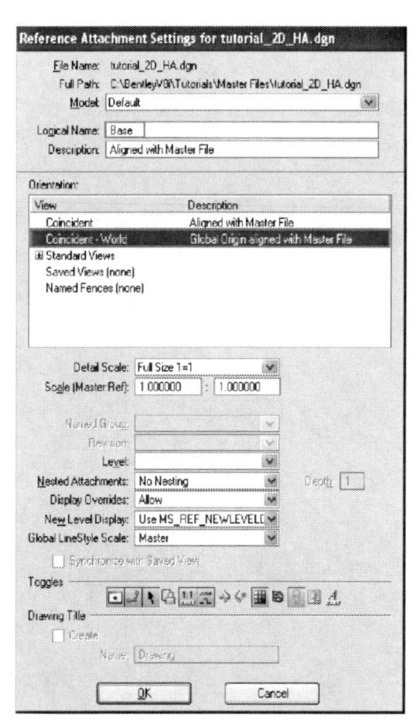

Figure 13-16 Reference Attachment Settings

2. Set a reference to the design file.
   a. Attach the **tutorial_2D_HA.dgn** file as a reference file from the **C:\Temp\Tutorial\Chp13\Master Files** directory to the newly created **KeyMap.dgn**.
   b. Select ***File → Reference → Tools → Attach***.
   c. Browse to the **C:\Temp\Tutorial\Chp13\Master Files** directory, select **tutorial_2D_HA.dgn**, and click **Open**.

      The Attach Reference Setting dialog box appears. Use the default settings.
   d. Enter **Base** for logical name and select **Coincident-World** for orientation. Click **OK** and **Close** the Reference dialog box (see Figure 13-16).
   e. To view the reference file on screen, click on the **FitView** icon.
3. Load the plan sheet clip boundary cell.
   a. Select ***MicroStation → Element → Cells*** and attach the Caltrans cell library.
   b. Select ***File → Attach File*** from the Reference Attachment Settings window. Locate the **Tutorialcellib.cel** or **ctcellib.cel** file, and click **Open**.

      A list of cells with name and description appears.
   c. From the list, select **CFFULL** and click **Placement**, and close the dialog box.
4. Place the clip boundary along the horizontal alignment.
   a. Go to ***Tools → Cells → Place Active Cell*** and place a boundary cell at the beginning of the Route 8 alignment.

b. Rotate it to the same direction as the alignment shown in Figure 13-17.

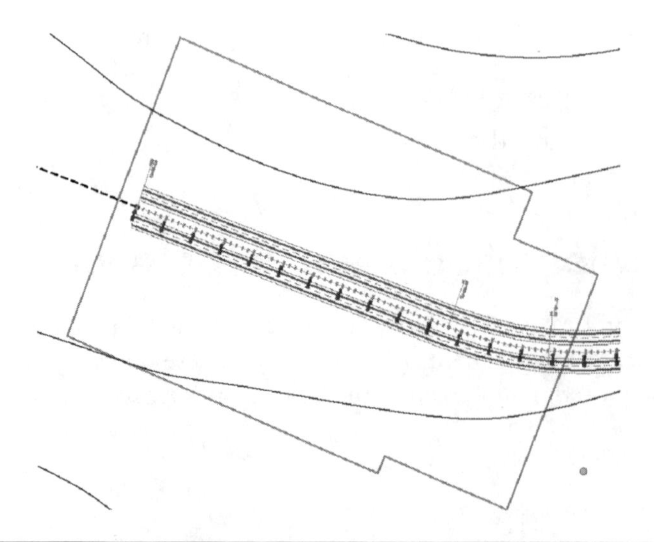

Figure 13-17 Clip Boundary Placed at the Beginning of Route 8

c. Use ***MicroStation → Copy/Rotate/Array*** tools to continue placing the clip boundary cell for the rest of the alignments.

Each should have a 100 ft. left margin and cover 1,000 ft. of the alignment. In other words, the midpoint on the left side of the clip boundary should be snapped at Stations 9+00, 19+00, 29+00, etc.

The completed key map is shown in Figure 13-18.

With the key map file **KeyMap.dgn** created, the key map and line index sheet can be developed by the procedure used for the title sheet.

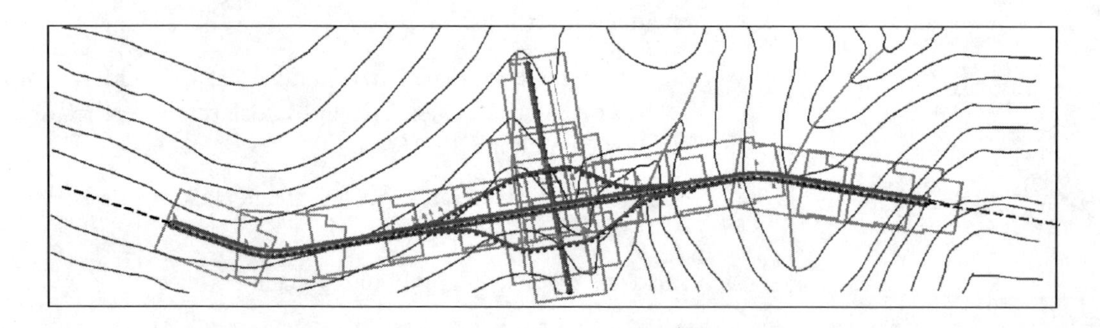

Figure 13-18 Clip Boundaries Covered the Entire Project Area

Step 1. Create a new file.

a. Open **MicroStation** if it is not open.

b. Create a key map sheet file **03_KeyMap.dgn** under the **C:\Temp\Tutorial\Chp13\Plan Sheets** directory.

c. Browse to the **C:\Temp\Tutorial\Chp13\Master Files** directory and select **V8eSeed.dgn** as the seed file. Click **Save** and **Open**.

Step 2. Set the reference to the master file.

a. Select ***MicroStation → File → References → Tools → Attach....***

b. Browse to the **C:\Temp\Tutorial\Chp13\Plan Sheets** directory and select **KeyMap.dgn** and Open.

c. From the **Reference Attachment Settings** window, enter **Base** for the logical name and select **Coincident- World ....**

d. Click **OK**. Use ***MicroStation → FitView*** to display the referenced **KeyMap** file.

e. Select ***MicroStation → File → References → Tools → Attach....***

f. Browse to the **C:\Temp\Tutorial\Chp13\Master Files** directory. Select **Tutorial_2D_HA.dgn** and **Open**.

g. From the **Reference Attachment Settings** window, enter **BaseHA** for the logical name and select **Coincident- World ....** Click **OK**.

h. Use ***MicroStation → FitView*** to display the referenced **KeyMap.dgn** and the **Tutorial_2D_HA.dgn** files.

Step 3. Load and place the plan sheet cells.

a. Use the **FULPLN** cell to insert a plan border.

b. Select the full clip frame cell **CFFULL** for the reference boundary clip.

c. For the same reason mentioned previously, set the X and Y scales to be **10:1**.

Step 4. Complete the key map and line index sheet.

a. Place notes and a design designation table.

b. Label station limits, a line index, and plan sheet numbers.

The completed key map and line index sheet are shown in Figure 13-19.

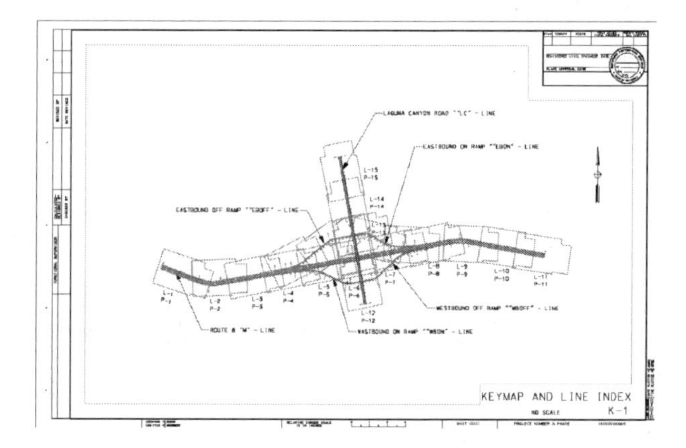

Figure 13-19 Key Map

13.1.4 Plan and Profile Sheets Preparation

There are three types of sheets (plan and profiles sheets, plan sheets, and profile sheets) that can be generated using InRoads. The plan and profile sheets are normally produced for small roadway improvement projects.

The plan sheets, normally for large projects, are set at the scale ratio of 1":100' (or 1":50') horizontal and 1":10' vertical. Each plan sheet includes existing and proposed roadway, drainage, utilities, and topographical information. It also includes construction notes to describe the location, type of work, and quantity of work. These notes are usually shown in the plan sheet's right margin or on a separate sheet if no space is available on the plan sheet.

The profile sheets present existing ground lines, proposed vertical alignments and grades, proposed and pertinent existing sewer profiles with appropriate grades and elevations, earthwork brackets, and other special information.

We will create both the plan sheets and the profile sheets for the tutorial project.

Plan Sheets or Layout Sheets

Follow these steps to generate the plan sheets or the layout sheets in InRoads:

Step 1. Open **MicroStation** and **InRoads** if they are not open.

a. Create a new file **04_Layout_Route8_.dgn** under the **C:\Temp\Tutorial\Chp13\Plan Sheets** directory.

b. Browse to the **C:\Temp\Tutorial\Chp13\Master Files** directory and select **V8eSeed.dgn** as the seed file. Click **Save** and **Open.**

Step 2. Load design files and cell library.

a. Select ***InRoads → Files → Open.***

b. Browse to the **C:\Temp\Tutorial\Chp13\Master Files** directory. Double click on the alignment file **Tutorials.alg** and the surface files **OG.dtm** and **Route 8.dtm.** Click **Close.**

c. Set **Route 8** as the active alignment.

d. Select ***MicroStation → Element***, select **Cells**, and attach the Caltrans cell library. Select **File → *Attach File***, locate the **Tutorialcellib.cel** or **ctcellib.cel** file from **C:\Temp\Tutorial\Chp13\Master Files** folder. Click **Open** and **Close.**

Step 3. Verify the global scale factor.

a. Select ***InRoads → Tools → Application Add-in → Check on Global Scale Factor → OK.***

b. Select ***Tools → Global Scale Factor*** and set the scale factors to **50, 1, 1,** for text, cell, and line style, respectively. Click **Apply.**

Step 4. Plan sheets generation.

1. Load the Plan and Profile Generator by selecting ***InRoads → Drafting → Plan and Profile Generator*** (see Figure 13-20).
2. Select the **Main** tab.
   a. Select Method: **Plan Only**
   b. Select Horizontal Alignment: **Route 8**
   c. Select **Use Station Limits** for Plan Views.
   d. Select **Generate Sheets** and **VDF Information** and **Host Files.**

   Use Default Station settings for Start and Stop Stations Limit:

Start=10+00.00

Stop=117+05.85

e. Define the alignment length to be covered by each plan view: **1000**.

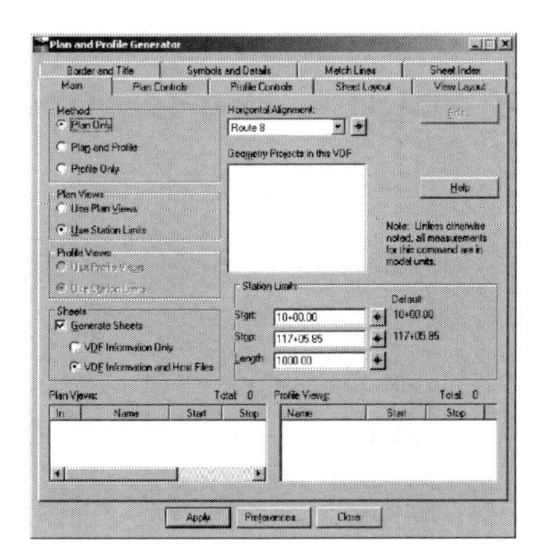

Figure 13-20 Main Tab Settings

3. Select the **Plan Controls** tab (see Figure 13-21).

   a. Rename Seed View Name as **Route8**.

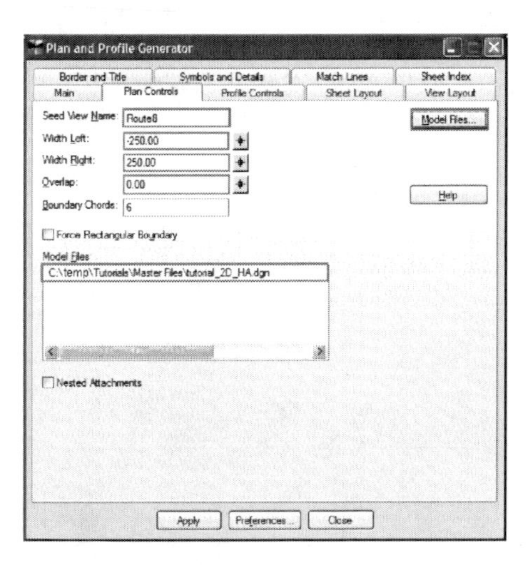

Figure 13-21 Plan Controls Tab Settings

   b. Define alignment view width to the left or right of the centerline:

Width Left:	**–250.00**
Width Right:	**250.00**
Overlap:	**0.00**

c. Click on **Model Files....** Browse to the **C:\Temp\Tutorial\Chp13\Master Files** directory.

d. Select **tutorial_2D_HA.dgn**. Click **Open.**

4. Select **Sheet Layout** tab.

a. Load the host file **C:\Temp\Tutorial\Chp13\Plan Sheets\04_Layout_Route8_.dgn** that was created in Step 1.

b. Load the seed host file **V8eSeed.dgn** from the **C:\Temp\Tutorial\Chp13\Master Files** directory.

5. Select **View Layout** tab.

a. Set Plan Location at **X=4.5, Y=10.5.**

b. Set Scale at **1:50.**

6. Select **Border and Title** tab (see Figure 13-22).

a. Under Border, check **Cell**, select **FULPLN** from name, and check **Retain Cell Levels for Each Sheet.** Set Scale to **1.00.**

b. Under **Symbology**, check only horizontal alignment, sheet number, and total sheets.

c. Select **Horizontal Alignment.** Change location in paper units to **X=850, Y=400.**

d. Select **Sheet Number** and change location in paper units to **X=1526, Y=1022.**

e. Select **Total Sheets** and change location in paper units to **X=1561, Y=1022.**

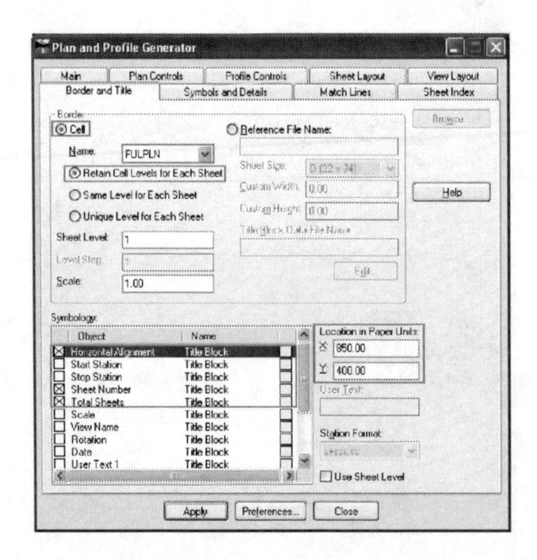

Figure 13-22 Border and Title Tab Setting

7. Select the **Symbols and Details** tab.

a. Under the north arrow, check **Attach**, select **NARR** from name, and check **Retain Cell Levels for Each Sheet.**

b. Set the Scale to **1.00.**

c. Set the Location in paper units to **X=24, Y=18.**

8. Select **Match Lines** tab.
   a. Use the default settings.
9. Select **Sheet Index** tab.
   a. Name the View Definition File (VDF) as **Route8_Plan_Only** and click **New**.
   b. Select Clipping Boundary Mode as **Calculate**.
   c. Click **Apply**.

   After executing the commands, the following 11 plan sheets are created under the **C:\Temp\Tutorial\Chp13\Plan Sheets** folder:

   04_Layout_Route8_01.dgn
   04_Layout_Route8_02.dgn
   04_Layout_Route8_03.dgn
   04_Layout_Route8_04.dgn
   04_Layout_Route8_05.dgn
   04_Layout_Route8_06.dgn
   04_Layout_Route8_07.dgn
   04_Layout_Route8_08.dgn
   04_Layout_Route8_09.dgn
   04_Layout_Route8_10.dgn
   04_Layout_Route8_11.dgn

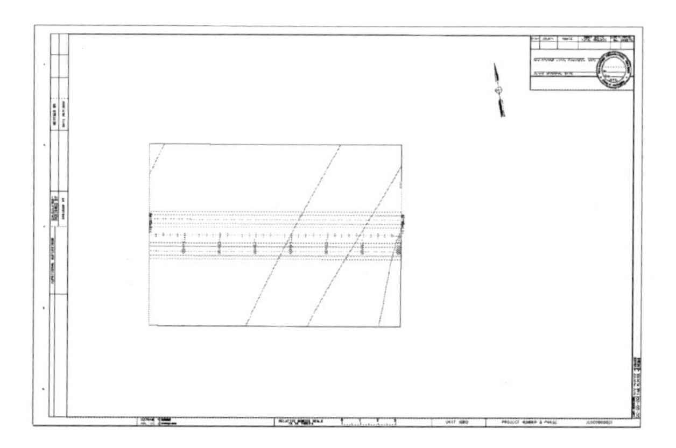

Figure 13-23 Generated Plan Sheet

   The view window displays the last plan. Figure 13-23 is the result plan of the **04_Layout_Route8_11.dgn** file. (If you cannot see the alignments, go to the **Troubleshooting Notes** later in this chapter.)

10. Save the VDF file and the preference file by doing the following:
    a. Select ***InRoads → Drafting → Plan and Profile Generator → Sheet Index Tab → Save as….*** Save **Route8_Plan_Only.vdf** file under the **C:\Temp\Tutorial\Chp13\Plan Sheets** folder so that sheet definitions can be recalled and modified if necessary.
    b. Under the Plan and Profile Generator dialog box, select ***Preferences… → Save as….***
    c. Name the preference file as **Route8_Plan_Only**, then click **OK** and **Close**.

    With the preference file, the control settings on the Plan and Profile dialog box can be loaded at a later time (see Figure 13-24).

Repeat the same procedures Step 1 to 4 for the remaining alignments.

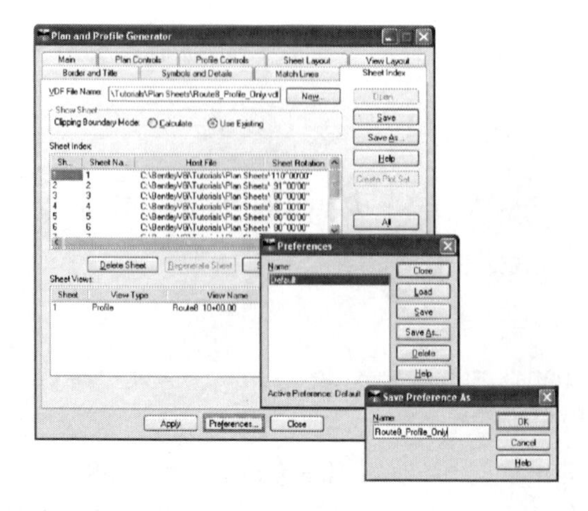

Figure 13-24 Saving VDF and Preference Files

» Create **04_Layout_Laguna_Canyon_Road_.dgn** and generate layouts for the Laguna Canyon Road alignment.

Figure 13-25 shows one of the layouts for Laguna Canyon Road.

» Create **04_Layout_EBOFF_.dgn** and generate layouts for the EBOFF alignment.

Figure 13-26 shows one of the layouts for the EBOFF ramp.

» Create **04_Layout_EBON_.dgn** and generate layouts for the EBON alignment.

Figure 13-27 shows one of the layouts for the EBON ramp.

» Create **04_Layout_WBOFF_.dgn** and generate layouts for the WBOFF alignment.

Figure 13-28 shows one of the layouts for the WBOFF ramp.

» Create **04_Layout_WBON_.dgn** and generate layouts for the WBON alignment.

Figure 13-29 shows one of the layouts for the WBON ramp.

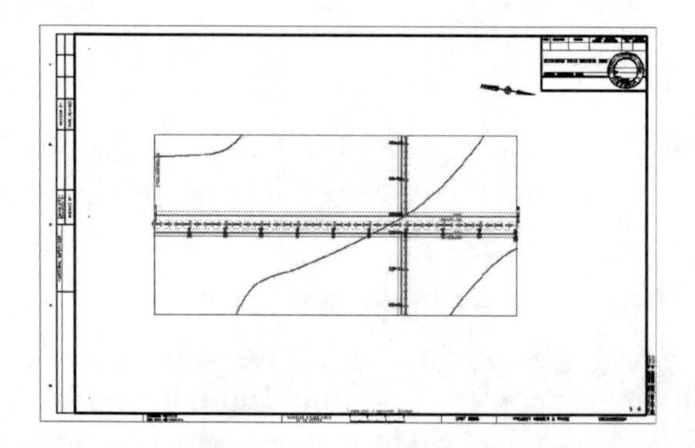

Figure 13-25 First Plan Layout of Laguna Canyon Road

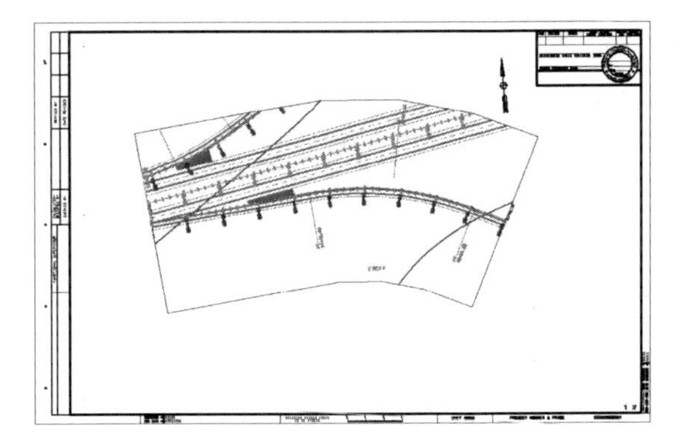

Figure 13-26 First Plan Layout of the EBOFF Ramp

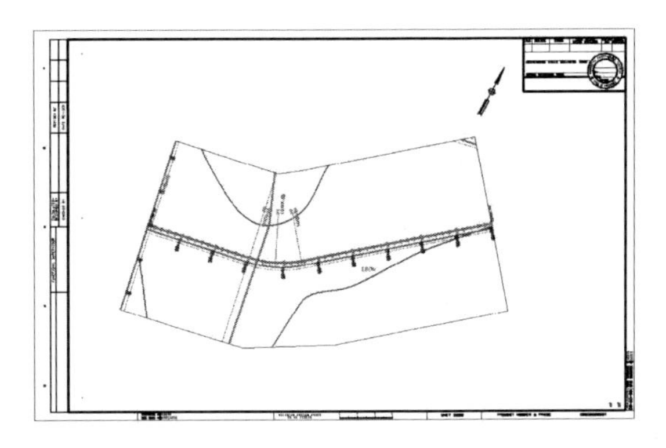

Figure 13-27 First Plan Layout of the EBON Ramp

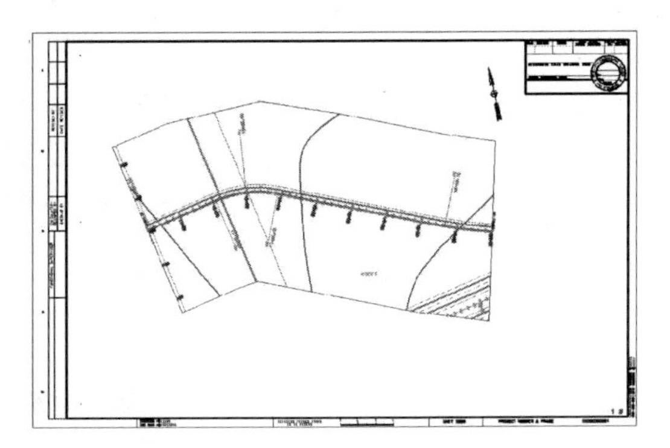

Figure 13-28 First Plan Layout of the WBOFF Ramp

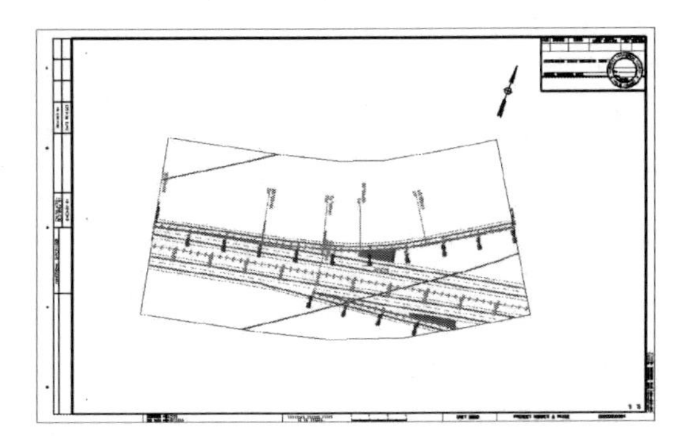

Figure 13-29 First Plan Layout of the WBON Ramp

Troubleshooting Note

Due to file compatibility issues, alignment may not be referenced correctly. The resulting plan of **04_Layout_Route8_11.dgn** file does not show alignments within the clipping boundary (see Figure 13-30). The following extra steps are necessary to correctly display the alignments:

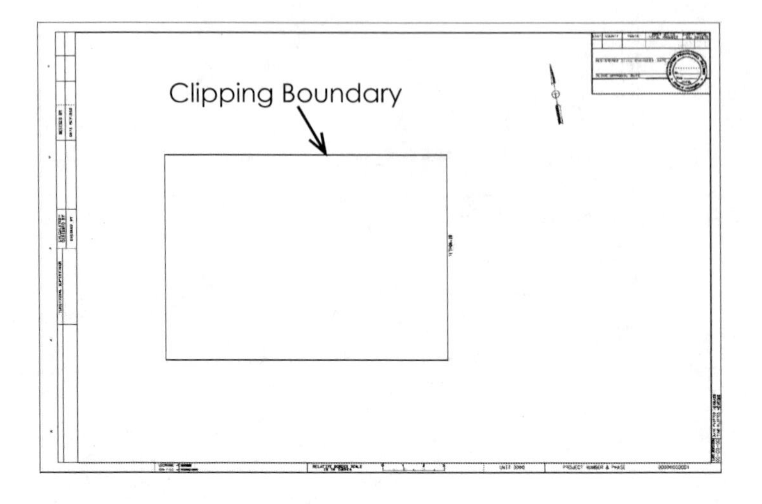

Figure 13-30 Generated Plan Sheet without Alignment

1. Go to ***MicroStation*** → ***References*** → ***tutorial_2D_HA.dgn***.
2. Right-click the empty field below **Orientation** (see Figure 13-31). Select ***Standard Views*** → ***Top*** and then select **Coincident–World ….**

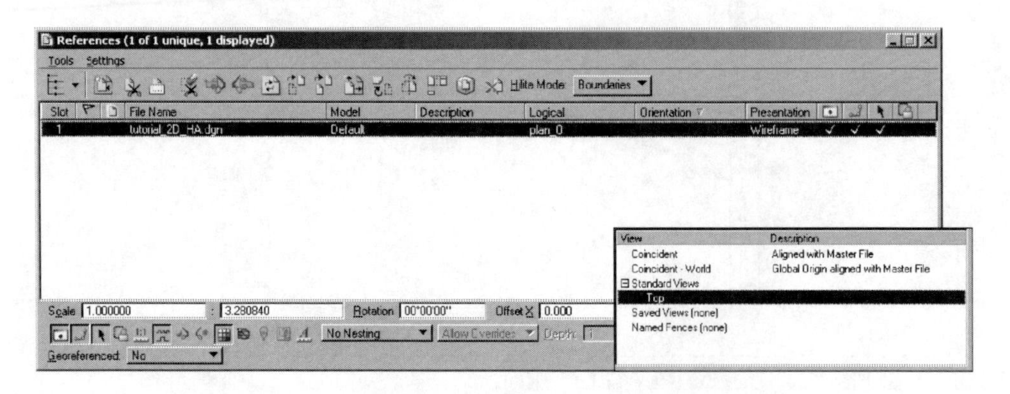

Figure 13-31 Set the Orientation for the 04_Layout_Route8_11.dgn file

3. Once the referenced **tutorial_2D_HA.dgn** file appears, select ***References*** → ***Tools*** → ***Clip Boundary***, then select the **clipping boundary**.

The clipped plan appears (see Figure 13-32).

4. Select ***MicroStation*** → ***Save*** to save the **04_Layout_Route8_11.dgn** file.
5. Open each generated layout file from the **C:\Temp\Tutorial\Chp13\Plan Sheets** folder and follow the same clipping process for each layout file:

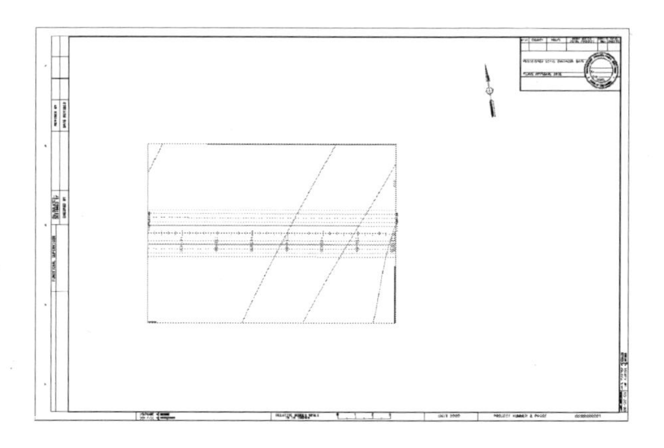

Figure 13-32 Clipped Plan for the 04_Layout_Route8_11.dgn file

04_Layout_Route8_1.dgn
04_Layout_Route8_2.dgn
04_Layout_Route8_3.dgn
04_Layout_Route8_4.dgn
04_Layout_Route8_5.dgn
04_Layout_Route8_6.dgn
04_Layout_Route8_7.dgn
04_Layout_Route8_8.dgn
04_Layout_Route8_9.dgn
04_Layout_Route8_10.dgn

Step 5. Complete plans.

1. Open each generated layout file from the **C:\Temp\Tutorial\Chp13\Plan Sheets** folder.

Note that we skip the next step because we do not have information about the construction notes.

2. Place notes using ***MicroStation → Text tools or InRoads → Drafting → Place Plan Note***.

Profile Sheets

Follow these steps to generate profile sheets in InRoads:

Step 1. Create the host file.

1. Open MicroStation and InRoads if they are not open.
2. Create a new file **05_Profile_Route8_.dgn** under the **C:\Temp\Tutorial\Chp13\Plan Sheets** directory.
3. Browse to the **C:\Temp\Tutorial\Chp13\Master Files** directory and select **V8eSeed.dgn** as the seed file. Click **Save** and **Open**.

Step 2. Load the profile border reference file (**FULPROF.dgn**).

1. Locate this file in the **C:/Temp/Book/Chp13/Master Files** folder or download the file from the book website (www.csupomona.edu/~xjia/GeometricDesign/Chp13/Master Files/).
2. Place **FULPROF.dgn** in the working project directory **C:\Temp\Tutorial\Chp13\Master Files**.

Step 3. Load the design files and cell library.

1. Select ***InRoads → Files → Open***.
2. Browse to the **C:\Temp\Tutorial\Chp13\Master Files** directory. Double-click the alignment file **Tutorials.alg** and the surface files **OG.dtm** and **Route 8.dtm**. Click **Close**.

3. Set Route 8 as the active alignment.

Step 4. Verify the global scale factor.

1. On the InRoads Main Menu Bar, select ***Tools* → *Application Add-in* → *Check on Global Scale Factor* → *OK*.**
2. Select ***Tools* → *Global Scale Factor*** and set the scale factors to **50, 1, 1**. Click **Apply**.

Step 5. Create a profile preference file.

1. The default setting of exaggeration for vertical is **10.00**, and for horizontal is **1.000**. Use **5.000** for vertical exaggeration for the tutorial project. Select ***InRoads* → *Evaluation* → *Profile* → *Create Profile***.
2. In the Create Profile dialog box, change vertical exaggeration to **5.000**.
3. Select **Grid**. From the **Symbology** List, uncheck **Major Horizontal** and **Major Vertical** (see Figure 13-33).

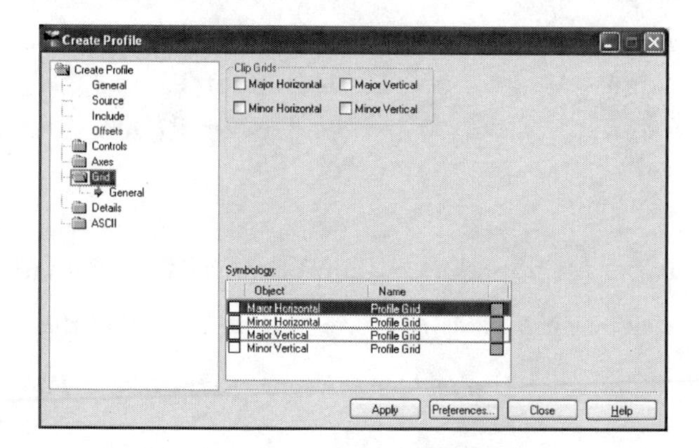

Figure 13-33 Profile Grid Settings

4. Expand **Axes**, select **Bottom**, change major ticks position to **Inside**, and click on **Preferences....**

The Preferences dialog box appears.

5. Select **Save As...** (see Figure 13-34).
6. Define the preference file name as **Profile_Route8.** Click **OK** and **Close**.
7. Select **Close** on the Create Profile dialog box.

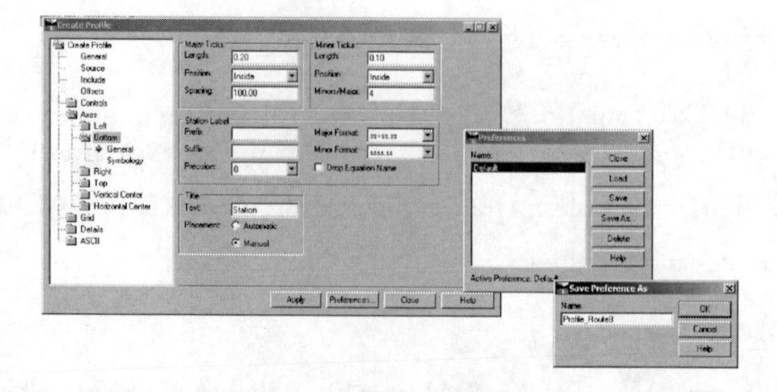

Figure 13-34 Profile Axes Setting and Create Preference File

Step 6. Profile sheets generation.

1. Load the Plan and Profile Generator by selecting ***Drafting → Plan and Profile Generator***.
2. Select the **Main** tab (see Figure 13-35).
   a. Set Method: **Profile Only**
   b. Select Horizontal Alignment: **Route 8**
   c. Select **Use Station Limits** for Profile Views.
   d. Select **Generate Sheets** and **VDF Information and Host Files**.
   e. Use **Default** Stations for Route 8.
   f. Alignment Length to be shown on each plan view: Use **1000** ft.
3. Select **Profile Control** tab (see Figure 13-36).
   a. Seed View Name: **Route8**
   b. Set Name: **Route 8**
   c. Profile Preference: **Profile_Route8**
   d. Vertical Alignment: **Route 8**
   e. Select **OG** and **Route 8** for surfaces.
   f. Select **Shift at Major Stations** for profile elevation shifts.
   g. Define profile height at **70.00**.

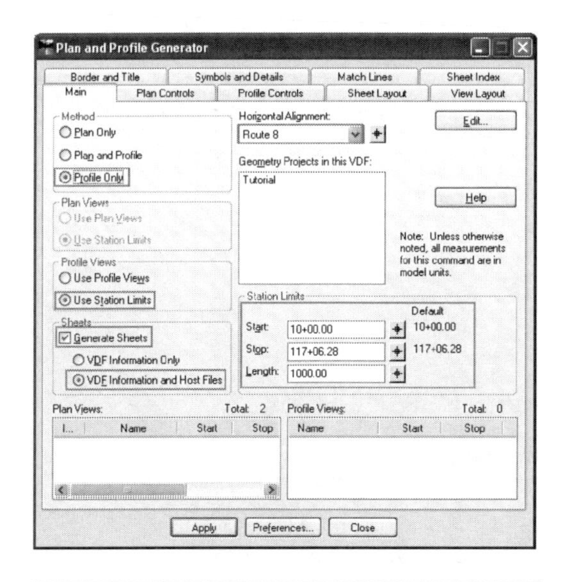

Figure 13-35 Profile Only Main Tab Settings

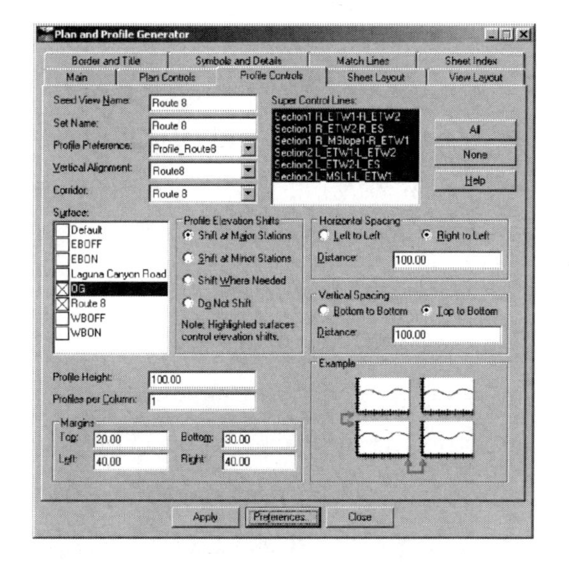

Figure 13-36 Profile Only Profile Controls Tab Settings

4. Select the **Sheet Layout** tab.
   a. Load the host file **05_Profile_Route8_.dgn** that was created in Step 1.
   b. Load seed host file **V8eSeed.dgn** from the project **C:\Temp\Tutorial\Chp13\Master files** directory.
5. Select the **View Layout** tab.

a. Set profile location at **X=2.80, Y=4.00.**

b. Set scale at **1:50.**

6. Select the **Border and Title** tab (see Figure 13-37).

a. Check **Reference File Name.** Click on the textbox and select **Browse....**

b. Load the **FULPROF.dgn** file from the **C:\Temp\Tutorial\Chp13\Master Files** directory. Set scale to **0.80.**

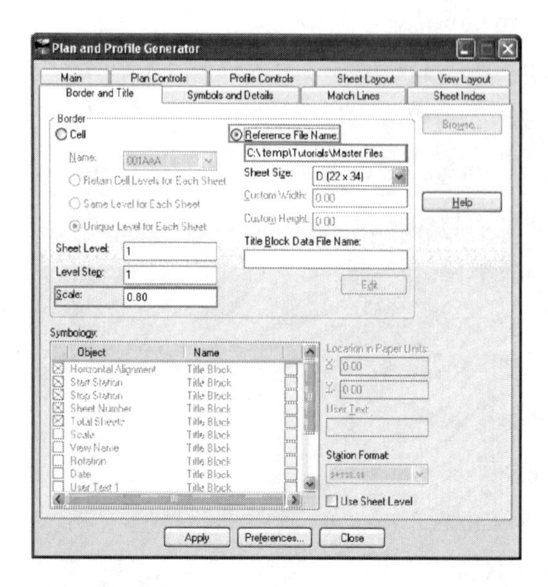

Figure 13-37 Profile Only Border and Title Tab Settings

7. Select the **Symbols and Details** tab.

a. Under North Arrow, uncheck **Attach.**

8. Select **Match Lines** tab.

a. Use the default settings.

9. Select the **Sheet Index** tab.

a. Name the View Definition File (VDF) as **Route8_Profile_Only**. Click **New.**

b. Select the Clipping Boundary Mode as **Calculate.** Click **Apply.**

c. Click on any point on the View Window.

After executing the commands, the following 11 profile sheets are created under the **C:\Temp\Tutorial\Chp13\Plan Sheets** folder:

05_Profile_Route8_01.dgn
05_Profile_Route8_02.dgn
05_Profile_Route8_03.dgn
05_Profile_Route8_04.dgn
05_Profile_Route8_05.dgn
05_Profile_Route8_06.dgn
05_Profile_Route8_07.dgn
05_Profile_Route8_08.dgn
05_Profile_Route8_09.dgn
05_Profile_Route8_10.dgn
05_Profile_Route8_11.dgn

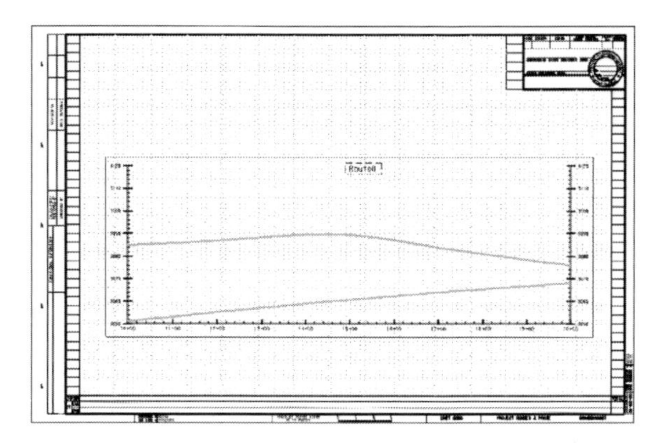

Figure 13-38 First Profile of Route 8

The view window displays the last plan. Above is the result plan of the **05_Profile _Route8_1. dgn** file (see Figure 13-38).

10. Save the VDF and preference files.

    a. Under InRoads Menu Bar, select ***Drafting → Plan and Profile Generator → Sheet Index Tab → Save as….***

    b. Save the **Route8_Profile_Only.vdf** file under the **C:\Temp\Tutorial\CHp13\Plan Sheets** folder, so that sheet definitions can be recalled and modified if necessary.

    c. Under the Plan and Profile Generator Dialog Box, select ***Preferences… → Save as….***

    d. Name the preference file as **Route8_Profile-Only**, then click **OK** and **Close**.

    With the preference file, the control settings on the Plan and Profile dialog box can be loaded at a later time.

Repeat the same procedures Step 1 to 6 for the remaining alignments.

» Create **05_Project_Laguna_Canyon_Road_.dgn** and generate profiles for the Laguna Canyon Road alignment.

Figure 13-39 shows one of the profiles for Laguna Canyon Road.

» Create **05_Profile_EBOFF_.dgn** and generate profiles for the EBOFF alignment.

Figure 13-40 shows one of the profiles for the EBOFF ramp.

» Create **05_Profile_EBON_.dgn** and generate profiles for the EBON alignment.

Figure 13-41 shows one of the profiles for the EBON ramp.

» Create **05_Profile_WBOFF_.dgn** and generate profiles for the WBOFF alignment.

Figure 13-42 shows one of the profiles for the WBOFF ramp.

» Create **05_Profile_WBON_.dgn** and generate profiles for the WBON alignment.

Figure 13-43 shows the one of the profiles for the WBON ramp.

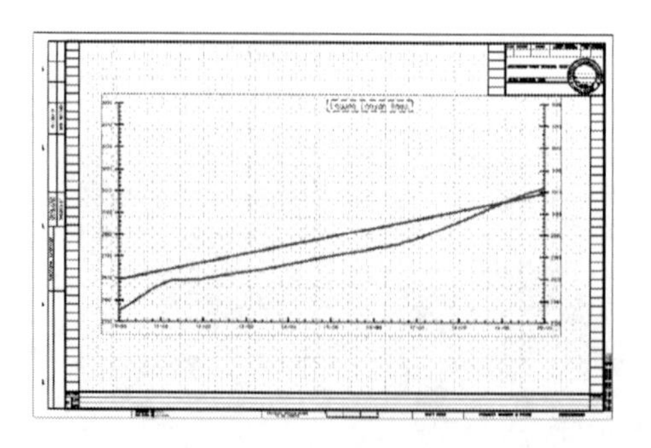

Figure 13-39 First Profile of Laguna Canyon Road

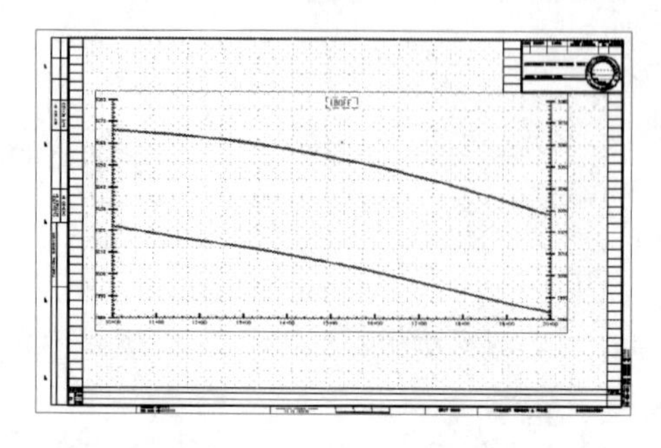

Figure 13-40 First Profile of EBOFF Ramp

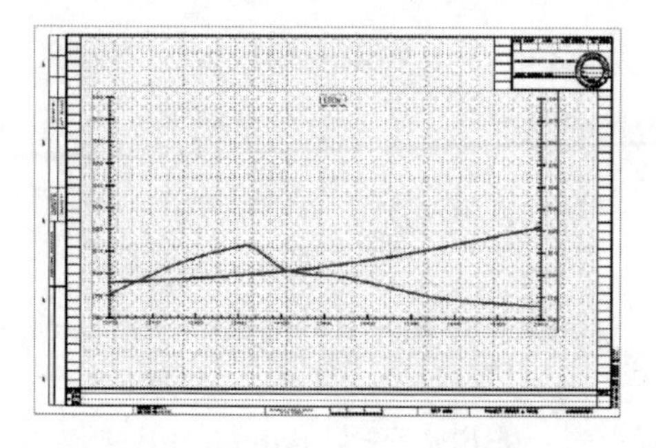

Figure 13-41 First Profile of EBON Ramp

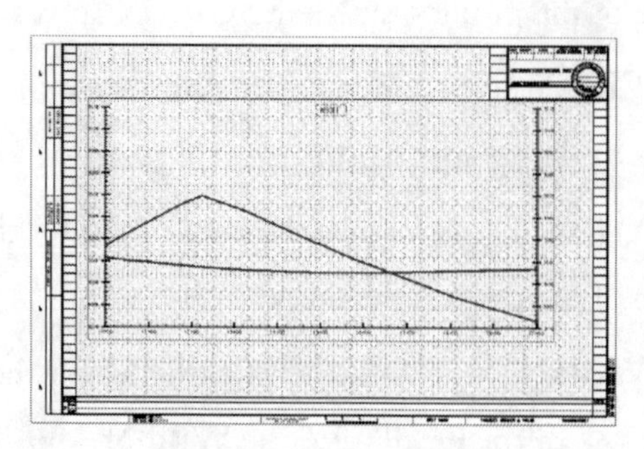

Figure 13-42 First Profile of WBOFF Ramp

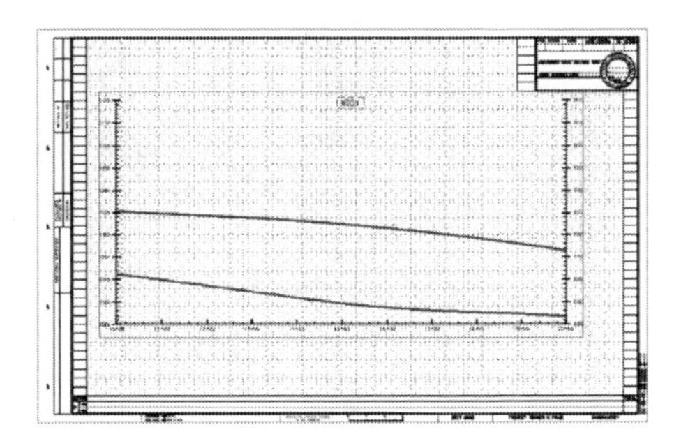

Figure 13-43 First Profile of WBON Ramp

Step 7. Complete profiles.

We skip this task due to the lack of construction notes.

1. Open each generated profile file from the **C:\Temp\Tutorial\Chp13\Plan Sheets** folder.
2. Place notes in the files using ***MicroStation* → *Text*** tools or ***InRoads* → *Drafting* → *Place Profile Note***.

13.2 Preparation of Cost Estimates

An engineering cost estimate is an approximation of a project's total cost based on design and quantity, material and labor, inflation, historical bid data, and other factors. Figure 13-44 illustrates the three stages of cost estimates.

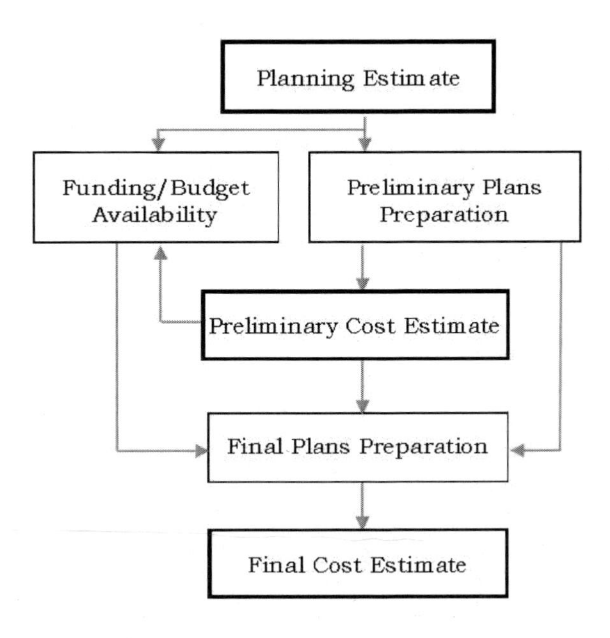

Figure 13-44 Project Cost Estimate Process

1. **Planning Estimate.** This stage involves a rough approximation of project costs. The approximation provides information for project funding assessment.
2. **Preliminary Cost Estimate.** This stage involves a preliminary approximation based on well-defined cost data and design information. The cost estimate is used to determine if there is enough funding available for project construction.
3. **Final Cost Estimates.** This stage involves an approximation based on preliminary cost estimates and information from the specifications used and the changes made since the preliminary cost estimate stage.

The procedure discussed in this chapter applies to both preliminary and final cost estimate stages. It includes the following four steps:

Step 1. Determine bid items.

In practice, project construction work should be divided into bid items to quantify and estimate project costs. Bid items should closely match the construction notes written on the final project plans. In this step, highway designers should develop a comprehensive list of bid items so contractors can make reasonably accurate estimates on the project construction costs.

Step 2. Identify bid items.

This step identifies bid items (number, name/description, and measurement unit) from the Caltrans website at www.dot.ca.gov/hq/esc/oe/awards.

For example, if **Bid Item Class 2 Aggregate Subbase** is designated in Step 1, its item description and item code (**250201**) can be found on the standard contract item list.

The measurement units for a bid item can be found in the *Contract Item Cost Data Book*, which is available online in PDF format at the website www.dot.ca.gov/hq/esc/oe/awards/.

For example, the unit for **Item 250201 Class 2 Aggregate Subbase** is in **Cubic Yard**.

Step 3. Determine item unit cost.

The unit cost of bid items can be extracted from the bid history of similar projects. For Caltrans projects, the unit cost of standard contract items can be found online on this website: sv08data.dot.ca.gov/contractcost.

For example, the unit cost for **Class 2 Aggregate Subbase** is **$26.29 per Cubic Yard** when the following information is provided to the website:

Item Code:	**250201**
District:	**District 8**
Years:	**2008, 2009, 2010, 2011**
Measurement Unit:	**CY**

Step 4. Estimate quantities of each item.

After the unit cost and the measurement of bid items are identified, the costs for the bid items can be estimated. You can use MicroStation to determine the quantities of bid items, while InRoads can be used to calculate earthwork quantities as discussed in Chapter 12.

13.3 Preparation of Specifications and Contract Document

There are two standard specifications available for projects located in California: 1) American Public Works Association (APWA) *Green Book* and 2) Caltrans standard specifications. The APWA *Green Book* is used for federal-aid projects unless otherwise requested by agencies. Caltrans standard specifications are used for state-funded projects unless otherwise requested by agencies.

Most public agencies have adopted either APWA standards or Caltrans standards for projects within their jurisdiction. Project engineers responsible for preparation of specifications and contract documents should check with agencies to verify which standard specifications to be used.

Name____ Date______________

13.3 Questions

Q13.1 What is PS & E?

Q13.2 List the steps to obtain the unit price of **8" thermoplastic traffic stripe** (in linear feet) from the Caltrans website.

Q13.3 How do you find the item code for the bid item **Install Roadside Sign** on the Caltrans website?

Q13.4 What are the two most commonly used standard specifications for roadway improvement projects in California?

Q13.5 How do you import the registered Civil Engineer stamp cell **CESEAL.cel** from the Caltrans website on a plan in MicroStation?

Q13.6 Review the Caltrans *Plans Preparation Manual* (*PPM*) and generate a set of **Combined Plan and Profile** sheets for the tutorial project using the **Plan and Profile Generator** in InRoads.